Nathan Zuntz

Nathan Zuntz

His life and work in the fields of high altitude physiology and aviation medicine

Hanns-Christian Gunga

AMSTERDAM • BOSTON • HEIDELBERG • LONDON • NEW YORK • OXFORD
PARIS • SAN DIEGO • SAN FRANCISCO • SINGAPORE • SYDNEY • TOKYO
Academic Press is an imprint of Elsevier

Academic Press is an imprint of Elsevier
30 Corporate Drive, Suite 400, Burlington, MA 01803, USA
32 Jamestown Road, London NW1 7BY, UK
525 B Street, Suite 1900, San Diego, CA 92101-4495, USA
360 Park Avenue South, New York, NY 10010-1710, USA

First edition 2009

Library of Congress Cataloging-in-Publication Data
A catalog record for this book is available from the Library of Congress

British Library Cataloguing in Publication Data
A catalogue record for this book is available from the British Library

ISBN: 978-0-12-374740-2

For information on all Academic Press publications
visit our website at www.elsevierdirect.com

Typeset by Charon Tec Ltd., A Macmillan Company (www.macmillansolutions.com)

Printed and bound in the United Kingdom

Transferred to Digital Print 2011

This book is dedicated to
my wife Luise,
and my children Leonard, Maxim and Arthur

This book is dedicated to
my wife Janet,
and my children Leonard, Marion and Arthur

Contents

Preface

This book deals with the life and work of Nathan Zuntz and specifically addresses the contribution he made to high altitude physiology and aviation medicine. While some of the material on this topic has already been published in English (Gunga and Kirsch, 1995a, 1995b), the overwhelming majority of it is only available in German (Gunga, 1989). Recently, new and interesting sources on the life and work of the Berlin physiologist have come to light. That was why the author gladly accepted the invitation from the History Book Committee of the American Physiological Society in 2005 to write a biography of Nathan Zuntz. The result is a completely new, revised version of the book which is based on more than twenty years of study on the life and work of this physiologist.

This biography could not have been written without the help of numerous people and institutions. First, I would like to thank Dr. John West, the current Chairman of the APS History Book Committee, who personally committed himself so firmly to making this project possible from the very first day.

I am very grateful to the Free University, the Charité, the DLR (German Aerospace Center) in Cologne and the management in Bonn-Oberkassel – namely Dr. Horst Binnenbruck, Dr. Peter Preu, Prof. Dr. Günter Ruyters, Dr. Hans-Ulrich Hoffmann, and Dr. Peter Gräf – and the sponsors Klaus-Peter Ludwig (EADS/EADS-Astrium), Dr. Gerd Bräunig (Kayser-Threde), Dr. Roman Skoblo (IFBL), and Dr. Bernd Mueller (Bayer-Schering Pharma), who made this book possible by providing financial and logistic support. If the *Zentrum für Weltraummedizin Berlin* (ZWMB) (Center of Space Medicine Berlin) had not been founded at the Free University of Berlin in 2000 and if these sponsors had not contributed to creating the "Nathan Zuntz" Endowed Professorship in *Weltraummedizin und extreme Umwelten* (Space Medicine and Extreme Environments) at the Charité Campus Benjamin Franklin (CBF) in 2003, I could not have realized this project. Thanks are also due to those institutions, departments and their staff who took interest in and lent their support to these historical studies – several national and international archives and libraries, the Department of the History of Medicine Charité Universitaetsmedizin Berlin, the historical working group of the DGLRM (German Society for Aviation and Space Medicine), the Life Science Working Group of ESA (European Space Agency), the ISGP (International Society of Gravitational Physiology) as well as the ÖGAHM (Austrian Society of Alpine and High Altitude Medicine).

My special thanks also go to my tireless and always very critical translator, Carmen v. Schöning, and her team. Without the months of very close cooperation and constant enthusiasm for the topic, this book would have been impossible to realize. Brigitte Bünsch, Anja Haller, Bärbel Himmelsbach-Wegner,

Eveline Hofmann, Darren Lipnicki, Thilo Noack, Thomas Schlabs, Mathias Steinach, Lutz-Rainer Weiss, Dr. Andreas Werner, and Dr. Eberhard Koralewski were particularly supportive in the research and the preparation of the manuscript and in dealing with electronic media. I thank them as well for their constant assistance and open minds.

Personal thanks are also due to all my colleagues at the Center of Space Medicine Berlin and the colleagues at the Department of Physiology, namely Prof. Dr. Axel R. Pries and his team, and Dr. Clarence Alfrey, Prof. Dr. Hans-Georg Baumgarten, Prof. Dr. Claus Behn, Prof. Dr. Gilles Clément, Prof. Dr. Rupert Gerzer, Prof. Dr. Bruno Günther, Prof. Dr. Gerda Horneck, Prof. Dr. Alan Hargens, Dr. Reed Hoyt, Prof. Dr. Egon Humpeler, Prof. Dr. Léon-Velarde, Prof. Dr. Dag Linnarsson, Dr. Günther Reitz, Surgeon General German Air Force Dr. Erich Rödig, Prof. Dr. Wolfgang Schobersberger, Dr. Peter Wittels, Prof. Lothar Röcker, Prof. Dr. Hans-Volkhart Ulmer, Prof. Dr. Li-Fan Zhang, Prof. Dr. Fengyuan Zhuang, and the late Carlos Monge and Peter Hochachka for many enlightening and enjoyable discussions. Last but not least, Prof. Dr. Karl Kirsch was largely responsible for teaching me the physiology of humans in extreme environments and calling for the creation of the Endowed Professorship. I would like to thank him expressly for the exciting years we spent working together, the conversations we shared, and the discussions we had about the different working philosophies in our field, the place of science in contemporary society, and on many other topics as well.

Finally, I wish to thank my dear wife, Luise, and my children Leonard, Maxim, and Arthur for their understanding those innumerable times I told them I was "going to be a little bit late."

Berlin, January 2008

Introduction

The year 2000 saw the foundation of the Zentrum für Weltraummedizin Berlin (ZWMB, Berlin Center of Space Medicine) at the Freie Universität Berlin (Free University of Berlin). The Nathan Zuntz Professorship for Space Medicine and Extreme Environments was established three years later. This book focuses on the very early days of high altitude physiology and aviation medicine in Berlin, a city that has played an important role in the history of these sciences, and on the accomplishments of Nathan Zuntz (1847–1920), who was a professor of animal physiology at the Königliche Landwirtschaftliche Hochschule (Royal Agricultural College). As Zuntz died in 1920, this book will not deal with the darkest chapter in the history of high altitude and aviation medicine in Germany: the involvement of physiologists, medical staff, and research institutions, such as the Deutsche Forschungsgemeinschaft, in human experiments during the Nazi regime. This topic has become, and continues to be, the subject of research in recent decades (Bower, 1987; Hunt, 1991; Harsch, 2000, 2001, 2004; Eckart, 2006, 2008). On the other hand, this study will illustrate how the life and work of Zuntz, as well as the technical and logistical capabilities available in Berlin before the 1930s, exemplified the leading role that Germany took in scientific research in comparison to the work done in England, France, and the United States. Immediately after the end of the war, from May/June 1945 onwards, under the codenames "Project Paperclip" and its successor operation "Operation Overcast," the Pentagon had already made efforts to seek out leading aviation scientists in the ruins of Germany and to recruit them for American research institutes (Gimbel, 1986; Bower, 1987: 203–272). Some of these scientists first gathered in Heidelberg at the Kaiser Wilhelm Gesellschaft (Kaiser Wilhelm Society) to give an overview of the knowledge on aviation that existed in Germany; they continued to do so for the next two years (*German Aviation in World War II*, 1950; Report from Heidelberg, unpaginated). German specialists in aviation medicine were among the first scientists ever to be brought to the United States in this way. In the years to come, they decisively shaped the development of aviation and space medicine in America (Winau, 1987: 348–350). Among them was Otto H. Gauer (1909–1979), whose life and work will have to be dealt with separately. He was a young scientist, educated in the physical sciences and medicine, who took an interest in gravitational physiology. Gauer returned to Germany from the United States in the early 1960s and was made Chair of Cardiovascular Physiology at the Department of Physiology at the Free University of Berlin in 1961.

In the early 1980s, Karl Kirsch (1938–), a student of O. H. Gauer and the late Rolf Winau,[1] visited some of the German scientists who were still living in the United States to interview them (oral history); a study which was funded by the Deutsche Forschungsgemeinschaft (DFG, German Research Foundation) (Kirsch and Winau, 1986: 633–635). In the course of their studies on the history of aviation and space medicine, they discovered the early work of Nathan Zuntz (1847–1920). His publications *Höhenklima und Bergwanderungen in ihrer Wirkung auf den Menschen* (The Effect of High Altitude Climate and Hiking on Human Beings, 1906) and *Zur Physiologie und Hygiene der Luftfahrt* (On the Physiology and Hygiene of Aviation, 1912), in particular, sparked their interest in the life and work of this scholar, especially since no comprehensive account and scientific evaluation of his life and work had yet been published. Of course, Jokl (1967: 321–328) and Wittke (1970: 295–298) had previously noted the importance of his work, above all in their history of high altitude physiology. However, the breadth of Zuntz's research and its influence on the development of aviation medicine, as well as his fundamentally new thoughts and methodical structures, had at best only begun to be explored.

In 1985 I was asked to discuss the life and work of Zuntz as part of a medical thesis. The work was completed in 1989 (Gunga, 1989), and excerpts of this thesis were later published in English (Gunga and Kirsch, 1995: 168–171, 172–176).

The following chapters will deal with the reconstruction of the biography of Zuntz, of his scientific work, and of the influence these disciplines had on the early development of high altitude physiology and aviation medicine. The following questions will be addressed:

- Were the results of Zuntz's research in Berlin his achievement alone?
- Where did his work rank both in a national context and abroad?
- Was there already a clear awareness of the problems requiring research on aviation medicine in Germany before 1930, on the basis of which this branch of research was able to unfold over the following years?
- If so, what form did this awareness take, and what part did Zuntz play in its development?

These are the principal questions guiding the present book.

Notes

1. The medical historian Rolf Winau was born on February 25, 1937. Under his leadership, the Institut für Medizingeschichte (Institute for the History of Medicine) at the Freie Universität Berlin (Free University of Berlin) became one of the largest institutes in Germany in this field. After being awarded the degrees of Doctor of History and Doctor of Medicine, Winau was appointed Professor at the Free University of Berlin in 1976. Together with Gerhard Baader and other members of his staff, he wrote many books and a large number of articles on the history of pharmacology, biologism, and medicine under National Socialism, as well as on the history of medicine in Berlin. He published a total of 219 books and articles between 1965 and 2005. After the Berlin Wall fell in 1989, Winau actively participated

in restructuring the different university institutions in Berlin. In particular, he was involved in founding the Zentrum für Human- und Gesundheitswissenschaften der Berliner Hochschulmedizin (ZHGB, Berlin Center for Humanities and Health Science). He was appointed as its founding rector in 1997 (Winau, 2005). R. Winau retired in March 2006 and died on July 15, 2006 in Berlin. His funeral, attended by many friends and companions and reported on widely in the media, was held on July 28, 2006, at the cemetery of St Matthias Church.

1 Biography

1.1 Genealogical overview

Nathan Zuntz was born in Bonn on October 6, 1847 (City Archives, Bonn). His father was the Bonn merchant, Leopold Zuntz. His mother Julie, née Katzenstein, had been born and raised in Kassel.

Nathan Zuntz's forefathers originated from Zons am Rhein, a city located on the River Rhine formerly known as Zuns (Figure 1.1). It was located on the left shore of the Rhine, about 20 km northwest of Cologne. Deeds record that, starting in 1372, the village of Zuns – originally under the control of the archbishop-elector of Cologne – was authorized to levy a toll on ships transporting cargo on the river, referred to as the *Rheinzoll* (Rhine toll), and it was granted its town charter in 1373. Zons was incorporated into Dormagen in 1975 as the city district *Feste Zons* (Zons Fortress).

Towards the end of the fifteenth century, the Jew Pesach von Zons (†1488) settled in Frankfurt am Main together with his wife and son, Michel. At that time, it was not unusual within Jewish families that the name of the home town be added to the first name. However, the name "Zons" left room for ambiguity, given that it existed in a number of variations and diverse spellings. Thus, the generations succeeding Pesach were recorded under the name of Zunz instead of Zons. Additional confusion results from the fact that the spellings Zunz and Zuntz were sometimes even used in parallel.[1]

Figure 1.1 Historical picture of the town Zons anno 1575.
(Blüm-Spieker, 1984, p. 7)

The Zunz family remained settled in Frankfurt, where they earned their livelihood primarily as small merchants. Like all Jews in Germany at this time, they were subject to various restrictions and regimentations affecting their choice of occupation so that hardly any professional opportunities outside of merchant trading and banking were open to them.

In 1813, Nathan Zuntz's grandparents, the merchant Amschel Zuntz (1778–1814) and Rechel Hess (1787–1872), were married in Frankfurt. Rechel came from Bonn, where her father had operated a coffee and colonial goods shop since 1783. The marriage was brief: just a few days after the birth of his son, Leopold, in September of 1814, Amschel Zuntz died. His young widow took her child and moved back to her parents in Bonn. Rechel Zuntz was described as a "fanatically pious Jewess," and thus it was important to her that her son be taught Hebrew at an early age and become familiar with the Talmud. Consequently, Jewish rules and customs were also strictly observed in the Zuntz household – habits which Leopold retained throughout his entire life, even though he distanced himself from the Jewish religion in later years. Despite the fact that Leopold Zuntz was interested in literature and philosophy even as a youth, and expressed the desire to become a scholar, his assertive mother demanded that he become a merchant, and thus he completed a corresponding apprenticeship. After the death of her father Rechel took over his shop in 1837, establishing, with her son (Lehmann-Brune, 1997, pp. 179–180), the company "A. Zuntz seel. Wb." This name was later slightly altered to A. Zuntz sel. Wwe. (*des seligen Amschel Zuntz Witwe* – widow of the blessed Amschel Zuntz) to reflect modern spelling.

By the time Leopold married Julie Katzenstein in 1846, he had already been entrusted with much of the management of the business. At this point in time, coffee-bean roasting had already become the main focus of the company's business. Although the company later grew in size and gained quite a reputation, its economic situation during the first decades of its existence was always precarious. The remarkable rise of the company did not begin until the 1870s, when Leopold Zuntz's sons and grandchildren entered the company's management. The first store in Berlin was opened in 1879, and ten years later another branch was inaugurated in Hamburg. "Zuntz Kaffee" developed into a well-known brand, and in 1893 the company was awarded the coveted title of a supplier of the Imperial Court (*kaiserlicher Hoflieferant*). Following the Berlin Industrial Exhibition of 1896, at which the Zuntz pavilion – an interesting structure in terms of its architectural design – offered a cup of coffee with pastry for ten *Pfennig*, the company became very popular in Berlin. Since the idea of the "cheap cup of coffee" had enjoyed such success, the first coffee sales outlet with a coffee bar was opened at Berlin's Spittelmarkt; this was similar to the "coffee to go" outlets found throughout the world today. Additional Zuntz coffee specialty shops with coffee bars were rapidly opened in all major streets of Berlin, Potsdam, Dresden, Hanover, and Cologne. At the same time, business operations had expanded, and the company was importing tea from India and Ceylon.

Following World War I, the company grew by leaps and bounds: by the end of the 1920s, "A. Zuntz sel. Wwe." was a flourishing company with 800 employees, numerous stores and its own cafés. In 1925 the famous company emblem was also developed – the *Biedermeier* woman with the wide-brimmed bonnet tied down around her face (Figure 1.2). However, the rise to power of the National Socialists in 1933 brought far-reaching changes: in the course of the discrimination against Jews in Germany and their economic disempowerment, the business was forcefully "arianized" and the last principal stockholder, August Zuntz, emigrated to England.

After the end of World War II, a member of the Zuntz family once again assumed management of the company, bringing about its gradual revival. In the early 1960s, the company employed a staff of 300 in Berlin and 80 at the Bonn branch. On October 26, 1962, the company's 125th anniversary was celebrated in grand style and the coffee house "Zuntz im Zentrum" was opened next to the famous Café Kranzler on Berlin's Kurfürstendamm. In the late 1970s, A. Zuntz sel. Wwe. was sold to the companies Dallmayr and Darboven (Lehmann-Brune, 1997, pp. 182–197).

Figure 1.2 The famous company emblem "A. ZUNTZ SEL. WWE" showing a woman with the wide-brimmed bonnet tied down around her face (*ca.* 1925). (Lehmann-Brune, 1997, p. 197)

1.2 Childhood and youth

When Leopold Zuntz married, his mother Rechel extended her influence to include his family life, living in the same house as her son and his quickly growing family. Between the years 1849 and 1863, Nathan's ten siblings – Albert, Carolina, Jeanette, Emma, Simonetta, Anna, Joseph, David, Mathilde, and Siegmund Richard – were born (Figure 1.3). All the children were raised in strict observance of religious principles, not least out of consideration for Leopold's mother. She was the one who gave Nathan Hebrew lessons when he was only four years old, and who ensured that Jewish rules were followed precisely. The degree to which the atmosphere in the Zuntz house was influenced by the pious and stringent Rechel is particularly vivid in the complaints voiced by Julie Zuntz, who felt compliance with Jewish dietary laws to be quite a challenge. In contrast to his mother, Leopold Zuntz is described as a weak and sickly person. He was also plagued by financial worries, as the profits generated by the coffee-bean roasting shop were meager and only sufficed for a very simple and frugal life for his large family.

Nathan, the first-born child, was of a delicate constitution and consequently did not begin school until seven years of age, attending the grammar school in Bonn in 1854. Nevertheless, he was a good pupil, thirsting for knowledge so strongly that Leopold allowed his talented son to take additional Latin lessons. In 1858, Nathan Zuntz enrolled in the sixth grade (*quinta*) at the local high school (*Gymnasium*). His father approved of this step and always supported his son in his desire to learn, yet he couldn't protect Nathan against the initial difficulties he experienced at the new school. In contrast to his well-off classmates, he often had to help out with household chores and in the store after school; in addition, he was shy and often ill. His daughter Emma Zuntz noted:

> *Nathan celebrated his Bar Mitzvah at thirteen years of age. A comprehensive instruction in the Jewish [theological] teachings preceded the ceremony ... This Bar Mitzvah of the first-born son was a large family celebration.*

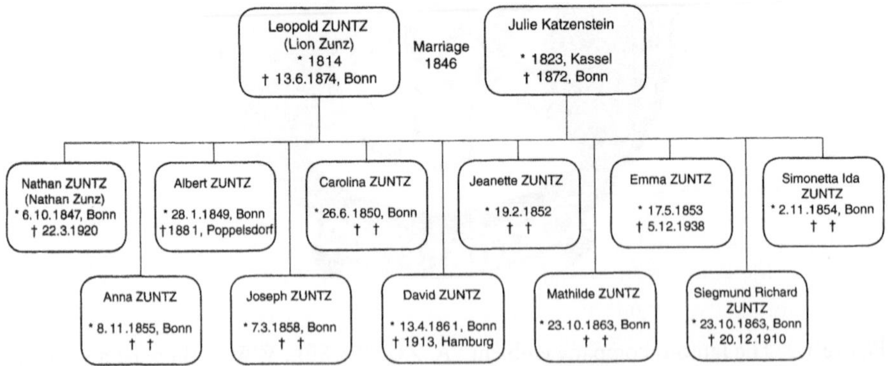

Figure 1.3 The siblings of Leopold Zuntz (1814–1874) and Julie Katzenstein (1823–1872).

The boy thanks his honored teacher in a long speech held in Hebrew. That was the future professor's first oratorical success.

(Emma (Sarah) Zuntz, undated, pp. 9–10)[2]

In 1863 – shortly before passing his *Abitur* (final high school examination) – Zuntz stopped attending high school and began an apprenticeship as a banker in Bonn. The reasons behind this decision remain a mystery, and may possibly never be resolved. According to von der Heide (1918, p. 332), the decision was made at the urging of relatives. They would have taken it as a matter of course that the oldest son would complete a merchant apprenticeship and later assume the family business. However, Zuntz's training at the Bonn bank was brief; after only a few weeks he was forced to give up the apprenticeship position because "he behaved poorly by spilling an inkwell over the record book and was subsequently released on grounds of inaptitude." Apparently his father stood steadfastly by his side in the matter, and did everything within his power to ensure that his son could return to the *Gymnasium*. Later, it was Nathan's younger brother, Albert, two years his junior, who was trained as a merchant and appointed by his father as successor to the company.

After returning to high school in Bonn, Zuntz passed his *Abitur* in the autumn of 1864 at the age of seventeen. He did not mention the brief interruption of his high school period either in the *curriculum vitae* set out in the appendix of his dissertation or in his handwritten resumé of 1870.

1.3 Professional and scientific development

1.3.1 The years in Bonn until 1880

After having obtained the necessary qualifications to attend a higher institute of learning, Zuntz enrolled at the University of Bonn,

and soon after was accepted as a student by the then rector magnificus Prof. Bauerband and registered by the Dean of the local faculty, Prof. Pflüger, in his album of students. During the following 8 semesters he studied medicine. His teachers were Messrs. Binz, Busch, Doutrelepont, Finkelnburg, Greef, Hildebrand, Landolt, v. Lavalette, Naumann, Obernier, Pflüger, Pluecker, Preyer, Rindfleisch, Ruehle, Saemisch, Schaafhausen, Max Schultze, Schroeder, Veit, M. I. Weber.

(Curriculum vitae, *N. Zuntz, July 6, 1870, Archives of the Rheinische Friedrich-Wilhelm-University Bonn)*

October 6, 1864, can be determined as the date on which Zuntz enrolled as a student – his eighteenth birthday – and August 13, 1868 as the date on which he left university.[3] It has not been possible to find any other documents pertaining to Zuntz's studies at the university (for example, on the seminars or courses he attended), and nor is any information available on his degree except for the certificate recording that N. Zuntz was awarded the degree of Doctor of Medicine.[4] Therefore, any biographer must depend on the *curriculum vitae* Zuntz wrote in 1870 and on the undated handwritten notes of his daughter Emma (Sarah) Zuntz, which she wrote for her children who wanted to know more about their grandfather, for information on this phase of his life. She noted:

> *[His] time as a student at university was different from the happy and carefree [times] we experienced, who did not know the troubles of life. But it was a wonderful time of growth and development. His father was very concerned about how his delicate, sensitive son was going to deal with the frights of anatomy. On the advice of more experienced people, he demanded that Nathan learn to smoke, to which he had an insurmountable aversion. As it turned out, he could tolerate the smell of corpses if only he could stop smoking. As early as the second semester, the young student became an anatomist, earning the tidy [sum] of 100 Marks per semester, and was laying the foundations of his precise knowledge of the human body. [His] father took great and loving interest in his work and influenced his development. During the phase in which young doctors, intoxicated with their new insights, could easily become materialists focussing only [on one thing], he read Spinoza's writings with his son and broadened his perspective from the small details to look at life's grand and eternal issues. Until his death, he remained the young scholar's best friend, his shining example.*
>
> *(Emma (Sarah) Zuntz, undated, pp. 10–11)*

Further on, Emma (Sarah) Zuntz continued:

> *At the time, universities and colleges were not yet overrun by students and individual students did not disappear in the masses. The anatomist Schulze and the clinician Ruehle soon noticed the extraordinarily gifted young Zuntz. It was Schulze who arranged for him to make the acquaintance of Pflüger, who was later to become his brilliant master and teacher. Schulze sent the young student to the* Physiologisches Institut *[physiological institute] with an abnormal, excellently dissected brain, and there the first of many stimulating discussions began, which later informed so much of both of the men's work. Zuntz was also popular among his fellow students, in spite of the fact that he could not be a part of their jolly high jinks because of his lack of means. He always found hiking companions for his trips into the Siebengebirge mountains, even though he never stopped at a tavern. He considered himself to be extraordinarily blessed that he was allowed to study and to do research while Albert (his brother) was preparing to support [his] father in their grandfather's firm, [and felt this so strongly] that he would have considered it sinful to spend even one penny on something as frivolous as his personal pleasure. As a youth among his youthful friends, he had already developed that peculiarity that I was always aware of as his child. He, to whom the friendship and trust of people came so easily – he did not need it. They were the ones to seek him out, they could not do without his calm and rational thinking, his selflessness. He found his closest relationships in the Levi family. There he, the most unmusical of all people, became an intimate friend of the famous cellist, Jacques Rendsberg.*
>
> *(Emma (Sarah) Zuntz, undated, pp. 11–12)*

Zuntz was made Schultze's[5] assistant in the course of the fourth semester of his studies. He particularly emphasized the guidance Schultze gave him on working with the microscope during this period. Zuntz writes that during his final four semesters he spent most of his free time in the university's physiological laboratory. Pflüger "was so very kind" to help him there (*Curriculum vitae*, N. Zuntz, July 6, 1870). Later, Zuntz dedicated a copy of his dissertation he gave to his teacher "with respect and gratitude" (Zuntz, dissertation 1868), as he did with his book on *Höhenklima und Bergwanderungen in ihrer Wirkung auf den Menschen* (The Effect of High Altitude Climate and Mountain Hiking on Human Beings) (Zuntz *et al.*, 1906), which he gave to Pflüger on the occasion of the latter's fiftieth anniversary of his doctoral degree. In the course of his final two semesters, Zuntz worked as an *Unterarzt* (resident physician) under Ruehle at the *Medizinische Klinik* (Medical Faculty of the University Hospital) in Bonn. He wrote that, in this position, he had "more opportunities to work with the practical aspects of our science" (*Curriculum vitae*, N. Zuntz, July 6, 1870).

After passing his *Rigorosum* (oral examination for the doctorate) with the grade *summa cum laude* (N. Zuntz, Certificate of Doctor of Medicine, July 31, 1868, Archives of the Rheinische Friedrich-Wilhelm-University Bonn), Zuntz was awarded the degree of *Doctor Medicinae et Chirurgiae* from the University of Bonn by Schultze on July 31, 1868. His inaugural dissertation was published under the title *Beitraege zur Physiologie des Blutes* (Contributions on the Physiology of Blood) (N. Zuntz, Certificate of Doctor of Medicine, July 31, 1868). In December 1868, Zuntz passed the state examination and was licensed to practice as a physician (von der Heide, 1918, p. 332). Immediately after having received the license, Zuntz substituted for a fellow doctor who had fallen ill, working in a large rural medical practice near Bonn.[6] Von der Heide states this practice was located in Oberpleiss am Siebengebirge (*Curriculum vitae*, N. Zuntz, July 6, 1870).

Emma (Sarah) Zuntz described the situation for Zuntz as follows:

> *Ruehle and Pflüger both offered the position of assistant to him. After a prolonged inner struggle, he decided to work for Pflüger, but did not commit himself right away because he first wanted to spend one winter in Berlin to continue his professional training. He earned the money for this by [working] as a country doctor. Grandfather often told me about his first practice and made fun of [himself as] the young doctor. The roads across country were long and any trip was an arduous undertaking, so that in some cases, the farmers would send a horse. He had never ridden [before] and he felt as unsure of himself on the back of a horse as he did in the practice. In the city, he had been able to ask for advice from his teachers in any difficult case, [but] here he had to depend on his own knowledge and his own initiative. However, his farmers were very pleased with him. He performed complex operations under extremely difficult conditions. To cite but one example, a cesarean operation took place in a barn by the light of an oil lamp and without any help in administering anesthesia. Even when he had reached old age, he still shuddered at the thought of the responsibility he had been given, at an*

age when our boys were just leaving school. But the difficult time passed and the young doctor traveled to Berlin.

(Emma (Sarah) Zuntz, undated, p. 13)

Zuntz practiced in Oberpleiss am Siebengebirge until September 1869, when he terminated his activities as a country doctor and went to the Charité and other institutions in Berlin for the winter semester.[7] There, he attended clinical courses and seminars held by Graefe, Frerichs, Virchow, Westphal and Traube. He writes that he spent time in their institutes and pursued mathematical studies (*Curriculum vitae*, N. Zuntz, July 6, 1870). According to Emma (Sarah) Zuntz, he was impressed by Traube but a little bit disappointed by Virchow:

Of [all of] the famous professors of medicine, Traube made the greatest impression on Nathan. He attended his lectures each day and was allowed to take part in all of his ward rounds. On the other hand, Virchow did not fully do justice to the expectations that he had associated with this great name. But while studying with Virchow, he did become acquainted with Senator, with whom he worked on fever curves. He felt his most enjoyable times in Berlin were spent in the company of his aspiring fellow students of medicine. The young doctors, nearly all of whom made a name for themselves and became famous, founded an association. They met in the small rooms they had rented, sitting on the host's bed, table and chair. One of them would present the recently published papers, or describe his plans for the future. These meetings were the breeding ground for the Berliner Physiologische Gesellschaft *[Berlin Physiological Society]; grandfather was its chairman for a long time. These scientific evenings were very serious and dignified. But youthful cheer also came into its own in the raisonneurs' bar (and tavern) society. While the same people were in attendance, it was wit and jokes that ruled the evenings, and if anyone had any weaknesses he was mercilessly teased. The food was very simple. Everyone brought their own supper, in their coat pocket, and then they shared and traded. One evening Zuntz saw beautiful apples lying in a basement close to the meeting place. Oh, he thought, I will bring dessert; the others will have brought enough staples. But as they all emptied their pockets, nothing but apples rolled in every direction that evening. Everyone had had the same brilliant idea.*

(Emma (Sarah) Zuntz, undated, pp. 15–16)

When he returned from Berlin, Zuntz began working as Pflüger's assistant in Bonn on April 1, 1870 (*Curriculum vitae*, N. Zuntz, July 6, 1870). It can safely be said that, by his work in neurophysiology and metabolic physiology, Pflüger[8] had the strongest influence on Zuntz as a young man. He opened the doors for Zuntz to a scientific career at the university, and likely found Zuntz to be one of his most talented and versatile students. When Zuntz completed his medical studies in 1868, the *Chemisch-physikalische und physiologische Laboratorium der Königlich Landwirthschaftlichen Akademie Poppelsdorf* (Chemico-Physical and Physiological Laboratory of the Royal Agricultural Academy of Poppelsdorf) near Bonn had just been established in the previous year, as had the Institute for Chemistry run by Kekulé (Figure 1.4).[9]

Figure 1.4 *Die Königliche Landwirthschaftliche Akademie Poppelsdorf* (The Royal Agricultural Academy Poppelsdorf) around 1868.
(Archives of the Rheinische Friedrich-Wilhelm-University, Bonn)

In 1872, the Anatomical Institute was moved from the building at the *Hofgarten* castle grounds (which now houses the *Akademisches Kunstmuseum*) to Poppelsdorf.[10] In Bonn the subjects Anatomy and Physiology had been taught separately since 1859, and Pflüger had been appointed to the newly-created chair for Physiology while still a young scholar.[11] This was how, in the course of the next two decades, a center for natural science developed in Poppelsdorf near Bonn from the simplest of beginnings.

Following his return from Berlin, Zuntz became Pflüger's assistant on April 1, 1870, after he had already published several scientific treatises on the acid–base balance of the blood – some of them together with Pflüger.[12]

Only two months after beginning work as an assistant, on July 4, 1870, Zuntz filed an application with Beseler, the University Curator, for permission to begin the *Habilitation* procedure – in which several formal requirements must be met in order to become a university lecturer – in the field of physiology.[13] It is very likely that Zuntz would have been supported by his doctoral advisor Pflüger, who held his former student and current assistant in high esteem regarding his scientific qualifications. In a recommendation in support of Zuntz's *Habilitation* project, he wrote:

> *I know Dr. Zuntz very well, who has worked at the Physiological Institute for four years. He is a very hard-working, knowledgeable and talented young man who has already made findings and gained insights that are* widely *recognized as valuable facts and laws and have been included in all physiology textbooks. In promoting him, we will gain a force that we will be sure to value.*[14]

On July 7, 1870, Zuntz's application was accepted by the University Curator (Letter from the University Curator to N. Zuntz, July 7, 1870, Archives of the Rheinische Friedrich-Wilhelm-University Bonn). On July 9, 1870, Zuntz addressed a letter to the Dean of the Faculty of Medicine, Rindfleisch, requesting "most humbly that you recommend my admission to the *Habilitation* procedure in order to become a university lecturer in the high Faculty of Medicine" (Letter from Zuntz to the Dean of the Medical Faculty, July 9, 1870, Archives of the Rheinische Friedrich-Wilhelm-University Bonn). Zuntz's letter to the Dean closed with the statement that he intended to study physiology and physiological chemistry, and that he wished to lecture in these subjects.[15] His inaugural lecture was given before the faculty that same month, on July 29, on the subject "*Die Ursache der Atembewegungen*" (The Causes of Respiratory Movements) (Letter from Dean Rindfleisch to the University Curator Beseler, October 1, 1870, Archives of the Rheinische Friedrich-Wilhelm-University Bonn). Following the lecture, a colloquium was held "to the greatest satisfaction of the faculty" (Letter from Dean Rindfleisch to the University Curator Beseler, October 1, 1870), "so that Dr. Zuntz was unanimously admitted as a lecturer."

The outbreak of the Franco-Prussian War of 1870/1871 caused an interruption to the young scientist's career. According to von der Heide, Zuntz volunteered for medical service in the Bonn military hospital (von der Heide, 1918, p. 332). He had most certainly not forgotten his practical medical skills in the meantime, since he had been working as a consultant doctor alongside his work at the university following his return to Bonn from Berlin (Loewy, 1922, p. 3). The years that followed must have placed Zuntz under extraordinary strain in terms of his career and family. Emma (Sarah) Zuntz noted:

> *She [his wife] was not unaware of the problems her husband was having at work; her quick mind helped her understand the situation. Pflüger took great pleasure [in the company of] his favorite student's wife and enjoyed sharing his plans and thoughts with her. It was a merry company in the small university town and Frieda (his wife) enjoyed it to the full. For her husband, whose scientific research and large practice kept him breathlessly busy, it was almost too much of a good thing. Consequently, he fell asleep in a corner one evening while his wife was dancing, until Pflüger woke him and teased him about his lack of wakefulness and jealousy. Nathan was far from having such thoughts; how he enjoyed the cheerfulness of his young wife. The household was indeed run in an original way: everything the young professor earned he threw into a money box, and the housewife took what she needed. And if they were ever low on funds, then they dined on* Blutwurst *[blood sausage] and potatoes without the jolly mood suffering.*
>
> *(Emma (Sarah) Zuntz, undated, p. 26)*

In late 1872, Zuntz took a position as a teacher at the *Königliche Landwirtschaftliche Akademie Poppelsdorf* (Royal Agricultural Academy Poppelsdorf).[16] His mother died the same year; and his father, Leopold Zuntz, died two years later in 1874.[17] This meant that Zuntz was responsible for his younger siblings[18] along with his own family, having married Friederike/Frieda Bing in 1874 (L. Zuntz, 1926, pp. 201–202) (Figure 1.5). Furthermore, Pflüger

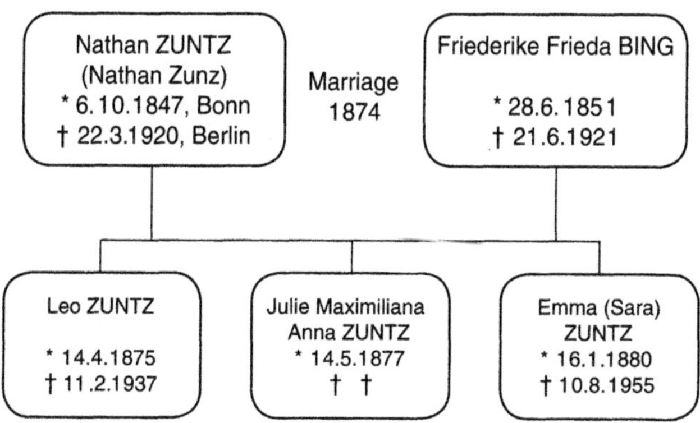

Figure 1.5 The siblings of Nathan Zuntz (1847–1920) and Friederike Frieda Bing (1851–1921).

had petitioned that Zuntz be promoted to *ausserordentlicher Professor* (full professor without tenure) at the Physiology Department of the Medical Faculty in spring of 1874 (Letter from Pflüger to the faculty in Bonn, February 7, 1874, Archives of the Rheinische Friedrich-Wilhelm-University Bonn). While, in his opinion on Zuntz written in 1870, Pflüger had focused on his assistant's scientific qualities, this letter to the department is an interesting source for understanding how Pflüger judged Zuntz's personal traits and character:

> *Dr. ZUNTZ is a talented, knowledgeable scholar who submits solidly researched work. He has proven himself to be a useful and valuable member of our faculty, which is why I believe that he deserves to be recognized and encouraged by us. Considering that his [family] situation has forced him to run a time-consuming medical practice to earn his living, and that in his role as the 1st Assistant he has provided guidance to the physiology [department's] students in preparing their scientific papers with great sacrifice and intelligent care, and moreover that without his contribution my research protocols could not have been prepared, then one must acknowledge that in their number and as regards their scientific value, Dr. ZUNTZ's publications are worth double, in particular since he has written them in only a few years' time.*
>
> *(Letter from Pflüger to the faculty in Bonn, February 7, 1874)*

The opinion sets out a list of Zuntz's publications and those term papers to which he contributed "a commendable share" (Letter from Pflüger to the faculty in Bonn, February 7, 1874). It bears noting that Pflüger referred to Zuntz's dissertation as a "work that has become famous."

> *Finally, where Dr. ZUNTZ's abilities as a teacher are concerned, he has explored almost all fields of physiology and is much appreciated by the students.*
>
> *Without any doubt, his activities as a lecturer would play an even greater role at the department if the students here had more time to attend lectures*

> *that are not directly required for completing [their] medical studies at university.*
>
> *For these reasons and in view of the fact that Dr. ZUNTZ has now been a lecturer without tenure for more than three years, and has furthermore constantly distinguished himself by his truly scientific mind, by the purity of his character and his loyalty to his colleagues, I hereby petition that the high faculty support and favorably advance the promotion of our lecturer and my first assistant of many years' standing to* professor extraordinarius *with his Excellency, the* Minister fuer geistliche, Unterrichts- und Medicinal-Angelegenheiten *[Minister for Intellectual, Instructional and Medical Matters].*
>
> *(Letter from Pflüger to the faculty in Bonn, February 7, 1874)*

His colleagues at the university approved Pflüger's petition. Rindfleisch added a commentary to the letter, stating that Zuntz was "a true asset" of the faculty and to science in general.

Five months later, on July 29, 1874, the Medical Faculty was informed that Zuntz had been appointed *ausserordentlicher Professor* (full professor without tenure) by the responsible minister (Letter from the *Minister fuer geistliche, Unterrichts- und Akademische Angelegenheiten* addressed to the Faculty of Medicine in Bonn, July 29, 1874, Archives of the Rheinische Friedrich-Wilhelm-University Bonn). Zuntz's appointment as Prosector at the Anatomical Institute of the University of Bonn followed in the fall of 1874 (von der Heide, 1918, p. 333).

During these years, Zuntz had issues of his own that he had to deal with:

> *The friend (of his wife) had a brilliant, but unhappy nature. She and her husband (Oppenheim family) were skeptics. In their manner of thinking, they negated and picked apart all existing values. [It was] a dangerous game that they sometimes even played in real life. Nathan, whose life so solidly rested on duty, work and his love of mankind would have dearly preferred to keep his wife away from these dangerous influences. But he himself was unable to completely evade the dangerous charm that the Oppenheim couple exuded. This was the* Gruenderzeit *[the historical "period of promoterism," of rapid industrial expansion in Germany between 1871 and 1873]. Oppenheim wanted to become rich quick, and all of his friends were to share his luck with him. Later Nathan was not sure whether he himself had believed in his friend's plans – most people thought he was nothing but a brilliant swindler – in any case Oppenheim fraudulently fell into bankruptcy, and Nathan lost a share of his assets just shortly before his second child was born. He earned enough to recover from the financial loss soon enough, but the damage this did to his faith in others was serious to someone who thus far had blindly trusted his fellow man. Add to this his great concern for the poor wife Anna (Oppenheim), who, in her distress and panic resorted to morphine, from which she suffered an unhappy death. These tragic events certainly were a factor affecting Frieda's (Nathan's wife) health. Though she pulled herself together in her husband's presence, her agitated nature and the volatility of her character increased more and more. She struggled against this, not wanting to be pulled down into the depths of the black night, but fate took its*

> *course. The heavy thoughts took her in their vise and Nathan had to commit his beloved wife to a mental institution in Nassau with the sad knowledge that she was carrying his child.*
>
> *(Emma (Sarah) Zuntz, undated, p. 27)*

Much later, he gave his daughter the following explanation as to how he was able to overcome all these difficulties: "As a growing girl I once asked father what made his life so happy despite all the difficulties. He answered, work and love" (Emma (Sarah) Zuntz, undated, p. 31).

Zuntz's son Leo was born in the spring of 1875 (Birth certificate no. 340, April 14, 1874, City Archives, Bonn), and was later to become a physiologist and doctor like his father. Zuntz's two daughters Julie Maximiliana Anna (Birth certificate no. 449, May 14, 1877, City Archives, Bonn) and Emma (Sarah) were born in the following years in Bonn (Figure 1.5), before the family moved to Berlin. Emma (Sarah) Zuntz noted:

> *Nathan was not someone who liked to occupy himself with small children. He had become a father at such a young age; he knew we would be well taken care of by Aunt Lisa, and so he entered into a close, personal relationship with us relatively late [in life]. In my early childhood he stood over me like an avenging god. I often teased him and told him that I was always relieved to hear the front door close behind him when he went on one of his frequent trips. This memory shows me how badly he suffered under my mother's fate. We were more of a burden to him than a joy because the fear always tormented him that one of us might have to accept her unfortunate heritage. Our entire relationship was to serve to strengthen our nerves.*
>
> *(Emma (Sarah) Zuntz, undated, p. 39)*

Zuntz's continued professional development in Bonn can be summarized as follows. In the fall of 1872, Zuntz began to expand the experimental station at the *Königliche Landwirtschaftliche Akademie* (Royal Agricultural Academy), which had been founded in 1856, for his zoo-physiological experiments while continuing his scientific work with Pflüger at the university.[19] The laboratory and offices – which later became Pflüger's Physiological Institute in 1878, and which some colleagues at the university felt was too large[20] – were where Zuntz pursued his early research in his home town. Because of his work as a teacher of zoo physiology at the Royal Agricultural Academy in Poppelsdorf and his simultaneous function as Prosector for Anatomy at the University of Bonn, Zuntz acted as the link connecting the research at the university with its scientific focus and the practical orientation of the Academy.

Nonetheless, everyone at the Poppelsdorf Academy was aware that the university had a positive and encouraging influence on it. Von der Goltz, who acted as director of the Agricultural Academy in Poppelsdorf for a time, later remarked in a ceremonial address:

> *The Academy always felt it was a very special advantage that it was so closely tied, both in the actual sense of the term and in its metaphoric sense, with the Rhenish Friedrich-Wilhelms-Universitaet. All of us who are familiar with the*

> *circumstances are fully aware of the many great advantages and benefits, the intellectual stimulation and promotion that the teachers and students of the Academy enjoyed.*
>
> *(von der Goltz, 1898, p. 250)*

It had been a well-known fact, at least since the 1860s, when Liebig had published his vehement criticism of the educational standards existing at the agricultural academies,[21] that universities and agricultural academies in Germany urgently needed to collaborate. Historical events show that Zuntz successfully took it upon himself to act as mediator in this difficult situation.

When Thiel,[22] the provisional director and *Geheime Regierungsrat* (privy higher executive officer) of the newly founded *Landwirtschaftliche Hochschule* (Agricultural College) in Berlin, had to suggest a candidate for the chair in the field of veterinary physiology to the ministry responsible, he chose Zuntz. Thiel's decision may have been strongly influenced by the fact that Zuntz had previously advocated combining scientific research and practical experience, and that the tasks and problems extant in Berlin had a similar structure, requiring that the professor appointed to that chair be able to cooperate with others in an interdisciplinary manner.

The administrative part of this procedure in terms of university policy indicates that, in founding an independent Institute of Veterinary Physiology at the Agricultural College, the political circles responsible understood that veterinary physiology and the research done in this field were very important and needed to be promoted in every possible way (Wittmack, 1906, p. 160). In the course of his last years in Bonn, Zuntz published six further papers and studies that essentially dealt with the physiology of blood, of circulation and of the respiratory system. His *Beitraege zur Kenntniss der Einwirkungen der Athmung auf den Kreislauf* (Notes on the Effects that Respiration has on Circulation) (Zuntz, 1878d, pp. 374–412) are particularly noteworthy in this regard. They bear witness to the fact that, even at a very early stage, Zuntz took an interest in matters of respiratory physiology under varying circumstances in terms of barometric pressure and oxygen content. Zuntz stated that the first experiments he performed in the field of respiratory physiology were done in collaboration with Pflüger in Bonn (Zuntz, 1878d, pp. 374–412), just after Waldenburg had published his first studies on the pneumatic therapy of lung disease (Waldenburg, 1873, pp. 465–469). Around 1860, it became known in Germany that increasing barometric pressure in pneumatic cabinets had a therapeutic effect on people suffering from lung disease (Liebig, 1885, p. 292). These first indications were only dealt with in France in the early 1870s, when Bert began his experiments in pneumatic cabinets, which he summarized in the work that made him famous, *La pression barométrique* (Barometric pressure) (Bert, 1878).

The scientific community in Germany entered into a heated dispute as to whether or not the therapy developed by Waldenburg was effective.[23] Zuntz made himself heard in this debate. For him, the search for a method that could be used to therapeutic effect in clinical practice was the primary reason for studying this field (Zuntz, 1878d, pp. 374–412). Just before he had concluded his experiments on the effects that modifying air pressure had on the organism, the studies by Drosdoff and Botschetschkaroff were published on this matter (Drosdoff and Botschetschkaroff, 1875, pp. 65–67). Zuntz saw his criticism of Waldenburg's thesis vindicated by their work and therefore decided not to publish his own experimental results, as he wrote later (Zuntz, 1878d, pp. 374–412).

In the meantime, the fall of 1880 had seen the appointment negotiations for the newly founded *Landwirtschaftliche Hochschule* (Agricultural College) in Berlin enter into full swing. On September 22, 1880, Thiel submitted his petition to Lucius, the *Staatsminister fuer Landwirtschaft, Domaenen und Forsten* (Minister of Agriculture, State-owned Domains and Forests), that Zuntz be appointed to the Chair of Veterinary Physiology (Letter from Thiel to Lucius, September 22, 1880, No. 20075, fol. 45, Secret Central Archives).

After Thiel had paid several visits to Zuntz in Bonn in order to convince him to move to Berlin (von der Heide, 1918, p. 333), Zuntz declared in a letter dated December 20, 1880, that he was prepared to accept the appointment to the chair provided the following demands were met (Letter from Zuntz to Thiel, December 20, 1880, No. 20075, fol. 65–66, Secret Central Archives):

1. Wages of 7500 Marks per annum
2. Subsidization of his rent
3. Payment of the costs of his move
4. One assistant
5. Provisional workspaces in Berlin (apparently, his actual laboratory and office were not yet ready for occupancy).

On December 27, 1880, Zuntz received a preliminary notice from Lucius regarding his being appointed to the Chair of Veterinary Physiology. In this letter, the ministry accepted the demands that Zuntz had formulated and stated that he was to take office on April 1, 1881 (Letter from Lucius to Zuntz, December 27, 1880, No. 20075, fol. 67, Secret Central Archives). Zuntz referred to this letter from Lucius and accepted the appointment to the chair on December 31, 1880 (Letter from Zuntz to Lucius, December 31, 1880, No. 20075, fol. 68, Secret Central Archives). At this stage, the notice of his appointment to the chair could only have been issued as a preliminary notice, since it had not yet been established whether the university in Bonn was going to release Zuntz as per April 1, 1881 (Letter from Lucius to Zuntz, December 27, 1880). The negotiations as to that matter must have taken place in the course of January 1881, because Zuntz confirmed to Minister Lucius on February 5, 1881, that he finally and conclusively accepted the appointment to the chair of the Agricultural College in Berlin (Letter from Zuntz to Lucius dated February 5, 1881, No. 20075, fol. 82, Secret Central Archives).

It should be noted that – at least as far as the documents archived in the *Geheimes Staatsarchiv Preußischer Kulturbesitz* (Secret Central Archives) show – Zuntz's Jewish faith (*Umlagenrolle* of the Bonn Synagogue Congregation 1880/1881, current no. 23 – Nathan Zuntz) at the time did not constitute any reason that would have prevented his appointment to the chair of Veterinary Physiology, as was the case in later appointment procedures involving Jewish scholars.[24] Schulte states (Schulte, 1976, p. 549) that Zuntz left the Jewish faith some time after 1880, but the source he cites in this regard cannot be corroborated (Letter from the *Nordrhein-Westfälisches Hauptstaatsarchiv* to the author, dated August 19, 1987). Since the Jewish Congregation in Berlin also does not have any documents in this regard, it is unclear at what point in time Zuntz left the Jewish faith (Letter from the Jewish Congregation in Berlin to the author, dated August 12, 1987). The fact is that many years later, on October 23, 1889, Zuntz was baptized as a Christian,[25] and according to Emma (Sarah) Zuntz she obviously played a central role in this decision:

> *It was Poske and Stein who familiarized Nathan ever more with the fundamental moral strengths of Christianity ... He had deliberated for a long time whether or not he should have his children baptized. He considered Christianity to be a continuation of the Jewish belief and simultaneously a strong element of the German culture, to which he felt he owed his best [qualities]. He felt he was a German and thought the combination of [Judaism] with the Germanic was beneficial for both sides. Poske, who wanted to see his wife become a Christian, encouraged him in his ideas. But it was a terribly difficult decision for him. As a Jew, he had achieved everything he could; he was holding the position that made him happy. Yet his Jewish friends would hold his conversion against him at best, they would consider it to be a concession to the tyranny of the state. He wrestled with the thought and struggled for a long time. There were several aspects that finally convinced him. If he just stayed Jewish for the sake of his friends, then he would be giving in for the sake of man. What moved him more strongly was my deep inclination to Christianity. I had read [about] the life of Jesus in a Catholic children's Bible when I was seven years old, and since that time his life and teaching have determined my being, have ruled my feelings and thoughts. I so much wanted to become a Christian. When I was eleven, he suggested I be baptized. [I responded,] "I can only do that if you do it too. Nothing can separate me from you." Poske brought Nathan to the preacher Kirms, a liberal and highly educated man who considered the dogmas to be nothing but a form, albeit one hallowed by long-standing traditions. After long discussions, it was decided to take the big step. I [can still] see all of us riding to church in the carriage. When I saw how moved my father was, I knew that he was suffering, and realized dimly that it was for my sake. Our baptismal text was, "Blessed are the meek: for they shall inherit the earth." I could not but apply the words to father; it was as if it had been written for him. Elfriede approached my father after the baptism and said softly, "It was the most difficult for the two of us." Christianity is a doubly holy legacy for me because father decided to convert for our sake. He was himself a genuine Christ figure, a mediator between God and man.*
>
> *(Emma (Sarah) Zuntz, undated, pp. 37–39)*

1.3.2 The Berlin years until 1920

When Zuntz finally gave in to Thiel's urging, the course was set for a new era in veterinary physiology research in Berlin. In this context, von der Heide notes:

> *When Zuntz founded his institute for veterinary physiology, it was new and unusual for physiologists and medical specialists to acquire their knowledge from agricultural institutions. Today, we view all these disciplines as working intimately with each other. For the* Landwirtschaftliche Hochschule *[College of Agriculture] itself, the establishment of the veterinary physiological institute was of far-reaching significance. Formerly, all essential knowledge on animal care, breeding, nutrition, and so on, had been acquired through courses in animal husbandry and scientific zoology. Any relationship between practical agriculture and veterinary physiology was nonexistent, forcing students of agriculture to obtain all their information from physiology schools. Thus, Zuntz's appointment by* Geheimrat *[Privy Councilor] Thiel meant a reform of revolutionary proportions for the College of Agriculture and ensured direct cooperation between agricultural practice and scientific physiology.*
>
> *(Von der Heide, 1918, p. 333)*

In the spring of 1881, Zuntz moved to Berlin with his family (Figure 1.6). During his early years at the *Königliche Landwirtschaftliche Hochschule*,[26]

Figure 1.6 Nathan Zuntz and his family (undated photograph, *ca.* 1900). Nathan Zuntz is standing in the second line far right while the person on the far left may be his son Leo. The other persons in the picture cannot be identified.
(Private collection, G. Zuntz, Cambridge, UK)

Figure 1.7 The main building of the *Königliche Landwirtschaftliche Hochschule* (Royal Agricultural College) located in Invalidenstrasse in Berlin (1906). (Wittmack, 1906, p. 27)

he was able to continue his research only under the most trying and primitive conditions. He was allowed access to no more than a few rooms in the main building for research purposes (Figures 1.7 and 1.8) (Zuntz, 1909b, p. 473). During these initial years in Berlin, Zuntz, in his new position as director of the Institute for Veterinary Physiology, was primarily occupied with supervising its establishment and growth. Many years later, Plesch, a long-time colleague of Zuntz, portrayed the latter's laboratory as a "chaotic stable" (Plesch, 1949, p. 62) in his biography:

> *The Institute itself displayed a decrepit shabbiness beyond compare, and nevertheless, everything needed was there – all you needed was a flair for improvisation. I cannot recall a single complete device in the Institute; anything needed first had to be brainstormed and then built. The floor was littered with cracks as thick as one's finger and filled with spilt mercury; the animal stalls were in the work rooms; dismantled equipment threatened to fall on one's head. Cleaning up was totally out of the question – no one would have dared touch the delicate, fragile equipment!*
>
> *(Plesch, 1949, p. 62)*

Emma Zuntz noted:

> *He had dedicated his life fully to [his] work. He certainly was extremely lucky in being able to establish his science from its earliest beginnings. Not only did he design it as if he were an architect, he also shaped the individual components and consulted others. He was teacher and leader for a large group of students. Mehring was often upset that many of father's ideas were sailing under false colors: He would give a student an idea; he guided him in his composition,*

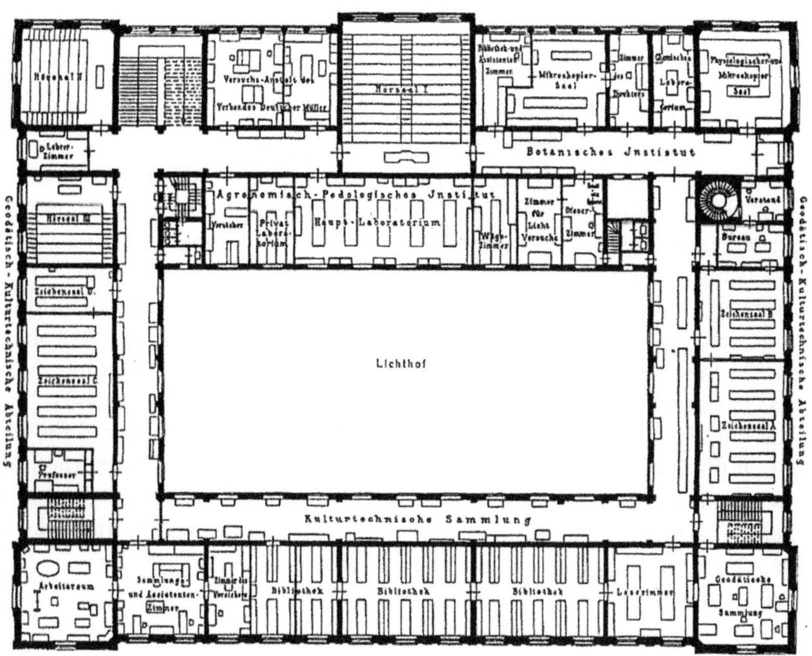

Figure 1.8 Floor plan of the second storey of the *Landwirtschaftliche Hochschule* (College of Agriculture), with the circular stairwell (upper left corner of the image) in which Zuntz, Schumburg and Loewy performed climbing experiments in addition to their field studies in the Alps and Tenerife (Monte Rosa).
(Wittmack, 1906, p. 29)

> *improved and modified the paper with a fine linguistic instinct, and, in the end, admired the other's work. And that is exactly what I appreciated so much in father's way of guiding others. But he was even happier when he found a highly talented young man. How he rejoiced at the successes Professor Neubig and the Viennese Durig achieved. When he was given his first institute at thirty years of age, it consisted of two rooms. Every year he expanded it, [and every year] the funding that the state approved became greater. In 1910, he and his staff moved into the large institute, which was constructed according to his plans [and] was the best equipped in the entire German Reich. On the one hand, this was owed to the importance of his scientific research, on the other, however, to his talent of putting the insights gained to use for agricultural practice.*
>
> *(Emma (Sarah) Zuntz, undated, pp. 43–44)*

Despite all the improvisation required, Zuntz was able to begin his experiments on the physiology of respiration and metabolism (Zuntz, 1882a, pp. 1–162; Zuntz and Geppert, 1886, pp. 337–338; Geppert and Zuntz, 1888, pp. 189–245) in the early 1880s in collaboration with Geppert.[27] These studies ultimately proved to be fundamental to the field. Since respiration came into play in all cases when testing metabolism, Zuntz and Geppert felt it would be sensible first to address the still obscure mechanisms that regulated breathing (Geppert and Zuntz, 1888, p. 189). In this context, two theories existed: one assumed direct nervous stimulation of the respiratory center; the other interpreted the accumulation of metabolic products in the bloodstream to be the stimulus for increased respiratory activity during physical exertion (Geppert and Zuntz, 1888, p. 189). In the course of this investigation, the two scientists invented the famous "Zuntz–Geppert respiratory apparatus"[28] (Figure 1.9) and formalized the equation to determine oxygen consumption during rest and exercise using expiratory (or inspiratory) volume measurements only, so that, according to Dempsey and Whipp (2003, p. 148), "metabolic rate could now be accurately quantified, allowing the relationship between ventilation and O_2 uptake, for example, to be established as a basis for understanding pulmonary gas exchange and the regulation of blood-gas and acid–base status."

In the initial stages of their studies, Zuntz and Geppert formulated the interesting hypothesis that when exertion begins, nervous impulses from the motor centers travel not only to the muscles but also simultaneously to the respiratory center (*zentrale Mitinnervation*, central co-innervation).[29] After concluding their tests, however, they came to the conclusion that respiration during exertion was not regulated by the nervous system but instead occurred as a

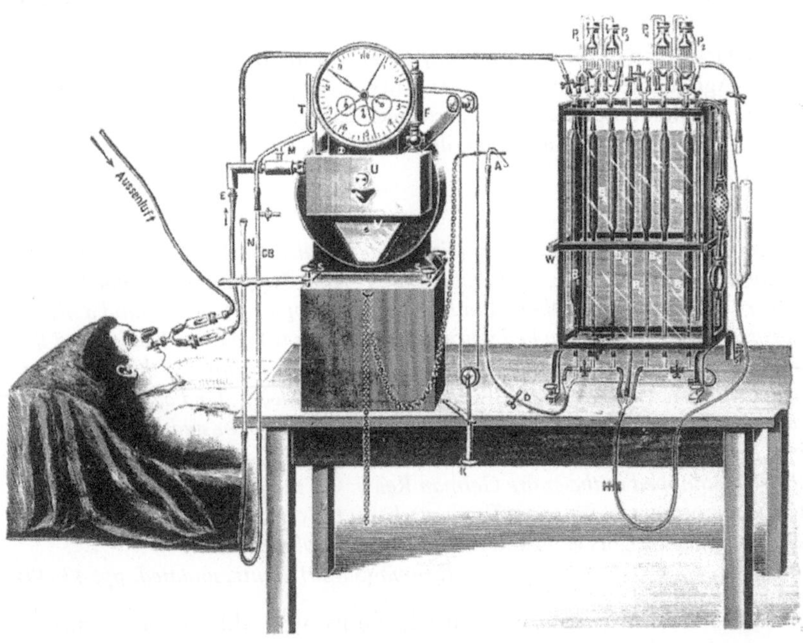

Figure 1.9 The Zuntz–Geppert respiratory apparatus in a typical experimental set-up (experiment with subject at rest): mouthpiece, valve mechanism for separating inhaled and exhaled air, gas meter, analysis apparatus.
(Zuntz *et al.*, 1906, p. 159)

result of "the blood absorbing unknown substances from the contracting muscles during exertion, this in turn stimulating the respiratory center" (Geppert and Zuntz, 1888, p. 244).

The neurophysiological aspects of this research topic obviously interested Sigmund Freud (1856–1939) when he visited the laboratories of Zuntz und Munk (1839–1912) on March 23 and 25, 1886. Although Freud felt warmly and honorably welcomed, he nevertheless was rather disappointed to see only "more or less blind dogs" (Tögel, 2006, p. 23).

During this same period, Zuntz and von Mering completed the investigations into the physiology of digestion that they had performed together in 1877 (Zuntz and Mering, 1883, pp. 173–221). Zuntz's introduction of the term "digestive process" to physiology can be considered a direct result of these studies (Zuntz and Mering, 1883, p. 221; Loewy, 1922, p. 8). It was as a result of his studies on respiratory, metabolic and nutritional physiology that Zuntz started to be interested in the physiology of strain and performance, which would later become the focal point of his research in Berlin. On closer examination, indications of a characteristic pattern of thinking become apparent, which can be found throughout his later scientific work. Zuntz's goal was to explore and record the parameters limiting the physiological capabilities of the organism while at the same time searching for solutions for improving the ability of both man and animal to perform.

With each passing year, the Institute for Veterinary Physiology, which was in the process of being established as a division of the College of Agriculture in Berlin (Figure 1.10), provided increasing opportunities for clarifying these issues.

Figure 1.10 Zuntz's Tierphysiologisches Institut (Institute of Veterinary Physiology) in 1906. (Illustrierte *Landwirtschaftliche Zeitung* 26 (1906), p. 54)

It is quite apparent that this research had pragmatic and economic significance, and Zuntz was well aware of this and in fact demanded it be part of his institute's considerations. In this regard, he and his co-author wrote in one of the first comprehensive publications to appear on this topic:

> *During our investigations on the metabolism of the horse which will be described here in after, we were guided by two motives. First, we wished to shed some light on questions of general physiological interest but also, by experimenting with one of our most valuable farm animals, we hoped to obtain usable results that could be applied in practical animal husbandry.*
> *(Zuntz and Lehmann, 1889, p. 1)*

Zuntz later emphasized the economic significance of his research on many occasions (Zuntz, 1899a, p. 3) and felt his findings were confirmed by the experimental results achieved by the *Compagnie des Voitures* (Coach Society) in France, among other scientific institutes (Zuntz, 1908c, p. 31). The scientific conclusions in his *Untersuchungen über den Stoffwechsel des Pferdes bei Ruhe und Arbeit* (Studies on the Metabolism of the Horse at Rest and at Work) (Figure 1.11), which he continued in the following years, were so fundamental to the field that they are held in high esteem even today (Kolb, 1974, p. 32; Klingeberg-Kraus, 2001, pp. 217, 219, 225) and have been adopted by standard textbooks of veterinary physiology (Scheunert and Trautmann, 1965, p. 543) and books dealing with the history of medicine and physiology.[30]

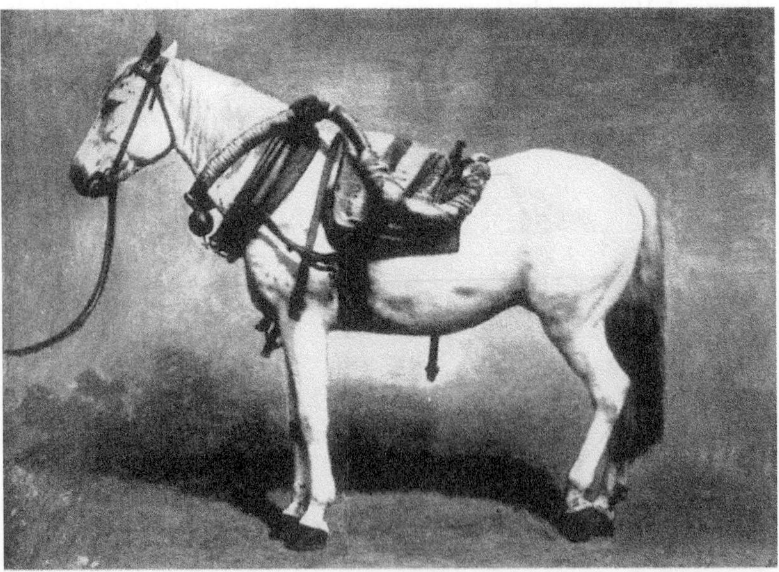

Figure 1.11 Zuntz's experimental horse "Barnabas" with its tracheal canula, tube, expiration valve, and saddle weighted with lead plates.
(Zuntz and Hagemann, 1898, Table IV)

As is evident from the celebratory speeches Zuntz gave on the occasion of the Kaiser's birthday, his research was not only driven by the economic considerations cited above but also, to a not inconsiderable degree, steeped in patriotism. Zuntz felt that science bore the responsibility of working ceaselessly for the prosperity of the people:

> *On this day that bestowed upon us the ruler and protector of our Fatherland, [the birthday of the Kaiser] we celebrate the stable unification of Germany under his powerful leadership, we rejoice in the beneficial and concerted efforts of all the forces springing up from German soil with one goal in common: achieving greatness, power and prosperity for our Fatherland. Naturally, we are well aware that the health of this great organism, this powerful empire, is only possible when all of its limbs and organs achieve that zenith of delight and purest joy in life which is the result and reward for faithfully performing our duty, for fully utilizing the strengths innate in mankind – and which spurs us on to repeatedly employ these strengths.*
>
> *We, the faculty of this university, are, in the eyes of many of our fellow aspirants, fortunate to have the opportunity to exert our powers along paths of our own choosing. Each one of us tills his most precious field, and would do the same even when following solely his inclination rather than the call of duty.*
>
> *Such a privileged position, however, results in the double obligation of applying ourselves body and soul to the task for which we have been selected to perform in accordance with our ruler's petition and the faith of his advisors.*
>
> *(Zuntz, 1899a, p. 3)*
>
> *(...)*
>
> *Even if my duty is performed as faithfully as possible, my work is but a small leaf; indeed, the activity of our entire college will seem but a tiny twig in the wreath woven by the united efforts of German science and German art, German agriculture, German industry and German trade for the prosperity of the Fatherland and to the honor of our ruler, who has ushered in peace, and who is called upon to guide [our nation's] destiny and whose birthday we celebrate today. Let us express our efforts and vows to devote our entire strength to the prosperity of the Fatherland and the fame of its ruler, by cheering His Majesty the Kaiser, our most gracious king and ruler, long live the King, long live the King!*
>
> *(Zuntz, 1899a, p. 16)*

One of Zuntz's outstanding qualities characterizing his work was the close relationship between theory and practice. This was apparent in the approaches he took during his early years in Bonn, and over the course of the 1890s this trait developed fully in a multitude of fields. The spectrum of his activities then ranged from the physiology of blood, circulation, and muscles (Zuntz, 1890a, pp. 337–341; Tangel and Zuntz, 1898, pp. 544–558; Zuntz and Hagemann, 1898, p. 1–338) to that of nutrition and digestion, animal breeding and starvation,[31] embryology,[32] and onwards to problems of climate and high altitude physiology (Schumburg and Zuntz, 1896, pp. 461–494). In the course of his studies, especially those on horses, Zuntz made direct measurements of the blood pressure in the pulmonary artery and aorta during rest and exercise. These were among the first measurements published in the scientific literature (Rowell, 2003, p. 114), and the methods applied by Zuntz were later (1929) used by Campos and associates at Harvard University to quantify the energy expenditure of dogs running on a treadmill (Tipton, 2003b, p. 219). By using Fick's principle for determining cardiac output, Zuntz was able to make far-reaching discoveries regarding gas exchange and cardiac performance.[33] Much later, following the end of World War II, Forssmann is said by Cournand[34] to have laid claim to these ideas originally developed by Zuntz.[35]

The animal experiments essential for his investigations at the Royal Agricultural College were met with bitter protests by the public. The discourse on the necessity and moral/ethical responsibility of animal experiments, a debate that was rekindled in the 1970s,[36] peaked in the 1880s and 1890s (Bretschneider, 1962, pp. 44–104). It speaks for Zuntz's civil courage that he did not evade the problem, but instead became publicly involved in this discussion and stated his position:

> *Should my previous arguments have succeeded in proving that our animal husbandry and manner of keeping them have been enhanced in a variety of ways by means of our physiology work and that the services of this science will be of immense value to it in the future, then I would already feel certain you would approve of the application of its indispensable methods. The fact that I here address the indispensable aspect of these methods, especially those which cannot be performed without certain harm being done to the well-being of the animal subject, seems justified in view of the fierce protests directed nowadays against vivisection, that is, against that encroachment upon the well-being and comfort of animals undertaken for scientific purposes. Perhaps the principal field of investigation in my laboratory can demonstrate more directly than any other the justification of vivisection. The investigations as to the metabolism and nutrition of animals, our efforts to examine and become familiar with these processes in detail and independent of the development of all the organs involved, compel us to analyze the function of these organs in a living body … I am not even mentioning the fact here that the operations are conducted painlessly on animals that have been anesthetized, (…), instead, I only want to emphasize the fact that these experiments, though they are, to an extent, certainly cruel to animals, bring about a continuous improvement of our ability to feed animals rationally and to utilize all nutriments available to us for their consumption.*
>
> *(…)*
>
> *Thus, for many animals, who might otherwise perish or who could have a less bountiful yet still complete diet, the conditions for their very existence and healthy growth are created as a result of the harm done to a few of their fellow creatures.*
>
> *In the natural course of events, countless creatures are maimed and killed atrociously in the battle for the survival of the fittest. Compared to this endless amount of suffering, the pain consciously inflicted by scientific researchers on individual animals is a mere drop in the ocean. This small amount of pain, however, achieves more favorable conditions for a much larger number of*

animals to survive and, to a great extent, for numerous human beings as well; indeed, it makes it possible for some of them to survive at all ... It is in particular the branch of physiology previously described, to which we have dedicated our especial attention, that has developed life-saving surgical methods which would not have come into existence so soon, if at all, without the preparatory work in physiology ... The situation is very similar in another field of experimental research of immediate interest to us: studies on the origin of epidemics and infectious diseases. In this case as well, it can be said that if we find a method to prevent, heal or at least mitigate the course of an epizootic disease by sacrificing a hundred animals to suffer an experimentally induced infection, thus sparing an endlessly larger number from pain and suffering, then indeed, we will voluntarily expose a few of their comrades to the same illness.

(...)

In view of these facts, who would wish to stigmatize the experiments of scientifically reasoning, well-educated researchers, even if they inflict severe suffering on the animals, as an unnecessary act of cruelty?

(Zuntz, 1899a, pp. 13–15)

In 1906, Zuntz again publicly addressed the issue of animal experiments in physiological research and took the opponents of vivisection to task much more severely this time:

The feelings of love and hate, trust and fear, which govern our relationships with our fellow humans, also control our behavior toward the animal world ... We understand that an animal lover is put in a state of extreme agitation when the animal of his or her affection experiences pain ... Such sentiments are very human, and we can neither will them out of existence nor would we want to do so. To consider warding off such danger, when only the possibility [of such danger] is threatening, is the next step. Based on such an attitude, those people who kill animals, often in a cruel manner, merely to satisfy their thirst for knowledge, appear odious. The very thought that a beloved animal could fall victim to a physiologist's craving for knowledge suffices for impulsive natures to condemn the entire work of the physiologist as dangerous, to condemn it as barbarous even, and to be blind to the compelling grounds that require animal experiments and indeed make them our duty.

(...)

The statement heard so often, that animals are better than humans, and therefore they are more deserving of our love, reveals a sorry state of the soul. 'I have suffered such injustice at the hands of my fellow man, animals have never betrayed me; they were grateful and showed their appreciation for kindnesses rendered.' Such thoughts show us the inferiority of much of the so prevalent love of animals and in particular passionate emotions. Man must love; our being, our disposition only allows happiness through love, but the love that most of us can give is much too narrow-minded, as if it were a business transaction: for each act of love, return payment is due in the form of gratitude, dependency, adulation, which the animal performs voluntarily, in accordance with its natural disposition, or else swiftly by force. The human, on the other hand, is often unappreciative and doesn't always perceive the kindness intended by others. We think and act independently and this increases in relation to our level of importance. This does not suit the majority of do-gooders and so they keep pets. Pets never say no, never disapprove of their master's actions; they lick his or her hand – not only when fed, but also when hit.

(Zuntz, 1906g, pp. 297–298)

In his review of a publication on "animal ethics" by the district court councilar Bregenzer, Zuntz commented:

In this book, experimenting on animals for scientific ends is not rejected absolutely. So-called vivisection is, however, only permitted "when a state of urgency exists in which the protection of very significant opposing interests are involved, in particular legally protected rights, for example by imminent danger to physical health and life or given the high probability that the fundamental rights of a number of persons can only be protected in the foreseeable future by performing the experiments in question. The latter case then comes to bear namely when another expedient is out of the question ...." "Nor can it be said that an alternative remedy or an equivalent substitute is impossible. If dissection is unavoidable, then without a doubt the human body constitutes in every respect the most suitable material. That the prevailing attitude is nevertheless that it is better to torment innocent animals than guilty humans (criminals) clearly arises from an unreflecting, egotistical anthropocentricism. Incidentally, I do not wish to advocate the vivisection of felons or persons who volunteer themselves for money." This is how far off the mark this jurist goes in his analysis of the "moral and legal relationships between man and animal." A criminal, that unhappy person whose constitution and childhood led him or her down a path that certainly often inflicted no less pain on themselves than they on others, is valued less than a dog or cat.

(Zuntz, 1906g, pp. 298–299).

However, Zuntz did not limit himself to performing physiological experiments on animals. When he felt it necessary and reasonable, he conducted physiological tests on himself and other human subjects (*cf.* below). This was especially the case in his high altitude physiological work during the Berlin period. In 1878, when Zuntz published his first article on the influence of altered air pressure on human circulation, this subject and different variations of it had already been addressed in a number of studies by German authors.[37] In Berlin in the late 1880s, Zuntz returned to this set of topics by participating in the work of Katzenstein (Katzenstein, 1891, pp. 330–404) and Loewy (Loewy, 1891, pp. 405–422), who were addressing, among other things, the influence of muscle activity on metabolism when the supply of oxygen was reduced. For some of these experiments Zuntz volunteered himself as a test subject, on occasion pushing himself to his physical limits (Katzenstein, 1891, p. 358).

Loewy's investigations built on and complemented Katzenstein's results. Zuntz and Loewy had already begun working together in 1885 (Report published by the Landwirtschaftliche Hochschule, 1922, pp. 28–29) and their cooperation was of particular significance as Loewy[38] was to become Zuntz's closest and most trusted colleague in the years that followed. When referring to his relationship with Zuntz to Jokl,[39] one of the fathers of modern high altitude physiology, Loewy described it in retrospect as one of brothers.[40] Over a period of many years, Zuntz and Loewy worked closely together with Lazarus, the director of the Jewish Hospital in Berlin (Zuntz *et al.*, 1906, p. 90), who had a therapeutic pneumatic chamber at his disposal that had been built and assembled from plans designed by Zuntz and Loewy (Jokl, 1967, p. 323). Zuntz described the equipment as follows:

> *A pneumatic chamber is the term used for a chamber in which a person can be exposed to altered air pressure over a longer period of time for therapeutic purposes. Such an apparatus, built to perfection, is to be found in the hospital of Berlin's Jewish community. We are immensely grateful to the Kuratorium (board of trustees) of the hospital and its medical director, Professor Lazarus ... for their permission to conduct numerous experiments on their grounds, and were most impressed by the exemplary facilities managed by their excellent technical director, the master-mechanic Mr. Gast.*
>
> *(Zuntz* et al., *1906, p. 90)*

There are no direct sources any more that might have potentially recorded or shed light on the cooperation between Zuntz, Loewy and Lazarus.[41] The pneumatic chamber itself (Figure 1.12) was a copy of the one built by Bert in Paris (Bert, 1878, p. 631). As stated above, the chamber was used mainly for therapeutic

Figure 1.12 The pneumatic chamber at the Jewish Hospital in Berlin. (Zuntz *et al.*, 1906, p. 87)

purposes, but was also made available to scientists, balloonists and other researchers for scientific studies and experiments (Lazarus, 1895, pp. 672, 702–705). For example, in addition to Zuntz and Loewy, the physicians Müllerhoff and von Schroetter as well as the meteorologists Assmann (Schumburg and Zuntz, 1896, p. 464), Berson (von Schroetter, 1902b, p. 90), and Süring (Süring, 1909, p. 55) conducted research here. When the Jewish Hospital moved from Auguststrasse to Schulstrasse in 1914 (Winau, 1987, pp. 242, 294–298) not all of the hospital's pneumatic chambers (Loewy, 1920, p. 434) could be used, so Zuntz took one of them for his laboratory (Loewy, 1920, p. 434).

For Zuntz's high altitude research purposes, the pneumatic chamber of the Jewish Hospital was of exceptional significance. The preliminary trials conducted there were the laboratory counterpart, so to speak, to the field physiological studies begun in the mid-1890s by Zuntz in the Alps, on Tenerife and while ballooning. Zuntz believed it to be imperative that final conclusions not only be based on physiological data from a laboratory setting, but also stand up to a field-study control and vice versa.

Zuntz's methodical manner of procedure became abundantly clear in his *Studien zu einer Physiologie des Marsches* (Studies on the Physiology of Marching) (Zuntz and Schumburg, 1901), which are closely related to his high altitude studies.

Yet before Zuntz could begin these investigations in 1894, he was forced to interrupt his teaching and research activities at the Royal Agricultural College in Berlin in the spring of 1893. In a letter dated April 24, 1893, addressed to the *Ministerium für Landwirtschaft, Domänen und Forsten* (Ministry of Agriculture, State Domains and Forests), Zuntz asked to be released from his professional duties for several weeks as his wife was seriously ill (Zuntz's letter to the Ministry for Agriculture, State Domains and Forests, 1893, No. 20079, fol. 119, Secret Central Archives). This letter attests to his deep personal attachment to her, and to his compassionate character:[42]

> *As is evident from the enclosed medical certificate, my wife is seriously ill and her suffering would worsen greatly were I obliged to suddenly deprive her of the care I have provided thus far and the comfort and soothing effects of my presence. Moreover, tormented as I am by anxiety about my wife's condition, I feel I would have great difficulty in meeting my obligations as lecturer and head of the laboratory satisfactorily.*
>
> *(Zuntz's letter to the Ministerium für Landwirtschaft Domänen und Forsten, 1893, fol. 119, Secret Central Archives)*

Zuntz took his teaching duties extremely seriously: "Research and teaching are the inseparable tasks of the academic profession" (Zuntz, 1899a, p. 3). That he could no longer meet his "obligations as lecturer" must have been considered a great loss by the students. According to Munk, his lectures were highly regarded, and frequently attended by medical students as well (Munk, 1956, p. 103). In this context, Loewy regretted "that Zuntz never developed a relationship to the university, because the clarity of his lectures is academic in the best sense of the word and they would have served to satisfy a larger and more broadly educated audience" (Loewy, 1917, pp. 1271–1272). Caspari[43] added another aspect, stating that:

Whoever had the good fortune of becoming more closely acquainted with him and to experience the friendliness of his manner will not only pay tribute to this great scholar today, he will admire the mild judge of all human frailty that he is, the friend who was always ready to help, and the adviser in scientific difficulties and personal hardship, in short, a man with an unfailing kindness of heart.

(Caspari, 1917, p. 620)

In his obituary for Zuntz, written in 1920 and published in the *Berliner Tageblatt*, the Berlin physician Mamlock (Schwalbe, 1926–1927, p. 97) emphasized that his contribution as a teacher specifically was "one of Zuntz's greatest strengths" (Mamlock, 1920). Forbes reported on the success of courses Zuntz gave in America:

Zuntz spent the summer of 1908 in the United States, sharing the instruction of a course in biochemistry presented at the Third Session of the Conference on Agriculture of Cornell University at Ithaca, N.Y. Others participating in the instruction were: Dr. C. F. Langworthy, Dr. A. L. Winton, Dr. L. B. Mendel and Dr. H. P. Armsby. Zuntz presented five formal lectures and two seminars. He brought some of his respiration apparatus with him and demonstrated its use. It is reported that his lectures were of unusual interest and his visit of great value to scholars at the school because of the information and inspiration they provided.

(Forbes, 1955, p. 3)

Von der Heide went so far as to describe Zuntz as a "master of lecturing in a manner comprehensible for anyone" (von der Heide, 1918, p. 344), and described what he meant as follows:

Zuntz has spoken before a mixed audience on an astonishing number of topics, without ever compromising himself in the context of these popular speeches and always upholding the dignity of the academic sciences. He knows how to introduce the most difficult scientific problems to his audience, [who are] always captivated by his sage remarks, warmed by his modest and charming manner and enlivened by his quiet humor.

(Von der Heide, 1918, p. 344)

For some of his contemporary colleagues, this "dignity of the academic sciences" represented quite an obstacle when they were called upon to present scientific matters in a manner also comprehensible to a lay audience (Ebel and Wuehrs, 1988, p. 17). In this regard Zuntz displayed no misgivings, as indicated by his earlier commitment as a shareholder in the Urania (an institution in Berlin presenting lectures and films for the general public)[44] and where he also later presented lectures on a number of occasions (Zuntz, 1910a, p. 10).

Presumably Zuntz was able to resume his lectures at the College of Agriculture toward the end of May 1893 (Zuntz's letter to the Ministerium für Landwirtschaft, Domänen und Forsten, 1893, fol. 119, Secret Central Archives), and finally started, in 1894, the initial experiments on the physiology of marching in cooperation with Schumburg of the Friedrich Wilhelm Institute for Medical Surgery. As already mentioned, Zuntz believed it to be vital that conclusions not be based solely on data gathered in the laboratory. Instead, he believed they needed to be confirmed by control tests performed in the field, and vice versa. This methodical procedure is reflected most clearly in his *Studien zu einer Physiologie des Marsches* (Studies on the Physiology of Marching), which are very closely related to his investigations in the field of high altitude physiology. Some results and conclusions from these studies broke new ground, and continue to be cited in basic textbooks of exercise physiology even today.[45]

In the broadest sense, the *Studien zu einer Physiologie des Marsches* was work commissioned for the German War Department, which had ordered an investigation "serving to establish the required physiological attributes for the permissible load of soldiers on marches" in the spring of 1894 (Figure 1.13). This ministerial decision had been preceded by discussions in the army's upper echelons addressing precisely this issue. All involved had concurred that the load that soldiers were carrying at the time was too heavy. In addition, the improvement of military medical studies in the nineteenth century had given rise to the justified hope that new discoveries in science could be applied practically in order to maintain and sustain the army's fighting ability.

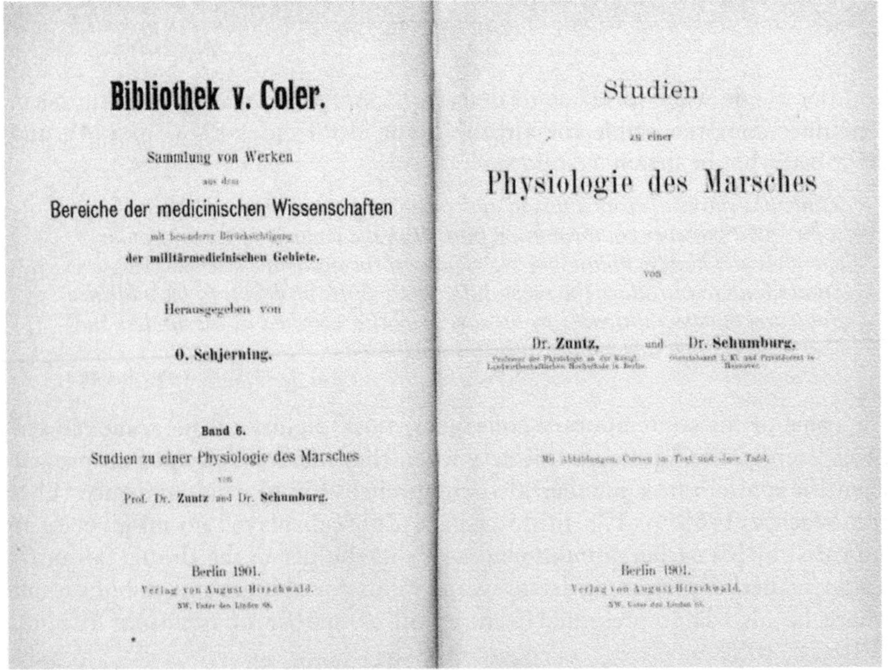

Bibliothek v. Coler.

Sammlung von Werken

Bereiche der medicinischen Wissenschaften

der militärmedicinischen Gebiete.

Herausgegeben von

O. Schjerning.

Band 6.

Studien zu einer Physiologie des Marsches

Prof. Dr. **Zuntz** und Dr. **Schumburg.**

Berlin 1901.

Verlag von August Hirschwald.

Studien

zu einer

Physiologie des Marsches

von

Dr. **Zuntz,** und Dr. **Schumburg.**

Berlin 1901.

Verlag von August Hirschwald.

Figure 1.13 Title page of *Studien zu einer Physiologie des Marsches* (Studies on the Physiology of Marching), published in 1901.

Investigations dealing with military marches in this context still represented a relatively new direction in research. Although in the 1890s various studies from France and Germany were published that dealt primarily with illnesses while marching and the treatment thereof (Kirchner, 1896), by and large these only reflected practical experience (marching hygiene). Scientific analysis of the topic had yet to be performed. As Zuntz and Schumburg state in their introduction to *Studien zu einer Physiologie des Marsches*, the answers to relevant questions had thus far been obtained only by means of "estimates, theoretical observation, subjective experience; the numerical foundation is missing" (Zuntz and Schumburg, 1901, p. 25).

For example, the question was still unresolved concerning the "limit of the soldiers' ability to march" (Zuntz and Schumburg, 1901, p. 22). While it was obvious that many factors played a role here, great significance was allotted to the baggage, the "millstone around the troops' neck as concerns their mobility" (Zuntz and Schumburg, 1901, p. 22). New research was required in order to provide clear data as to how much weight an average, healthy infantryman could carry "while marching at medium [speed] and during average weather conditions, without damage being done to his performance ability" (Zuntz and Schumburg, 1901, p. 23). It was hoped that by performing appropriate physiological experiments on people, those characteristics would be discovered that would "reveal the limit of permissible load [carried by] soldiers" (Zuntz and Schumburg, 1901, p. 25).

In light of the background discussions regarding possible alleviations, in the spring of 1894 the army command had load-carrying experiments performed on ten battalions of the army. While up until that point in time the clothing and baggage of infantrymen weighed a good 33 kg, it was found that this burden could be reduced by almost 7 kg by means of various modifications – for example, by the "use of less thick and heavy leather" for boots, or by doing away with the coat lining (Zuntz and Schumburg, 1901, p. 23).

The following statement of the military doctor Martin Kirchner illustrates the special priority allocated to the problem of the soldiers' marching ability within the War Department:

> *... in a future war, considerable marches will be required to reach what will presumably be very large battles ... where is the soldier to rest and restore his performance ability and the mental peace of mind and concentration, which he will need for his movements during combat and for the deft handling of firearms, if his strength is already exhausted from the march when he arrives at the destination?*
>
> *(Kirchner, 1893, p. 546)*

The *Felddienst-Ordnung* (field service ordinance) of January 1, 1900, also vividly illustrates the great importance that the march had for all things war-related and "the defense of the nation:"

> *303. By far the greatest part of war activity for troops is marching. The march forms the foundation of all operations, and the success of all undertakings*

> *is dependent on its solid execution. Often it is crucial that the army battalion arrive at the point prescribed at the correct time and that it is ready for battle.*
>
> *306. A troop accustomed to disciplined marching will only remain consistently and completely able to march if every strain not necessarily required for the purpose of the march is avoided intelligently.*
>
> *(Zuntz and Schumburg, 1901, p. 1)*

Presumably, Zuntz and his school had little reservations about mixing scholarly and military research, and thus Zuntz was able to begin initial studies on the physiology of marching soon afterwards. The results were first published in 1895 in the *Deutsche Militärärztliche Zeitschrift* (German Military Medicine Journal), and six years later they appeared in a more extensive form as volume 6 of the *Bibliothek v. Coler* (von Coler library), a collection of works from the fields of medical science with particular emphasis on military medical areas (Zuntz and Schumburg, 1901, pp. I–XVI).

The series of investigations themselves lasted from the end of April until the beginning of July 1894, and took place under the following aspects:

> *The intent was now to have the students, clothed in soldiers' uniforms, perform marches in order to discern the influence of various loads by means of investigations as diverse as possible into the behavior of the men both prior to and after the march.*
>
> *(Zuntz and Schumburg, 1895a, p. 50)*

Zuntz und Schumburg had found five students of the Friedrich Wilhelm Institute to act as subjects for the load-carrying experiments. Although they had offered their services voluntarily and with "sacrificial willingness," their active service had already taken place some time ago. When critics complained that the subjects of the experiment were not soldiers from the front, Zuntz countered that in the case of mobilization, primarily untrained reservists would be drafted. Furthermore, he stated, the students were chosen because the marching subjects had to be in a position to cooperate in the investigations by means of precise self-observation.

The experiment protocol envisaged that a constant route distance of 24.75 km be marched, and that baggage of varying weight (22, 27 and 31 kg) be carried. In order to emulate real marching conditions to the greatest extent possible, all subjects were "clothed completely in marching uniforms used in the field" (Zuntz and Schumburg, 1901, p. 27). The Guard Fusilier regiment had provided the appropriate clothing and equipment, though the loaded cartridges were replaced with pieces of lead.

The marches began at 7:00 am, sometimes as early as at 5:00 am, after various bodily functions had been measured in the Animal Physiology Institute. Further measurements occurred during the march, mostly during the two breaks, which lasted ten and thirty minutes, respectively. In addition, the physical status of each participant was recorded prior to the series of investigations and following it – in other words, four months apart.

Overall, an impressive twenty-eight marches were undertaken during this period, primarily due to the changing weather conditions. Since the weather drastically affected performance ability, the decision was made "to organize the highest number of marches and investigations possible; only by this means was it possible to achieve the purpose of comparing the influence that different loads had by evaluating two or more marches which were experienced under almost identical weather conditions but with varying loads" (Zuntz and Schumburg, 1901, p. 26).

The investigations and measurements performed on the subjects were diverse, and covered the functions of all organs whose activity might be influenced by the march and the load. In this context, Zuntz and Schumburg made the following assumptions:

> *The load carried by a marching soldier is to be regarded as acceptable as long as his performance ability while marching is not too greatly depreciated as a result of bearing the load, and as long as the physiological functions such as respiration, cardiovascular-, digestive system, muscle system, and nerve activity do not degenerate considerably.*
>
> *Thus, we first had to investigate what changes occur as a result of load-carrying while marching, and secondly, where the limits of the observed change lie in terms of their being acceptable. The latter can be achieved by measuring the individual functions after the march, should extreme states of exhaustion occur, and to interpret the changes obtained in this way as the limits of the soldiers' performance ability. This was successful many times.*
>
> *The damage to the physiological functions of the load-carrying soldiers was, however, only partially a consequence of the load alone. This was also caused by increased lengths of the march, thirdly by high temperatures and high moisture content in the air, and fourthly by individual states of exhaustion and often by only slight physical injury (i.e. sore feet).*
>
> *These four factors increase the demand on various organs, causing alterations in their activity, and resulting ultimately in the exhaustion of the soldier, in the worst case risking serious health repercussions or even the loss of life.*
>
> *This fact has been long known to anyone observing soldiers. Nevertheless, no numerical evidence exists for this long-known fact, especially the exact influence of each of the four individual factors remains unknown. Furthermore, we do not know which of the physiological functions (respiration, cardiovascular system, digestion, muscular system and nerve activity) are altered by each individual damaging factor and how.*
>
> *(Zuntz and Schumburg, 1895a, p. 50)*

At the end, results were available regarding the effects of marching, both on the individual organs as well as on the body's metabolism and energy exchange (which could be realized because of some new methodological approaches, such as a special valve mechanism for the marching soldier (Figure 1.14)). Furthermore, the *Studien zu einer Physiologie des Marsches* provides a multitude of annotations and practical suggestions, some of them in considerable detail, on issues of suitable clothing, advisable nutrition, etc.

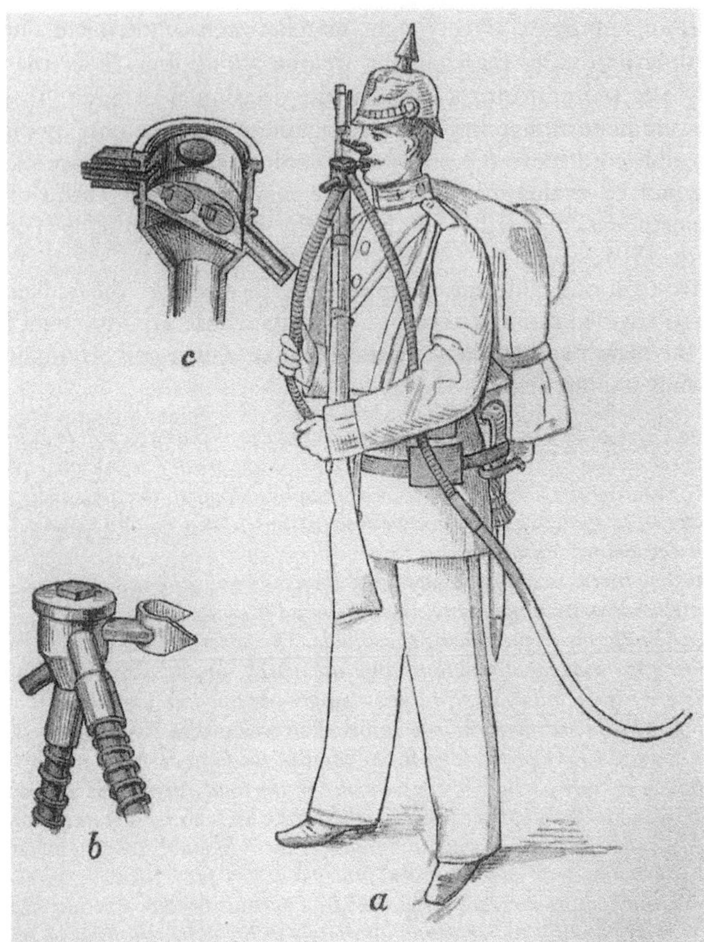

Figure 1.14 A special valve mechanism to measure the oxygen consumption of a fully equipped Prussian soldier during marching.
(Zuntz and Schumburg, 1901, p. 208)

Measurements of blood pressure, pulse and respiration frequency, as well as the vital capacity revealed that the highest degree of strain in particular (i.e. marches carrying 31 kg of baggage) led to considerable impairment and a clear drop in the performance ability of the respiratory apparatus. This the scientists attributed equally to the mechanical impediment resulting from the heavy pack and to general muscle fatigue (Zuntz and Schumburg, 1901, p. 113). In this context, Zuntz and Schumburg note that reservists should at first be prevented from taking long marches with heavy loads.

Furthermore, observations on the effects that physical activity, performed at various degrees of strain, has on the heart and liver were carried out. In this context, too, Zuntz reached the conclusion that increasing the load from 27 kg to 31 kg made a damaging dilatation of the heart during longer marches significantly more probable (Zuntz and Schumburg, 1901, p. 76). Furthermore, he found that the strain may cause the liver to swell (Zuntz and Schumburg, 1901, p. 87).

Along with the physical effects, Zuntz and Schumburg also looked at the influence that strenuous marches had on the mental faculties of the subjects. This was evaluated using reaction tests and by testing how well the subjects could remember numbers. By employing an appropriate experimental set-up, the reaction times were measured from the start of an electrical impulse to the forehead or fingertips to the interruption of this impulse by squeezing the right hand.

The results indicated that the subjects' performance ability was clearly reduced when preceded by strenuous physical exercise. It was also revealed that on the days following demanding marches, the reaction time was prolonged even when resting. Previous exertions thus appeared to reduce performance ability even prior to the start of new marches. In contrast, improved performance was recorded when the exertion had been moderate. Overall, however, taking the measurements turned out to be more difficult than expected because of the significant individual differences among the individual subjects, and also their respective daily form, and thus the influence of the load carried by the subjects on their reaction times was not always clear.

The effects on memory were investigated by having the subjects repeat a spoken number series, and here too performance was better in some instances than beforehand if the marches had not been too strenuous. Thus, the stimulatory effect of moderate physical exercise could also be confirmed in this context (Zuntz and Schumburg, 1901, p. 130).

Finally, the fatigue of muscle groups not directly required while marching was measured using ergographic methods. It was determined that over the course of the four-month experiment series, the muscle performance ability of all subjects increased from week to week – a fact ascribed to the increase of muscle tissue caused by marching. As had already been established in the previous experiments, it was confirmed that marching with light and moderately heavy packs led to improved performance ability, whereas excessive heat and a heavy pack load resulted in the subjects' performance deteriorating significantly (Zuntz and Schumburg, 1901, p. 140).

Alongside the influence of marching on individual bodily functions, Zuntz and Schumburg further emphasized the implications the marches had for the subjects' metabolism and energy exchange. The primary objective was to obtain indications as to the nutritional requirements of soldiers.

The most obvious effect was, of course, the changes in body weight that the five experimental subjects underwent. The measurements mentioned, taken before and after the total experiment time period, revealed a weight loss of between 1.5 and 3.5 kg, resulting primarily from the loss of fat tissue. Supporting this was the fact that the waist measurements had clearly decreased on four of the subjects, while the muscle mass has increased (Zuntz and

Schumburg, 1901, p. 29). Zuntz and Schumburg concluded "that the metabolism of the soldier marching under strained conditions is at the upper limit of what he can replenish by means of ingesting food. Every significant additional demand will thus be to the detriment of the organs' health" (Zuntz and Schumburg, 1895a, p. 53).

In order to arrive at exact figures and provide firm conclusions, Zuntz and Schumburg carried out a balance trial at the end of the marching series in which the intakes and excrements of two of the five subjects was controlled completely. This meant that the nutritional composition and quantities of food had to be recorded very precisely, as was the intake of food and excretion. The fact that Zuntz always kept in mind the person as a whole, both physically and emotionally, when describing individual detailed physiological processes is illustrated by the following introduction on nourishment during the metabolic experiments:

> *The daily nourishment was now standardized as follows:*
>
> *125 g lean beef,*
> *150 g summer sausage,*
> *350 g bread,*
> *50 g egg yolk,*
> *100 g butter,*
> *50 g orange marmalade,*
> *30 g sugar,*
> *1750 g beer,*
> *320 g coffee made from 13 g of coffee beans,*
> *20 g cognac.*
>
> *Both men partook of these food quantities equally, while Mr. Pochhammer, whose metabolism is higher and who weighed 5 kg more, additionally received*
>
> *50 g rice and*
> *300 ccm sterilized milk per day.*
>
> *It may be that the fairly significant beer quantities which we administered will raise eyebrows. But we discovered that it was impossible to provide the necessary quantity of nitrogen-free foods through, for example, increased amounts of bread, and we also feared that by increasing fat intake, digestive problems would occur with time. Particularly after marching, [the subject's] appetite was often reduced, as we had learned from the preliminary experiments; it would have been impossible to intake larger quantities of solid stuffs, while [in contrast] beer was particularly welcome, especially on marching days as a result of the great thirst they suffered. In order to prevent the risk of abnormal decomposition and digestive problems caused as a result, which would have ruined the entire experiment, a particularly high-quality export beer was chosen (Münchener Löwenbräu). In light of the digestive problems of the diet likely to occur due to the summer heat, the cognac was also ingested from the start.*
>
> *(Zuntz and Schumburg, 1895a, p. 60)*

Zuntz and Schumburg's objective was to determinine the nutritional needs of marching soldiers. Therefore, they first calculated the calories needed on days of rest during activities such as walking, standing, etc. The difference between energy intake and consumption would then, they assumed, provide information as to whether body fat was gained or lost. This calculation of theoretical weight change was, however, called into question by the measured results, which revealed that the actual gain was less than calculated even during the multi-day periods of rest. Weight loss was even recorded on days of rest during the marching periods, which Zuntz and Schumburg explained by theorizing that the water the body lost during a strenuous march was still not balanced out even after twenty-four hours (Zuntz and Schumburg, 1901, p. 178).

For marching days, the calories required were increased by about 1000 kcal, which meant a daily requirement of about 3600 kcal given the average weight of 65 kg for the infantrymen. When the actual amount of calories required was compared with the ration customarily allocated to soldiers, it was revealed that the food contained too little fat and not enough calories. In this context, Zuntz and Schumburg recommended that the fat and sugar content be increased in the soldiers' food; in contrast, they regarded the protein content to be sufficient (Zuntz and Schumburg, 1901, pp. 185–186).

Measurement of the respiratory gas exchange took place at the beginning and end of the marches. Essentially, the device was composed of the Zuntz–Geppert respiratory apparatus and a treadmill. The treadmill had been developed in the course of work done by Zuntz, Lehmann and Hagemann on the metabolism of the horses (Figure 1.15). Later, during their field studies in the Alps, Zuntz looked back on the treadmill marches in the animal physiology laboratory with some discomfort. On this topic, *Höhenklima und Bergwanderungen*

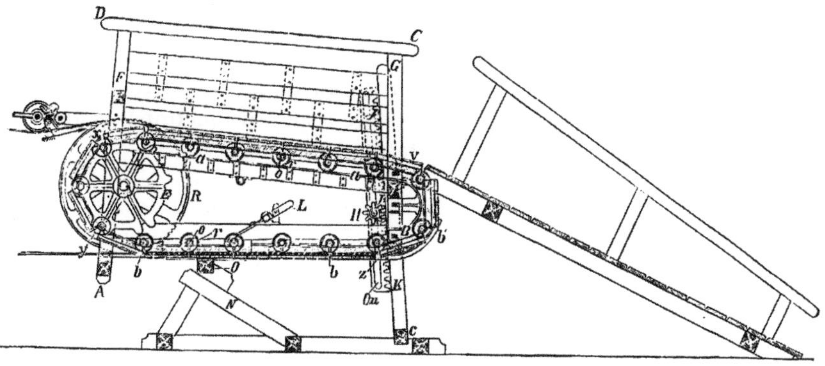

Figure 1.15 A technical description of the treadmill invented by Zuntz and Lehmann in 1889.
(Tigerstedt, 1911, p. 63)

(High Altitude Climate and Mountain Hiking on Human Beings) gives us the following insights regarding these experiments:

> *While climbing the well-tended fields, reveling in the magnificent view of the lake, the tiny village and the Faulhorn peak jutting up opposite us, we recalled, with a relieved aesthetic shudder, the Berlin treadmill experiments during which the bored eye could only wander up and down the gray walls of the laboratory building!*
>
> *(Zuntz* et al., *1906, p. 130)*

The results revealed that the respiratory quotient at the end of the marches was always reduced – in other words, that in place of the carbohydrates used up by work, fat was metabolized during long periods of sustained physical exercise. As demonstrated by further measurements, the body's carbohydrate stores, reduced by the march, had not been replaced even by the next day (Zuntz and Schumburg, 1901, p. 255).

Finally, Zuntz and Schumburg made several excursive comments on the clothing of the soldiers. Their studies of the clothes' properties in terms of heat conduction, air permeability and water-binding abilities showed how important it was to allow air to permeate to the underclothes. This result also meant that the uniform of the soldier in the summer was hardly suitable and

> *if aesthetic aspects are not considered and solely the requirements for the highest marching ability are at the fore, it would certainly be recommendable to carry the uniform jacket and wool pants in the knapsack and then to don them, if necessary, after having arrived at the quarters, but to execute the march itself in fatigue clothes or in the plain grey work tunic called* Litewka.
>
> *(Zuntz and Schumburg, 1901, p. 333)*

The plethora of "results not insignificant for practice" and "facts which the officer as well as the army doctor ... will happily use in further work" (Zuntz and Schumburg, 1901, p. 333) was summarized by the authors as cited below. However, they also refer to the partially provisional character of their results and provide an idea of further worthwhile research goals. Zuntz and Schumburg stated that more studies were needed that were to address this matter in greater depth, in order to investigate issues of functional distribution of baggage, sensible clothing, and the effects of training:

> 1. *Under a moderate load (up to 22 kg) and at not too high outdoor temperatures, no detrimental effects whatsoever will occur during a march that does not extend beyond 25 to 28 km. On the contrary, we found that a state of tiredness brought on by other causes and slight impairments to the functioning of individual organs, can be remedied by the march itself.*
>   *However, if the air is very hot or humid, a number of impairments of a minor sort have been proven (reduction of vital capacity, considerable fluid loss of the body, high pulse rate and breathing frequency, plethora). Still, these disappeared soon after the march and were in any case completely*

remedied the next day, meaning that no accumulation of impairments in marches taking place over many days in a row was observed.

2. *For the second level of load (27 kg) no impairment was noticeable given good weather and the same marching distance. In contrast, when this load was carried in hot weather, changes resulted which were not balanced out, not even on the next day. The second march thus took place under less favorable conditions than the first. In any case, a march of 25 to 28 km is the limit of what can still be favorably sustained in fairly hot weather.*
3. *A load of 31 kg encroached in a disruptive way on certain bodily functions even during cooler weather and the same marching distance.*
4. *As concerns becoming used to the baggage (training), it was observed that light baggage (up to 22 kg) no longer had an impairing effect after only a few marches if the exertion was gradually increased; on the other hand, the heavy (31 kg) load continued to have detrimental effects even after a longer period of practice.*

(Zuntz and Schumburg, 1895a, pp. 79–80)

All these experiments were definitely difficult to arrange and time-consuming. Fortunately, Emma Zuntz gives us a vivid description of how her father scheduled his days in the laboratory and at home:

I would like to describe his usual work day to you now. He got up early in the morning and generally was already sitting at his desk at half past six to continue the work of the night before. [He] prepared tables of figures obtained in experiments, designed new important apparatuses. He would look through the mail after breakfast and answer it immediately whenever possible. When he did not have to give a lecture in the morning, he went to the institute around half past nine where his students already awaited him. He went from one researcher to the next and his mind quickly found new ways of continuing projects that had come to a standstill. On some days, large-scale experiments had been prepared for him [and] took priority before everything else. Lunch was scheduled at home for four o'clock, but punctuality was not his strong point. How could he have been hungry when he was following the sequence of an experiment with feverish expectation, or when he had to dispel the doubts harbored by a fellow researcher? After lunch he stretched out on the sofa; the entire family gathered around him and we began telling him of our experiences [that day]. He would slowly doze off and fall fast asleep, and after around ten minutes, sometimes after half an hour, he awoke invigorated.

(Emma (Sarah) Zuntz, undated, p. 44)

In addition to those experiments on exercise physiology, in the summer of 1895, Loewy completed his famous work on respiratory gas exchange under normobaric and hypobaric conditions (Loewy, 1895); in parallel, Zuntz was already thinking about the use of increased atmospheric pressures (*hyperbaric oxygen therapy*) for medical therapy in case of rapid decompression (Zuntz, 1897c; Petersen, 2002, p. 39). These studies were conducted in Zuntz's laboratory as well as in the pneumatic chamber at the Jewish Hospital, and yielded early basic facts on respiratory and high altitude physiology (West, 1998, pp. 88–91).

They laid, together with the laboratory experiments in metabolism conducted over many years, the foundations for the field of physiological studies in high altitude (Zuntz *et al.*, 1906, p. XV), which commenced with Zuntz and Schumburg's first Monte Rosa expedition in August of 1895. Thus, Zuntz and his school embarked on research that was to lead to the work *Höhenklima und Bergwanderungen in ihrer Wirkung auf den Menschen* (The Effect of High Altitude Climate and Mountain Hiking on Human Beings) being published in 1906.

Chapter 3 discusses in detail the objectives, procedures, and results of Zuntz's physiological field expeditions in high mountain regions, which took place in the years 1895 to 1903 in the Monte Rosa massif and in 1910 on Pico del Teide in Tenerife. Zuntz personally participated in all of the expeditions except one. At this point, only the central scientific focus points of Zuntz's expeditions will initially be addressed, these being:

1. The effect of exercise on metabolism
2. The effect of a high altitude climate on metabolism
3. The effect of a high altitude climate on hematopoiesis
4. The etiology of "mountain sickness" (altitude sickness).

The expedition taken by Zuntz, Loewy and Schumburg in August of 1895 (Figure 1.16) lasted about two weeks, and was followed by their joint participation in the Third International Congress of Physiologists in Berne (Zuntz, 1895b, p. 863; Franklin, 1938, pp. 254–258), during which they met the most renowned high altitude physiologists of their time, Mosso[46] and Kronecker.[47]

Figure 1.16 Presumably, Schumburg, Adolf Loewy and Nathan Zuntz (left to right) in the Alps during their first high altitude studies in 1895.
(Private collection, G. Zuntz, Cambridge)

According to Zuntz, they also had the chance to relate and discuss their latest findings made during the Monte Rosa trip in private (Zuntz, 1895b, p. 863). In the years to come, a scientific dispute was to develop between Zuntz and Mosso that can justifiably be termed inspiring. It was based on the lifelong adherence of the latter to his hypocapnia theory on the etiology of "mountain sickness," as opposed to Zuntz's view that the reduced oxygen partial pressure was the major triggering factor. Nonetheless, their opposing scientific standpoints never affected their private relationship or personal interaction. Zuntz spoke of a "congenial relationship" (Zuntz, 1911m, p. 94) between his school of thought and that of Mosso, who until his death "supported [Zuntz's research] most unselfishly" (Zuntz, 1911m, p. 94). Jokl reports that Loewy, who had the opportunity to accompany Mosso on mountaineering expeditions in the 1890s, spoke in a similar manner about their friendship and work together (Jokl, 1967, p. 324).[48]

In 1896, A. Loewy, J. Loewy and the medical student Leo Zuntz, Nathan Zuntz's son, continued the high altitude physiological research undertaken on Monte Rosa, but now in the pneumatic chamber at the Jewish Hospital and in Zuntz's laboratory in Berlin (Loewy, Loewy and Zuntz, 1897, pp. 477–538). Issues concerning the influence of exertion on metabolic processes remained the focus of their work. Once again, for these research purposes, Lazarus granted them access to the pneumatic chamber of the Jewish Hospital "under the most liberal terms" (Loewy, Loewy and Zuntz, 1897, p. 479). In addition, they were able to continue the research they had just started on high altitude physiology under very favorable circumstances:

> *First, the test subjects had to be introduced and made familiar not only with the objective and purpose of the work but also with the relevant methods, since they were required not only to substitute for one another but also to take on certain tasks unassisted. Only through the cooperation of three individuals familiar with the problems to be addressed and sufficiently experienced in the methods in question, was it possible to accomplish the entire task in the short time available. The project necessitated considerable funds. These were provided by the* Kuratorium *[board of trustees] of the Countess Bose Foundation, to whom we express our sincere gratitude. Furthermore, we are exceedingly grateful to Professor Zuntz, who placed the necessary equipment at our disposal, to the extent it still existed after last year's expedition, and who provided continuous support by proffering his advice. And finally [we want to thank] Professor A. Mosso, who most considerately supported us with his counsel during our ascent from Gressoney la Trinité to the Gnifetti cabin.*
> *(Loewy, Loewy and Zuntz, 1897, p. 480)*

Despite such propitious conditions, this expedition also failed to achieve any conclusive breakthrough; the central problems of high altitude physiology remained unresolved (Loewy, Loewy and Zuntz, 1897, p. 538).

These unsatisfactory results prompted the decision to conduct a lengthier expedition with a larger number of participants, and so Zuntz's major expedition to Monte Rosa came about in the summer of 1901. Once again, metabolic experiments and the effects of oxygen deficiency on the organism were the primary questions investigated. It should be noted, however, that within the scope of this field study Zuntz and his colleagues pursued other questions as well. The areas examined were, among other things, observations of pulse and blood pressure changes, hematopoiesis, protein balance and thermal balance, perspiration, physiology of nutrition and blood gases, respiratory mechanics, balneology, clothing, and sports physiology. Under Zuntz's leadership, the expedition departed from Berlin for Monte Rosa on August 1, 1901 (Zuntz *et al.*, 1906, p. 127). The initial experiments on metabolism got under way on August 5, and all investigations were completed by September 10 of the same year (Zuntz *et al.*, 1906, p. 148). Immediately thereafter, Zuntz took his entire group[49] to Turino to attend the Fifth International Congress of Physiologists, taking place on September 16–21, 1901 (Franklin, 1938, p. 264).

After Mosso opened the congress on September 17 by appealing for an intensification of international high altitude research on Monte Rosa, Zuntz took the floor. He reported on the latest findings from his expedition, and demonstrated his new transportable gas meter (Figure 1.17).[50] On his recommendation, a

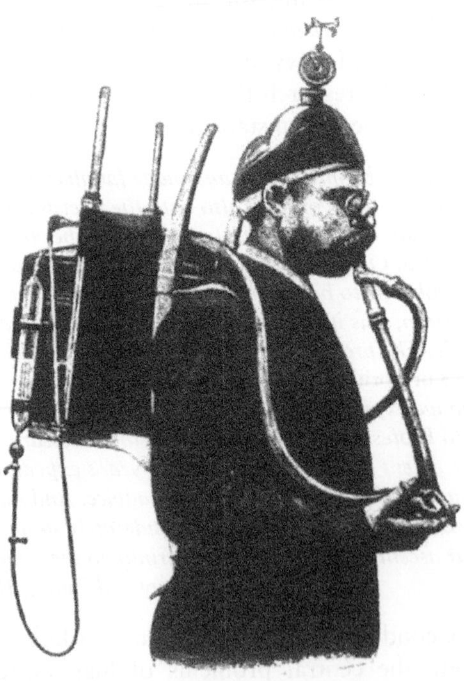

Figure 1.17 Zuntz's transportable dry gas meter being used during a marching experiment. On his head the subject is wearing an anemometer cap to measure wind speeds. (Zuntz *et al.*, 1906, p. 165)

letter of thanks was sent to the Queen of Italy and the Minister of Education, who had provided financial and political support for the high altitude experiments done on Monte Rosa (Franklin, 1938, p. 264).

Meanwhile, in Berlin during the summer of 1901, final preparations were under way for balloon ascents to be undertaken by von Schroetter[51] (Figure 1.18) along with Berson and Süring from the Meteorological Institute, also in that September (von Schroetter, 1901, p. 404; 1902b, p. 92). As previously mentioned, von Schroetter and his colleagues also conducted experiments in Lazarus' pneumatic chamber at the Jewish Hospital, simulating high altitude. At the Congress of Physiologists in Turino (von Schroetter, 1901–1902, pp. 86–87) in September, von Schroetter presented the initial results he had obtained in the physiological tests and demonstrated the design and use of an oxygen mask for high altitude excursions that he had constucted himself. At this point in time Zuntz and von Schroetter complemented one another's work excellently, which undoubtedly became apparent to both of them during the Congress at the very latest. Zuntz could draw on a long history of high altitude studies. Moreover, he had outstanding knowledge of mature, tried and

Figure 1.18 A portrait of Hermann von Schroetter (1870–1929) at the age of approximately thirty years.
(Hoernes, 1912, p. 275)

true methods, which permitted, for example, metabolic measurements of sufficient precision even in the confined space of a balloon basket. By contrast, the young von Schroetter had the advantage over the elder Zuntz of having gained personal practical experience on several balloon excursions. Nonetheless, because the methodology on which he was basing his work was flawed, his physiological observations in the balloon were meaningful and pertinent only to a limited degree. This anecdote shows why von Schroetter later referred to Zuntz as his "revered teacher" (von Schroetter, 1902c, p. 1423; 1906, p. 216, p. 229), since evidently Zuntz trained him in these scientific methods.

After Zuntz's return from Switzerland that year, he completed his *Studien zu einer Physiologie des Marsches* (Zuntz and Schumburg, 1901, pp. 333–335), which he had originally begun together with Schumburg by order of the War Ministry in the summer of 1894. After years of physiological field and laboratory tests, the authors were forced to admit that the core question regarding the permissible loads for soldiers while marching could not be answered as definitively as they had initially conceived. The physical and psychological condition of the individual subject varied too greatly. In addition, certain medical parameters, such as cardiac output, could not yet be determined with sufficient accuracy. Nonetheless, one of the remarkable features of their work, which renders it a classic in physiology, is the clearly recognizable integrative approach taken in its research: the simultaneous evaluation of multiple physiological parameters (respiration, circulation, metabolism, temperature regulation, etc.), the subtle descriptions of the external circumstances under which the experiment took place, and the implementation of the whole course of the experiment under both laboratory and field conditions have been a model in particular for integrative physiological research ever since. Zuntz applied this underlying concept from *Studien zu einer Physiologie des Marsches* to his physiological research on high altitudes and aviation.

In 1903 Zuntz set off once again for the Monte Rosa massif, this time accompanied by Durig[52] (Figure 1.19). This expedition was sponsored by the *Königlich Preussische Akademie* (Royal Prussian Academy) in Berlin and the Austrian *Ministerium für Kultur und Unterricht* (Ministry for Culture and Education). The objective was to expand and conclude the respiratory and metabolic experiments conducted during the major expedition of 1901 (Durig and Zuntz, 1904a, p. 418; 1904b, p. 1042). This Alpine expedition of 1903 was, however, clearly beyond the stamina and health of Zuntz, who was now over fifty years old. Years later, in his obituary for Zuntz, Durig remarked:

> *Images come to mind from the past when we lived, worked and cooked together during the weeks in the icy regions, in remote seclusion from mankind, and how you – with indescribable modesty – were content with even the worst of all I could offer you. You said to me then – seventeen years ago – at the highest frontier glacier, "I am so tired, just let me die." This must have been the first sign that your heart was beginning to fail.*
>
> *(Durig, 1920, pp. 344–345)*

Figure 1.19 A painting of Arnold Durig (1872–1961) by the Austrian Anton Filkuka (1888–1957) at the age of about forty years.
(Museum of the Association for the Preservation of the Homeland, Montafon, Estate of Arnold Durig)

During that journey, Zuntz and Durig spent almost three weeks in the peak regions of Monte Rosa. This was to be Zuntz's last major expedition in the Alps. The months that followed were devoted to sorting through and evaluating the extensive material collected, and the results of these studies made up the major portion of *Höhenklima und Bergwanderungen in ihren Wirkung auf den Menschen* (High Altitude Climate and Mountain Hiking on Human Beings) by Zuntz and colleagues, published in 1906 (Figure 1.20).

Finally, in May of 1902, the Third Conference of the International Commission on Scientific Aviation took place in Berlin (Minutes of the third assembly of the International Commission for Scientific Aerospace Travel dated May 20–25, 1902 in Berlin, Strassburg 1903, pp. 1–65). In the course of this conference, Zuntz and von Schroetter undertook their first balloon ascent together for physiological tests, followed by a further one in June of the same year (von Schroetter and Zuntz, 1902, pp. 479, 489).

In October of 1904, Podbielski, the Minister for Agriculture, State Domains and Forests,[53] submitted a proposal that Zuntz be conferred the title of *Geheimer Regierungsrat* (Privy Councilor) (Letter of Podbielski to Kaiser Wilhelm II, 1904, fol. 130 R, Secret Central Archives) in the light of his excellent research and teaching at the Royal Agricultural College. Podbielski's proposal was apparently approved by the Kaiser with retroactive effect, since there is a later notation that indicates that Zuntz had already been promoted

Figure 1.20 Title page of *Höhenklima und Bergwanderungen in ihrer Wirkung auf den Menschen* (The Effects of High Altitude Climate and Mountain Hiking on Human Beings), published in 1906.

on September 13, 1904 (Petition of the Minister for Agriculture, 1918, fol. 22 recto). Then, at the beginning of 1906, Zuntz was recommended by Podbielski for the position of rector at the *Königliche Landwirtschaftliche Hochschule* (Royal Agricultural College) for the term of office ending in 1908 (Letter from Podbielski to Kaiser Wilhelm II, 1906, fol. 141, Secret Central Archives).

As was customary in such cases, a prior nomination of Zuntz by the faculty of the college had taken place. The authorization by the Kaiser was given on January 31, 1906 (Letter from Kaiser Wilhelm II to Podbielski, 1906, fol. 141, Secret Central Archives). Accordingly, Zuntz took over from Orth as rector in April of 1906. It is worth noting in this context that later proposals submitted to Kaiser Wilhelm II for the rectors who succeeded Zuntz at the Royal Agricultural College (i.e. Börnstein, holding office from 1908–1910) explicitly emphasized that the candidates were Protestant (Letter from the responsible Minister for Agriculture, 1908, fol. 152, Secret Central Archives). In Podbielski's application on behalf of Zuntz, there is no mention of anything in this regard.

In the succeeding years as rector of the College of Agriculture in Berlin, Zuntz's scientific work was forced to take second place to his other duties. However, he was visited during this period by many famous scientists, such as Marie and August Krogh (of Copenhagen) and Joseph Barcroft (of Cambridge) in the summer of 1907, who had come to see him before the International Physiological Congress of 13–16 of August in Heidelberg.[54]

In the few publications of this period, Zuntz addressed issues of fisheries – in particular, the feeding and breeding of fish. Evidently out of interest for this subject, Zuntz participated in the International Agricultural Congress in Vienna in May 1907, at which the "national economic significance" (Zuntz, 1907) of fishery research was emphasized. It may be assumed that Zuntz contributed considerably to this view over the course of the congress, given that, after all, such practice-oriented and applied physiological research corresponded with the manner in which Zuntz thought and acted as a scientist.

At his institute, in cooperation with Knauthe and Cronheim, Zuntz had already conducted successful research into fisheries for a number of years – chiefly pertaining to the metabolism of fish.[55] Testimony to this is the gold medal awarded to Zuntz's Veterinary Physiological Institute at the World's Fair in St Louis in 1904 for an improved respiratory apparatus (Figure 1.21) and a number of curve graphs on the nutrition conditions of fish farms (*Jahresbericht der Königliche Landwirtschaftliche Hochschule*, 1905, pp. 35–36).

In a speech given before the *Deutsche Gesellschaft für volkstümliche Naturkunde* (German Society for Popular Natural Sciences) in 1905, Zuntz based some physiological considerations on the course and control of hibernation in endothermic animals on his work on the metabolism of ectothermic animals (Zuntz, 1905a, pp. 145–148) – a scientific topic which had much interested him since 1871 (Zuntz, 1871, p. 321). It is again characteristic of Zuntz to attempt to incorporate his physiological knowledge and findings across such seemingly remote branches of science.

In 1908, four years after the foundation of the *Deutsche Physiologische Gesellschaft* (German Physiological Society) in Breslau (Blasius, 1959, p. 3), Zuntz became a regular – although not lifelong – member of the Society (Deutsche Physiologische Gesellschaft, Mitgliederverzeichnis 1911, Zentralblatt für Physiologie 25, p. 1114).[56]

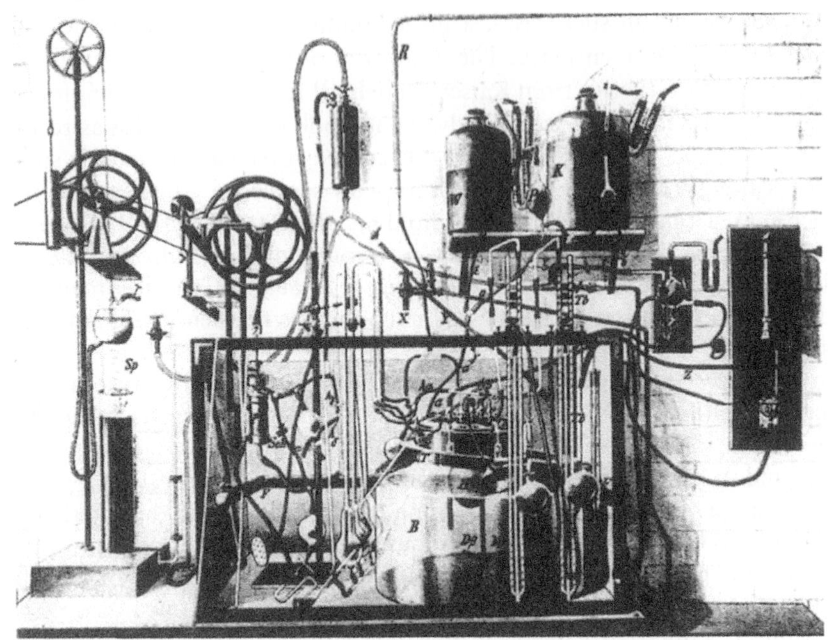

Figure 1.21 Side view of the respiratory apparatus serving to measure the metabolic rate in fishes, invented by Zuntz, which won the Gold Medal at the World's Fair in St Louis (USA) in 1904.

As already mentioned, Zuntz was unable to devote much time to his research during his term as rector. Nonetheless, this position had a definite advantage for his future scientific activity in Berlin, in that Zuntz was evidently able to use his position of authority to expedite the planning of the new building that was to provide space for many different research laboratories for the Institute for Veterinary Physiology. Once the property at Chauseestrasse 103 had been purchased, nothing stood in the way of Zuntz's plans, which, despite the "enthusiastic interest" (Zuntz, 1909b, p. 473) displayed by Thiel, had failed to be realized for years owing to lack of space (Zuntz, 1909b, p. 474). As early as 1909, it was possible to move into half of the new Institute in Chauseestrasse. Plesch, Zuntz's colleague of many years, relates the following anecdote about the start-up phase in the new laboratory, which once again shows the pleasant working atmosphere in Zuntz's institute:

> *He [Thiel] built Zuntz a new laboratory with all the bells and whistles and technical innovations – but with it vanished the charm, atmosphere and coziness. Laboratories are like dance bars in this respect: they only stay popular as long as the dancers are tripping over each others' feet and perching uncomfortably on stools.*
>
> *However, the spell of newness in the newly built Institute was soon broken by a happy accident. It takes a good deal of time to learn the ins and outs of*

a new research institution because everything is backwards, confusing and in the wrong place. The work routine is disturbed and it can turn one's head. And thus the unbelievable happened: I dropped a bottle of concentrated sulfuric acid on the brand-new varnished wooden floor. With lightning speed the acid ate into the floor and not only the puddle but each drop left its considerable traces.

I stood there in shock – horrified of having damaged the new institute – and shook before Zuntz, who had rushed over. I awaited the well-deserved storm to break over my sorry head. Instead – what happened? The good old guy clapped his hands with glee, jumped about beaming and shouted: "Thank God! Now we can finally work like normal people again and stop this damn being careful!" The accident was like waking from a nightmare and the atmosphere returned to being comfortable.

(Plesch, 1949, p. 63)

Toward the end of his term as rector, Zuntz received an invitation from some American scientists to visit the United States, which provides notable proof that Zuntz had garnered an international reputation over the years of his applied physiological studies on man and animal in the field and laboratory. In a letter to the Minister of Agriculture, State Domains and Forests Zuntz requested leave and support for the scholarly journey to the United States, summarizing his motives, objectives and tour plan as follows:

Your most obedient undersigned has been invited by Messrs. True and Bailey, the directors of a course taking place next July for officials at agricultural experiment stations in the United States, to participate in said course as a lecturer.

Subject to Your Excellency's permission, I have given my acceptance. I now wish to request most humbly that leave be granted me as of July 8, 1908 for this purpose.

I will make arrangements that my seminar is completed before my departure, and the practical course in veterinary physiology can be concluded very well by private lecturer Dr. Caspari together with assistant Paechtner.

The course, to be held in Ithaca in the state of New York, will be completed at the beginning of August, and my intention is to subsequently take a research trip through the United States.

During this trip, I would like to visit the most important agricultural experiment stations and physiological institutes in the country. In particular, I wish to work together with the renowned physiologist Jacques Loeb for an extended period in his laboratory in San Francisco in order to study the methods of artificial parthenogenesis that he discovered. In addition, I hope to spend some time in Boston to study the apparatus, moved there from Middletown, of the late Professor Atwater, and also some time at the State College of Pennsylvania to become acquainted with the working methods applied by Professor Armsby.

My hope is to become familiar with pieces of equipment and methods which could be used by my new institute through these studies.

Moreover, I would like to take the opportunity to enhance my studies of many years on the effects of high altitudes on humans. At Pikes Peak, 4300 meters high, a cog railway makes it possible to reach altitudes at which the typical effects of mountainous regions become strongly evident, without any physical exertion being required, thus eliminating the complications caused by strenuous muscle activity, which were unavoidable in my investigations thus far.

Partly based on my experience of the course at Ithaca, and partly during the rest of my journey, I would like to gather material to compare our educational methods with those in America, especially at agricultural secondary schools and colleges, but also in the teaching of science and medicine in general.

Specifically, it is the system developed in Cambridge, near Boston – learning through practical work, the success of which is manifest in terms of a well-rounded view of nature – which I would particularly like to look at more closely.

Furthermore, I intend to direct my attention to the students' way of life, the advantages and disadvantages of athletics, the role played by alcoholic beverages, as well as comparing the manner of preserving one's personal honor.

Although the American committee has consented to cover my travel expenses to and from Ithaca in addition to an honorarium of 100 dollars, the realization of the program outlined over a period of three to four months away from home will involve very considerable costs, which on the basis of inquiries made I estimate to be at least 4000 [presumably Reichsmark*].*

Therefore I wish to ask for a travel allowance of approximately half this sum.

(Letter from Zuntz to the Minister for Agriculture, 1908, fol. 192–193 recto, Secret Central Archives)[57]

The importance of his travels should not be underestimated, as can be seen by the remarks of Emma Zuntz:

The ability to devote himself to work was as strong as his ability to relax, to take pleasure in nature and to leave all thoughts of his work behind on his travels. [While living] in Berlin, it became a favorite habit of his to return home during his vacations and to see his siblings and the old homeland. Longer trips to Switzerland followed. There he, the man who did not have even a moment of [free] time to train his body in Berlin, became a competent mountain climber. He himself experienced how the wonderful mountain air can restore the spirits of anyone living in a big city. This is how the plan matured to study the effects of high altitude climates on humans systematically. After long and tedious preparations, he was able to take the wonderful [scientific research] trips to the Alps, which were what the highlight of his [scientific] work was based on: Grandfather summarized the results of his experiments on the mountains in a great book: Höhenklima und Bergwanderungen[in] ihre[r] Wirkung auf den Menschen. *If you read it, you will learn how much work, deliberation and self-restraint was required just for the preliminary experiments. Then in 1901, the first expedition took place, during which he first performed experiments on the Brienz [side of the] Rothorn [mountain], then on the Monte Rosa [mountain range]. Whenever he told stories about his life in the wonderful Alpine region, grandfather's eyes would just light up with joy. For him, this combination of work and the greatest pleasure he could take in nature left nothing to be desired. What luck for him, then, that so many questions remained unanswered after he had worked out the results, and so back to work he went on Monte Rosa with Durig, his favorite student, in 1903. Later, he went on another research trip to Tenerife, but he was not as happy there as in his beloved mountains. He also traveled to Denmark, Sweden, Norway and England, paying courtesy visits to his foreign colleagues who had come to see him in Berlin, and working as a guest of honor at the different institutes. He observed life in these foreign countries with great interest and he made many new friends. America made an overwhelming impression on him – it was really a new world for him. At the beginning, for the first months, he devoted his entire strength [to his work] as an exchange professor. Then he took a vacation and traveled the continent to California, where his student Jacques Loeb headed a large laboratory. He spoke to people wherever he went, and he made friends with the faith healer women and the students, who worked as cattle drivers to earn the money for their next year of study. He just had to see everything, even the plague laboratory despite all the warnings. He experienced a couple of terrible days when he fell ill with a high fever, but, thank God, it was only influenza. [However,] the wonderful time had a tragic close, because before returning home, he received the news that his wife had once again fallen seriously ill. In fact, she never recovered, nor returned to his house again.*

(Emma (Sarah) Zuntz, undated, p. 46)

The wish to visit Jacques Loeb in San Francisco goes back to an old relationship. Loeb,[58] who had been educated by Friedrich Leopold Goltz and Adolf Fick, had worked as Zuntz's assistant in 1885. Furthermore, as can be gathered from the sources mentioned above, Zuntz was not an introverted person, an academic who shut himself off from his surroundings. Rather, besides his purely scientific endeavors, Zuntz placed great value on the contemplation of the social framework within which daily research took place, extending to the question of the individual's "preservation of personal honor." It may have been this attitude, this characteristic trait, that moved his students, colleagues and friends to describe the working atmosphere at Zuntz's Berlin laboratory, and on the numerous expeditions, as harmonious – even unique:

Noble, helpful and good, those were the essential features of your nature. You could forego pleasures, live in want and modesty, and still help others with your advice. You would share the last bread crumb, the last warm spot by the stove, and likewise the hardest work in the burning sunshine, with your students. And in the same way, you would also share your entire wealth of knowledge and skills with everyone who came to you for counsel and assistance. Whether young novice or mature scholar, you would listen to all with the same patience. Never would you show your superiority, rather with kindness and in your patient, quiet manner you directed each of us down the right path. In the morning when you came into your laboratory and work was at a standstill at all the tables, it took just a few kind words and a clear hint, and things took off again at full steam, until the next time when things faltered and once again your master hand pointed us in the right direction ... You were like a father to us all. Everyone came to you with their matters of the heart and all could drink from your virtually inexhaustible fount of learning. You did numerous good works and helped so many to achieve the joys of home and hearth. How great was our love and admiration for you, how high the respect and esteem of your colleagues throughout the world.

(Durig, 1920, p. 344)

The fact that Zuntz's merits as scientist and teacher favored the formation of a special kind of school around his personality was an intrinsic consequence of his approach to research and his very nature. So often do we see a famous man shining at the center of a circle by virtue of his unsurpassed knowledge and triumphant success. His followers sit in amazement and admiration at the feet of the master; but it is easy for a certain coolness and reserve to separate them. This also means that the seeds sown by the master cannot come to full bloom, for they lack the warmth of personal contact. But Zuntz had built a school about him, a circle of people, that was united not only through the bond of common scientific interests but also in their devotion to a great man. Zuntz did not reign here as the Almighty, the Virtuoso, the Omniscient, not as the strict teacher and master, nor as an academic. No, Zuntz lived among his students as a kindred soul and friend. He lost himself in the crowd of his students, his spirit filled [the halls of the institute] and yet he had just the same ideals as his students, and was involved in the same problems. Wherever Zuntz was occupied, whether in the field or on expeditions, one never worked under him, no, one worked together with him, one lived together with him.

(von der Heide, 1918, pp. 343–344)

Loewy (Loewy, 1922, pp. 4–6) and Müller,[59] his other close colleagues in Berlin, spoke about the "working atmosphere" in Zuntz's laboratory in a very similar manner.

At the end of the summer in 1908 Zuntz returned from his research trip to the United States. During his travels many personal experiences with the living and working conditions in America had left a deep and lasting impression, which were decisively to influence his opinion and estimation of local developments and tendencies in Europe. Specifically, Zuntz took a critical look at popular athletics in a direct comparison of Europe and America (Zuntz, 1917a), a point that will be dealt with in greater detail later in this chapter. Scientifically speaking, the greatest impression made on Zuntz during his travels in America was the work of Loeb in Berkeley. In December 1908, soon after his return from the States, Zuntz presented a report of his experiences to the Physiological Society in Berlin (Zuntz, 1908a, pp. 708–709).

Once he had relinquished his post as rector at the Royal Agricultural College in the spring of 1908,[60] had spent the summer in the United States, and had initiated the construction of his new Institute for Veterinary Physiology, Zuntz could again devote more time to his scientific activities. The variety and quantity of publications in the ensuing years lead to the belief that Zuntz was now drawing upon a wealth of forty years' experience with all the energy he could muster. A key issue at this time was the cirumstances under which physiological circumstances lactate is produced. Zuntz believed that lactate was produced only under hypoxic conditions. As stated recently by Barnard and Holloszy, this was a point "on which Zuntz was mistaken, and his interpretation was very reasonable considering the complete lack of information regarding the regulation of glycogenolysis" (Barnard and Holloszy, 2003, p. 294). But Zuntz and his school were absolutely right when they concluded, based on their numerous studies on nutrition during the preceding decades, that both fat and carbohydrates mainly serve as substrates for energy during exercise in humans, and that the role of protein as an energy-providing substrate was negligible (Barnard and Holloszy, 2003, pp. 294–297; Brooks and Gladden, 2003, pp. 323, 330–331). It might be said that in these papers on nutrition Zuntz was leaving the classical field of nutritional physiology and taking a further step towards a biochemical understanding of living processes. At the end of 1908, two lengthier articles again dealt with the physiology of nutrition (lactation). The first was jointly published with Ostertag (Ostertag and Zuntz, 1908a, pp. 201–260), while the second, prepared in cooperation with Oppenheimer, appeared in *Biochemische Zeitschrift* (the Biochemical Journal) (Zuntz and Oppenheimer, 1908, pp. 361–368) – of which, not surprisingly, Zuntz was a co-editor – and reported on decisive technical improvements in the respiratory apparatus of Regnault and Reiser.

The next year, Oppenheimer gratefully dedicated his *Handbuch der Biochemie des Menschen und der Tiere* (Handbook of Human and Animal Biochemistry) to "his most honored teacher N. Zuntz" (Oppenheimer, 1911, front page). One of the many contributors to this work was Leo Zuntz, Nathan's eldest son (Oppenheimer, 1911, front page). At this time Leo Zuntz worked as a doctor in Moabit (Schwalbe, 1908, p. 137), and participated periodically in various research projects in his father's institute (Loewy, Loewy and Zuntz, 1897, pp. 477–538; Zuntz L., 1903, pp. 601–606, 631–634).

Finally, in 1909, the first edition of the *Lehrbuch der Physiologie des Menschen* (Textbook of Human Physiology) appeared under the mutual sponsorship of Zuntz and Loewy as chief editors (Zuntz and Loewy, 1909). In 1913 the second edition was published, and reviewed as follows by Fuerth in the *Zentralblatt für Physiologie* (Physiology Newspaper):

> *It was certainly no mean task to edit a work compiled for teaching purposes that had almost as many authors as chapters, and to do so in such a fashion that the unity of the entire work did not suffer as a result. Proof that this has been accomplished, and that the volume met with the approval of readers for whom it was intended, is most easily apparent in the fact that within just a few years the second edition is already available ... The clarity and expedient choice of illustrations as well as the book's presentation deserve a laudatory mention.*
>
> *(Fuerth, 1913, p. 742)*

One of the contributors to this work was Scheunert, who in 1921 would succeed Zuntz as chair of the Institute for Veterinary Physiology at the Royal Agricultural College.

During the years around 1910, Zuntz turned his attention with renewed interest to a research topic which had intensively captivated him several years before: climate physiology.[61] His interest in the corresponding issues had grown during the expeditions to the high mountains and during his studies on the physiology of marching, especially in conjunction with problems of clothing physiology and occupational health research.

Indeed, on June 14, 1908, Zuntz sent a letter to the *Geheime Medizinalrat* (privy medical councilor) Dr Loebker in Bochum, in which he described his concept for research on the work physiology of miners. Zuntz hoped to receive information from Loebker, who had contacts in the *Institut für Hygiene und Bakteriologie* (Institute for Hygiene and Bacteriology) in Gelsenkirchen, as to "whether this institute or any other authorities would be interested in conducting the investigations portrayed." Zuntz presumed that we

> *demand extreme exertion of humans in today's world for an entire series of professions, and thus it appears to be of great import to investigate more precisely the conditions under which this work can be carried out most favorably. Studies of this sort would first need to address the issue of how the work can be most economically carried out, that is, with the least expenditure of nourishment, and secondly, to what extent the same can be called for under a variety of special conditions without the workers' health being jeopardized.*
> *(Letter from Zuntz to Loebker, 1908, fol. 88–89, Staatsbibliothek Berlin)*

In order to clarify the practical benefits of a study of this sort, Zuntz referred to his own analyses of the physiology of marching, which he had performed with soldiers:

> *These investigations have had the practical result that the marching ability of soldiers has been increased by way of a certain reduction of baggage weight and by the clothing being better adapted to [meet] the needs created by heat production. Furthermore, these investigations also showed that consuming easily-digested foodstuffs in small quantities, especially of sugar, during exhausting marches is very beneficial. These experiments revealed that the greatest obstacle preventing high performance was excessive temperature ... The temperature at which the performance is considerably restricted depends on multiple factors, among which the most crucial are humidity and air movement. While experimenting on the marching soldiers, I was able to determine that the danger of heat stress given a certain temperature is lower to the same degree that the air moves and is dry. [We] were even successful, at that time, in exactly determining in numbers the amount of sweat that the body needed to excrete in order to avoid overheating at various temperatures, wind velocities and humidity levels.*
> *(Letter from Zuntz to Loebker, 1908, fol. 89–92, Staatsbibliothek Berlin)*

Since the opportunity existed, as Zuntz contended, in particular in mining operations, to reduce the negative effects of excessive temperatures through "pro-active ventilation and, to a certain degree, by using measures which counteract the saturation of air with water vapor," the practical benefits of a corresponding study which investigated "the work performance and its influence on the entire organism in connection with the special climate conditions which dominate in mines" was clear. In this context, Zuntz had both the economic benefits for the businessman as well as concrete improvements for miners in view:

> *If one considers the practical significance, for example, of creating a standard for the longest permissible shift length; how important the question may be, under certain circumstances, of whether the health of a miner is seriously jeopardized when certain conditions demand a longer stay [in the mine], and whether measures exist which can remedy any dangers without complications, and if one furthermore considers the significance the businessman attaches to his workers always having the highest possible energy levels and ability to perform, then it appears clear from the very start that the expenditures which would be required, while not inconsequential for such an investigation, if it is to be thorough and exhaustive, represent a completely economical capital expenditure which would be very well worth it.*
>
> *(Letter from Zuntz to Loebker, 1908, fol. 93–94, Staatsbibliothek Berlin)*

Similar to his study on the marching performance of soldiers, Zuntz's detailed plans included determining the amount of oxygen consumption during the normal exertion of a miner by studying miners working in the mine:

> *The apparatus, which I have already tested under other difficult conditions, for example while mountain climbing, represents the means [for conducting the experiment]. Likewise, the previously tested method for investigating the performance of the digestive system under the conditions given in the respective case would be applied. In addition, the conditions of thermoregulation, that is, the behavior of the body's temperature and the sweat excretion under the conditions of this exertion, would need to be researched and, finally, the influence of the same on the circulation [system] and the respiratory organs both during exertion as well as thereafter would need to be studied.*
>
> *(Letter from Zuntz to Loebker, 1908, fol. 94, Staatsbibliothek Berlin)*

Therefore, not surprisingly and not simply by chance, Zuntz also got deeply involved during these years in the evaluation process for the inauguration of the Kaiser-Wilhelm-Gesellschaft.

The Kaiser-Wilhelm-Gesellschaft (hereinafter referred to as the Kaiser Wilhelm Society) for promoting science and research (today's Max Planck Society) was established on January 11, 1911, as a registered association under the chairmanship of the Prussian Minister for Education August von Trott zu Solz (1855–1938) in Berlin. By founding this organization, Prussia followed an international trend of establishing research institutes that were independent of the state, much to the interest of Emperor Wilhelm II. He lent his name to the self-administrating institution, which was financed primarily by private patronage but was subject to state supervision. Its objective was the establishment and maintenance of non-university research institutes focussing primarily on the natural sciences, which was to complement the *Königlich Preussische Akademie der Wissenschaften* (Royal Prussian Academy of the Sciences). The first president of the Kaiser Wilhelm Society was the theologian, Adolf von Harnack.

Despite the fact that biology, in contrast to chemistry, was still not a generally recognized scientific discipline in 1910, the Kaiser Wilhelm Society proved itself determined to make short work of establishing a research institute for biology. The first preparatory step taken in the meeting of the *Verwaltungsausschuss* (administrative committee) on February 21, 1911, was to commission the corresponding expert reports (Archive: A 1 1218: *Vorbereitung der Gründung biologischer und medizinischer Institute* [Preparation for founding biological and medical institutes], February 3, 1910–November 18, 1911, fol. 34). The responsible departmental head in the Prussian Ministry of Education, Friedrich Schmidt (who changed his name, and was known after 1920 as Schmidt-Ott) created a list of thirty professional scientists, *sachverständige Gelehrte* (practiced scholars), who were to assume this task (Letter from Harnack to the Prussian Minister of Education dated February 25, 1911 [Archiv der Max-Planck-Gesellschaft, Archives of the Max Planck Society: A 1 1218, fol. 35]). The list of scientists printed in the annex of the meeting's record (Archives of the Max Planck Society: A 1 1218, 1911, fol. 76/77) reads like a Who's Who of the field's contemporary elite,[62] and can be seen as proof of the great esteem in which Zuntz was held, in that he, as a non-university researcher, was requested to state his position alongside such illustrious colleagues as *Wirklicher Geheimer Rat* (high privy councilor) Professor Dr Emil Fischer (Berlin), *Geheimer Medizinalrat* Professor Dr Waldeyer (Berlin), *Geheimer Medizinalrat* Professor Dr Rubner (Berlin), *Geheimer Medizinalrat* Professor Dr von Wassermann (Berlin), *Geheimer Regierungsrat* (privy higher executive officer) Professor Dr Max Delbrück (Berlin), and Professor Dr Svante Arrhenius (Stockholm, Sweden).

In order to spur on the planned questionnaire campaign, the President of the Kaiser Wilhelm Society himself composed a letter to the Minister of Education on February 25, 1911, in which he asked the "*hochgebietende*" ("illustrious and commanding") State Minister personally to appeal to the scientists listed to prepare and submit the relevant expert reports. Von Trott zu Solz complied with this request, and on April 18, 1911, the corresponding letter was sent to the selected scientists:

> *... Under today's altered conditions of study, a large number of disciplines are finding that the funds and working capacity of their university institutes are taken up to such a degree by instructional duties that the free development of a researcher is becoming increasingly constricted. Also, scientific advances have resulted in new special fields coming into existence which have not yet established a corresponding representation in the universities, since they either cannot be adjusted to fit the framework of the universities without additional effort, or, in light of the scope of funds and institutions which would be required to successfully make them a part of the universities, cannot expect to find such funds and institutional framework ...*
>
> *The Kaiser Wilhelm Society now has the task of deciding which research institutes to support. In the basic concept leading to the establishment of the society, biological research fields have been targeted as those subjects, in addition to chemistry, at which efforts and energy should first be directed ... The issue now is how to go about dealing with the multifarious fields of biology.*

In order to gain clarity on this topic, it would seem that hearing the viewpoints of competent experts representing various fields of biological sciences would be of great value. The Kaiser Wilhelm Society thus contacted me with the request to obtain expert reports on this issue from a number of practiced scholars. Thus, gracious gentlemen, I would be most grateful if you would be so kind to submit a detailed expert statement on the matter.

(Letter of the Minister of Education dated April 18, 1911. U I K. Archives of the Max Planck Society A 1 1219, fol. 69)

A summary of all applications for funds and suggestions which the ministry had received thus far was attached to this letter (this was marked as "Confidential," Archives of the Max Planck Society, 1911, fol. 71–75).

In this context, the *Denkschrift* (essay) published on January 26, 1910 by Otto Cohnheim, a leading German physiologist at the time, outlining his thoughts on the subject, is of particular import. From 1903 to 1913, Cohnheim was an associate professor of physiological chemistry at the University of Heidelberg. He had been a guest professor in 1904 in the United States, and in 1909 again visited a large number of scientific research institutes in New York. At the request of Schmidt-Ott, he summarized his impressions of his time in the USA, criticizing, in this context, the structures of German academia, which in his view compared unfavorably with the American system. His analysis referred to the theoretical medical fields of physiology, physiological chemistry, pharmacology, experimental medicine, bacteriology and ontogenesis based on experiments. His evaluation of German biological sciences led him to conclude that the Americans had "been leaders [in this area] for years. In comparison, there can unfortunately be no doubt that German physiology has hardly distinguished itself in the past few years" (Archives of the Max Planck Society: A 1 1218, fol. 1–5). Cohnheim believed an important reason for the increasing prominence of the Americans lay in the better financial situation of American research institutes, but he also listed causes tracing back to the organization of the scientific establishments, citing first and foremost the lecture fees charged by professors at the German universities:

... this system has resulted in two consequences:

1. *It makes the creation of new subjects, or the branching off of fields which have become or are becoming independent, more difficult. After all, when a professor is appointed and then told that while he will be earning only a meager salary, 10,000 Marks of supplementary income are connected with the position, then it would be unreasonable, presumably even legally impermissible, to deprive him of a portion of these 10,000 Marks by establishing a new professorship without compensating him for this loss. In fact, in the last 40 years, very few new independent professorships for medicine or for a natural science have been created in Germany. This means that all those who are creating a new specialist [field] or working in special [areas], in other words, the most progressive scientists of all, are getting short-changed ...*
2. *... There are exceptions, but the rule is generally that in German institutes for physiology, pharmacology, bacteriology and biology, directors have neither the time nor assistants to do the work necessary.*

(Archives of the Max Planck Society: A 1 1218, fol. 1–5)

In order to help abate the negative circumstances mentioned, Cohnheim proposed to "found research institutes for very specific tasks," such a one being a "nutrition institute which would unite the Pavlovsk Institute in Petersburg and the Boston Institute. An institute for hereditary and breeding issues, where animals could be observed over generations" (Archives of the Max Planck Society: A 1 1218, fol. 1–5). Furthermore, Cohnheim suggested creating a "fund for scientific work alongside the general institute funds" from which, for example, the very costly animal experiments could be paid for.

It may or may not be a direct consequence of Cohnheim's essay, but all scientists contacted by the Minister of Education were to state which areas of work listed in the annex "most" deserved consideration:

> *As a result of the fact that in the negotiations held prior to the founding of the Kaiser Wilhelm Society, the need was also contemplated, alongside the demand for new research institutes for chemistry and physics, to grant the biological sciences more space and funds than has been the case thus far, there is strong interest in establishing biological research institutes as a means of performing the tasks to be addressed by the Kaiser Wilhelm Society. This was expressed on the one hand by the fact that a series of endowments made to the society have been expressly earmarked for biological research purposes, while on the other hand, resulting from various recommendations from professional circles, attention has been drawn to a series of fields of study in biological sciences which could initially be considered for the establishment of special research institutes.*
>
> *(Archives of the Max Planck Society: A 1 1219, fol. 71–75)*

The fields of study referred to were experimental zoology, a biology facility (zoo), the chemical and physical mechanics of ontogenesis and genetics, researching the biological relationships of land and fresh-water mollusks, general pathology, pathological physiology, protistological research, experimental diagnostics and therapy, as well as experimental psychology.

It is particularly interesting that this summary submitted by the Minister of Education called for the establishment of a bio-ontological institute in Dahlem, outside of Berlin. This institute was to "combine all sciences which deal with organisms and their development," and was "intended to be a complex of connected, yet individual institutes" (Archives of the Max Planck Society: A 1 1219, fol. 71–75):

> *These should not be institutes intended to primarily serve research purposes alongside the university establishments. Instead, the university itself is to undergo restructuring in biological fields, which as a result of the more purposeful grouping of individual disciplines and the geographical location of the institutes would give research work greater freedom alongside teaching. Those fields of natural science, which are innately linked but worked on separately due to the practical requirements of instruction, are to be joined. Key areas of zoology such as comparative anatomy, embryology and physiology are to be released of the excessive demands placed on them for medical purposes, and*

areas which thus far have been considered remote such as palaeontology and anthropology are to be brought closer to the related biological disciplines.

(Archives of the Max Planck Society: A 1 1219, fol. 71–75)

On the topic of work physiology, the following was stated:

Two endowments totaling 400,000 Marks were donated to the Kaiser Wilhelm Society for the establishment of an occupational hygiene and physiological institute. *Even if one could assume that this would extend beyond the sphere of work the Kaiser Wilhelm Society has defined as its practical task, nonetheless, in light of the conditions stipulated here as elsewhere by the donations, said practical cannot be excluded, provided the overall operation of the institute, its methods and objectives are determined by scientific research. But doubts have also been raised as to whether it is necessary to found a special institute for this work, whether it is right to separate the science of hygiene and the physiology of the worker from general hygiene and physiology of humans, as would naturally occur as a result of its being allocated a special institute, and whether the problems arising in this special field could be successfully addressed with the same or even greater prospects by presenting generally oriented institutes with these issues as a particular task.*

(Archives of the Max Planck Society: A 1 1219, fol. 71–75)

The hopes in relation to nutrition physiology were summarized as follows:

Furthermore, attention has been drawn to certain problems of physiology *that specifically concern* investigations of nutrition and digestion. *It is noted that also in this context, research is almost entirely dependent on experimenting with living animals and that for such issues as, for example, how certain forms of nutrition affect the organism (vegetarianism, influence of alcohol, coffee, etc.), a reliable answer can normally only be anticipated if the experiments can be continued over a longer period of time and, if possible, be performed on several consecutive generations of animals.*

(Archives of the Max Planck Society: A 1 1219, fol. 71–75)

In his expert report dated May 15, 1911, Zuntz extensively addressed the latter two topics (Expert report by Zuntz dated May 15, 1911. Archives of the Max Planck Society: A 1 1220, September 8, 1911–October 6, 1911, fol. 144). To him, "more than anything, experimental biology in the broadest sense is in need of support, first and foremost." In this context, he referred to the *Biologische Versuchsanstalt* (Institute for Experimental Biology) in Vienna as a model institution,

... which covers both branches of biology: botany and zoology. The institute is furnished with exemplary equipment for experimental breeding experiments. It provides the possibility to subject generations of animals continuously to various climate factors and has already boosted our knowledge significantly regarding experimentally producible variations.

(Expert report by Zuntz dated May 15, 1911)

With particular explicitness, Zuntz noted that the Viennese institute was "maintained essentially from private funds" and was thus in a position to assume tasks which even the best-equipped zoological and physiological institutes of the universities could only perform to a modest extent:

> *Alongside the Viennese institute, zoological centers, first and foremost the one in Naples, would also serve as examples for the new institute that I am imagining. The institution would have to offer the same opportunities for studying land animals as the Naples institute does for ocean creatures. But it would need to provide the possibility, and this in a far larger scope because of the tasks involved in studying land animals, of keeping and observing the experimental subjects for years under all sorts of natural conditions imaginable, in other words,* **and particularly** *so, without a significant restriction of freedom.*
> *(Expert report by Zuntz dated May 15, 1911)*

Zuntz also acknowledged the very practical requirements that were to be taken into account when establishing such an institute:

> *Thus an institute of this sort can only thrive outside of a metropolitan area on spacious rural property. From this perspective, the Viennese establishments referred to above are much too cramped. On the other hand, and quite naturally so, the researchers at such an institution cannot continuously do without the mental stimulation offered them by scientific centers. Thus, convenient travel connections to a large city must exist.*
> *(Expert report by Zuntz dated May 15, 1911)*

An original thought that Zuntz put down on paper at this juncture, and which of course was connected to his own diverse research projects, was the idea of founding sister institutes in various regions under different climate conditions in order to logically enhance the work of a parent institute:

> *When an institute of this sort develops positively under competent direction, it will not be long before the need arises for sister institutes [located in various geographic areas and subject to] particular conditions [to be founded]. It will be desired, for example, to have a research institute on the ocean, on a larger lake, at a higher mountain altitude and finally both in the tropics and in the arctic. In this regard, permit me to call to mind the great approval with which the international research institute in the high mountains (Col d'Olen) was met, founded by the recently deceased Angelo Mosso. Also, the institute just mentioned is exemplary inasmuch as it reveals that at relatively modest cost, a large number of researchers are offered the opportunity [to carry out] important work year after year.*
> *(Expert report by Zuntz dated May 15, 1911)*

In contrast, he believed that the founding of a nutritional physiology institute was less urgent, since the corresponding establishments already existed in the United States, although in view of his direction of work the concept was of particular import to him:

> *Whether it would be worth while to set up such [an institute] in Europe, taking the excellently equipped Nutrition Laboratory in Boston as a model, seems doubtful to me. The institute, founded with Carnegie funds and abundantly*

> *outfitted with operational means, is directed so admirably and is so willing to open its doors to researchers of all countries that a second establishment of this sort in Europe does not seem to be urgently necessary. But if one considers setting up such an institute [regardless], then one would need to attend primarily to the comparative physiological field in our country, that is, studying the nutrition of animals, possibly all the way down to invertebrates. And if competent chemists and physiologists work hand in hand, great successes can be anticipated.*
>
> *(Expert report by Zuntz dated May 15, 1911)*

Finally, in his report, Zuntz stressed one last important factor pertaining to the organizational structures he envisioned for the newly founded institutes. In this context, he focused his attention mainly on future employees: in contrast to the universities, at which strict hierarchical conditions made independent work difficult for young researchers and associate professors, collegial cooperation should characterize the new institutes. He also championed strict limitation of the strain and time commitment that teaching duties imposed:

> *But it appears to be of particular importance to me to not forget the importance of drawing on competent and creative researchers, in any case. Naturally, it will be necessary to appoint tried and true persons to head the planned institute, that is, researchers who are no longer very young. Nevertheless, when selecting them, in addition to and indeed, outweighing their work performed thus far, it is vital to consider whether they have the ability and desire to allow younger and ambitious professional colleagues to work independently alongside them. Sometimes, it is precisely the most talented and creative minds who are little suited to teach; they cannot be deployed as teachers for their revolutionary ideas alone. Even if they have already proven themselves capable of turning out first-class work as researchers, one would still harbor concerns about allowing them to teach at educational institutes of the universities. This is why, above all, minds such as these should be given space to develop freely at the institutes that the Kaiser Wilhelm Society intends to establish.*
>
> *Thus I would consider it to be particularly important that in the various institutions to be newly founded, positions be reserved for independently acting researchers. Their work rooms would naturally first have to be furnished with the equipment and tools required in each case, and for this purpose the institutes would have to have corresponding funds available to them. One can say that even if 50 percent of the selections prove to be mistakes and the young people to whom positions are granted fall short of the hopes placed in them, the few who are allowed to pursue their independent way sufficiently justify such a measure as a good use of the funds available. No one can deny that in all scientific endeavors it is the people who are the essential for success, while resources are only of secondary import.*
>
> *(Expert report by Zuntz dated May 15, 1911)*

As can be seen already, several of the considerations touched on by Zuntz in his report were also discussed by Cohnheim, who had also been asked to submit this position. In his letter dated May 17, 1911, Cohnheim remarked that he advocated, “with utmost urgency, the creation of a biological facility revolving around a zoo." But he did not share Zuntz's opinion regarding a potential institute for nutritional physiology:

> *Secondly, but hardly less importantly, I feel an institute analyzing the physiology of digestion and nutrition is needed, which would also deal exclusively with experimenting on and observing living animals. In this case as well, above all sufficient space must exist in which the animals can move about freely. Only when the animals live under the most normal conditions possible can observations be continued over multiple generations.*
>
> *(Report by Cohnheim dated May 14, 1911. Archives of the Max Planck Society: A 1 1220, fol. 143)*

In this context, Cohnheim also expressed his opinion on the founding of an occupational hygiene and physiology institute, which in light of the earmarked donations already made was as good as guaranteed:

> *The work hygiene and work physiology institute might be connected with the institute for nutritional physiology. Its program would then have to be expanded and its projects would have to be defined in advance so that particular consideration would be given to the conditions of human nutrition. The investigations planned in the nutrition institutes on the influence of certain nutrition forms and poisons (vegetarianism, alcohol) throughout multiple generations would of course bear a direct relation to work hygiene. I would like to note that a connection between experimental-physiological investigations and studies about worker nutrition has already been carried out at two locations. In Brussels, the major industrialist Solvay founded two neighboring institutes, one physiological and one sociological. Recently, the two directors jointly published an interesting study about the nutrition of the Belgian worker. Most importantly, however, the deceased American physiologist Atwater directed an institute in which respiratory and calorimetric experiments were performed using the most modern technology. Based on this institute, he created an organization for the investigation of mass nutrition, which we have to thank for the best and most abundant discoveries and insights in this field. Atwater planned an international organization for the investigation of workers' nutrition; but the large Nutrition Laboratory in Boston, which the Carnegie Institution built for Atwater's successor, has already been designated in advance for both types of investigations, even though at the moment the work at this laboratory is mostly of an experimental nature.*
>
> *(Report by Cohnheim dated May 14, 1911)*

The Minister of Education announced on September 26, 1911, that he would have a paper prepared by a professional biologist "in which the material contained in the reports is presented in a summarized form" (Archives of the Max Planck Society: KWG Generalverwaltung: *Vorbereitung der Gründung biologischer und medizinischer Institute* [Preparation for founding biological and medical institutes], September 8, 1911–October 6, 1911: A 1 1220, 136). This *Zusammenfassender Bericht* (summarized report), which was published on November 12, 1911, by and large reflected the recommendations made by Zuntz and Cohnheim (Summarized Report on proposals submitted as to the establishment of a biology research institute of the Kaiser Wilhelm Society for the promotion of the sciences. Archives of the Max Planck Society: A 1 1221: *KWG Generalverwaltung: Vorbereitung der Gründung biologischer und medizinischer Institute* [Preparation for founding biological and medical institutes], October 12, 1911–October 18, 1911, fol. 155a).

Coincidently, Zuntz held a lecture in 1910 entitled *Climate and Man* in the Urania Lecture Hall in Berlin. A presentation at this venue, which was founded with the intention of informally educating the populace, shows that the topic was already of interest to the general public. Indeed, Zuntz was very interested in presenting the results of his research to the broader public, using language his listeners would understand. In contrast to many of his professional colleagues who displayed little willingness to make a scientific presentation oriented towards the audience, Zuntz had no reservations about conveying his knowledge in a manner generally comprehensible to an interested public. In addition, such engagement naturally went hand in hand with positive implications for the recognition and survival of a comparatively new professional branch such as climate physiology.

Zuntz collaborated extensively with the *Balneologische Gesellschaft* (Balneological Association), contributing regularly to its *Zeitschrift für Balneologie, Klimatologie und Kurort-Hygiene* (Journal for Balneology, Climatology and Health Resort Hygiene) and publishing many of his lectures in it. The *Zentralstelle für Balneologie* (Center for Balneology) stated its primary tasks to be the promotion of balneology in scientific and practical relations by researching issues of general interest and proliferating the results obtained through suitable publications, and, furthermore, enlightening and instructing the public as to the correct and hygienically proper design of the health resort and spa operations and the supporting spa administration (Excerpt from the statutes of the *Zentralstelle für Balneologie, 1916*, volume III, issue 11). It becomes clear at this point already to what extent scientific research in this field believed itself to be oriented by practice and applicability: researchers sought specific answers to specific questions, oriented by the practical needs of health resorts. In view of the constantly increasing number of health resorts, who referred explicitly to their special "climate effect," Zuntz's climate physiology research work obviously was of particular interest to health resort and spa administrations. He was well aware of this fact: "Accordingly, climate therapy plays a very large role among the remedies being applied today" (Zuntz, 1912g, p. 523). He outlined the function of the Center for Balneology as follows: "The main task of the Center for Balneology [consists of] solidifying the scientific foundation of balneology, that is, creating precise documents that can be used by health resorts for healing purposes" (Zuntz, 1912g, p. 523).

In this context, Zuntz commented repeatedly that the required scientific funding had been much too scant thus far, and was needed now "more than ever." Thus, his research in climate physiology met need. Zuntz set himself benchmarks and objectives for this new field by which to perform his application-oriented physiological research. And it is remarkable to note the approach he took, which corresponds to our modern-day practice, and which takes account of the practical impact of his work. To quote from a lecture Zuntz gave to the balneological society in Berlin in 1911:

> *Climate research is of course of extremely high interest to the forum you represent, inasmuch as it provides the basis for hygienic and therapeutic measures. But being aware of the relationships between climate and man is also of higher anthropological and economic significance. It gives us the basis for assessing the ability of man to perform in various climates. And to the extent that standard rules for remedying damaging climate effects are the outcome discoveries made, they will enable us to know how man can survive in extreme climate conditions.*
>
> *(Zuntz, 1911, p. 854).*

Later, Zuntz summarized the purpose and tasks of climate research for balneology as follows:

> *Recently, efforts have been made to put individual climate factors to use – high and low temperatures, intensive light, electrical charging of the atmosphere and its emanation content – as remedies. Normally, however, we will only be able to avail ourselves of the mixture of the many varied effects given in the climate of a location.*
>
> *Climate physiology, serving medical science, now has the task of researching and determining the effect of each of these individual factors on as many bodily functions as possible, and of determining to what extent, these effects support or mutually neutralize each other in the various combinations in which they may appear in different locations and their climates.*
>
> *It is easy to recognize the tremendous magnitude of this task and to understand that a series of various methods must be utilized in order to carry it out. On the one hand, every individual climate factor has to be applied in varying intensities to persons of various constitutions, healthy and ill, and the effect on the various organs and on the metabolic processes has to be studied. On the other hand, the climates as nature offers them to us have to be researched in their effects on individuals of various body build, various age and under various pathological strains. The second category of studies will always remain the most significant for the physiology of climate because it alone enables an impact lasting for weeks or even years.*
>
> *(Zuntz, 1912g, p. 524)*

Zuntz pursued the questions of climate physiology using tried and true methods and the techniques he had developed primarily in the context of his metabolism experiments. In accordance with his integrative approach Zuntz also worked empirically, yet compared the discoveries made in this manner with the results obtained during control trials in the laboratory. Only there could he isolate the individual climatic factors, such as heat, cold, humidity, etc., and perform a simplified analysis of climate effects (Zuntz, 1911j, pp. 643–644). In order to create such artificial climates, Zuntz resorted to the respiratory chamber in the veterinary physiology laboratory, which had recently been upgraded to an increased capacity (80 cbm) at the *Königliche Landwirtschaftliche Hochschule* (Royal Agricultural College) (Figure 1.22). This facility was perfectly suited for clearing up the questions he encountered in his field studies and significantly enhancing the climate studies performed. The temperature and air humidity factors could be changed as the researcher desired; it was also possible to expose the experimental subjects to various wind strengths using corresponding ventilators. The equipment of the new respiratory chamber included the treadmill and analysis devices already used for earlier investigations to determine oxygen consumption, and carbon dioxide and water release. In addition, Zuntz was now able to measure cardiac output in humans using an innovative method which he had published in 1911 with Markoff and Müller (Markoff, Müller and Zuntz, 1911). They introduced the nitric oxide rebreathing method in this paper, which was elaborated further by Krogh and Lindhard in Copenhagen (Fishman and Richards, 1964, p. 102). The foundation for these studies on cardiac ouput measurements during rest and exercise had been laid with the animal studies on horses as early as in 1898 (Zuntz and Hageman, 1898), described above in this chapter.

Zuntz had already performed extremely detailed investigations on high altitude climate during the numerous earlier expeditions to the mountains, and thus was able to use the data already collected.

As Zuntz explained in a lecture at the 5th International Congress on Thalassotherapy in Kolberg in 1911, his particular interest in the influence of ocean climate on the organism was aroused during a journey to the United States in 1908. He went on to state that this was a "field of thalassotherapy of vital

Figure 1.22 External view of the respiration chamber with (a) gas meter and (b) gas holder.
(von der Heide and Zuntz, 1913, Table IV)

importance for the future" (Zuntz, 1911d, p. 165), and that he felt the physiological effects of a stay on the open sea had only began to be researched thus far.

He stated in this lecture that, while crossing the Atlantic Ocean to New York, he had noticed that his food intake had increased significantly, but that he had not gained any weight. Since Zuntz furthermore found that the opportunities for physical exercise on board a ship were rather restricted, he believed this had to mean that the ocean climate led to an increase in the metabolic processes.

In order to corroborate these presumptions and thoughts dating from 1908 with exact data, Zuntz participated in a multi-week expedition led by Pannwitz to the Canary Island of Tenerife in the spring of 1910. A letter (greetings from the ocean liner) from Zuntz to Darmstaedter is signed by the members of the expedition, and provides evidence of the interest taken by many international researchers who took part in it (Figure 1.23). Zuntz stressed the point in this letter that they would like to complete their knowledge on the impact of climate on humans.

Some studies on sea climate had already been carried out by Zuntz and Durig on the crossing from Hamburg to St Cruz on Tenerife, March 14–19, 1910.

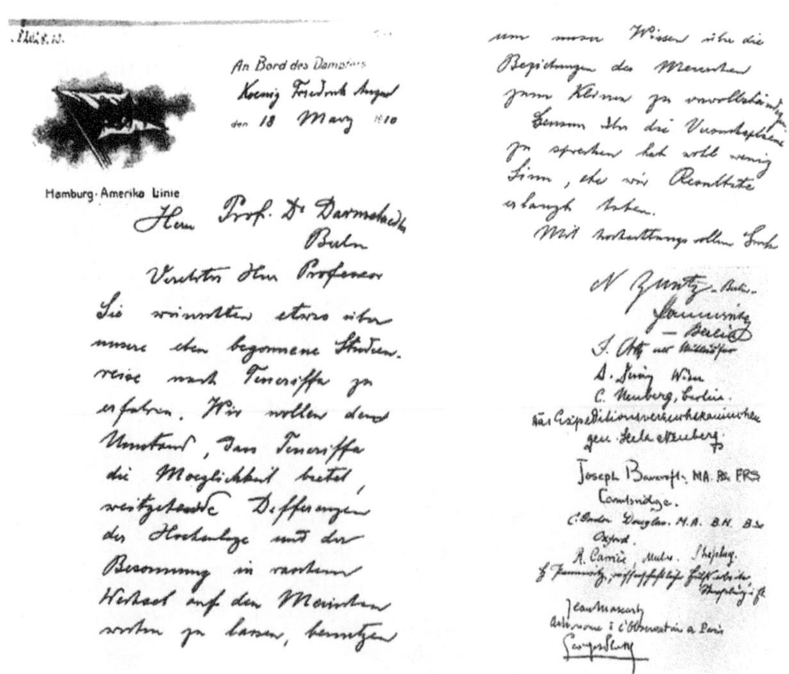

An Bord des Dampfers König Friedrich August den 18 März 1910

Hamburg-Amerika Linie

Herrn Prof. Dr Darmstaedter Berlin

Verehrter Herr Professor

Sie wünschten etwas über unsere eben begonnene Studienreise nach Teneriffa zu erfahren. Wir wollen den Umstand, dass Teneriffa die Möglichkeit bietet, weitgehende Differenzen der Hochlage und der Besonnung in [illegible] Wirkung auf den Menschen wirken zu lassen, benutzen um unser Wissen über die Beziehungen des Menschen zum Klima zu vervollständigen. [illegible] zu sprechen hat wohl wenig Sinn, ehe wir Resultate erlangt haben.

Mit hochachtungsvollem Gruß

N Zuntz Berlin
Pannwitz — Berlin
J. Ott [illegible]
A. Durig Wien
C. Neuberg, Berlin
[illegible]
Joseph Barcroft MA. BSc. FRS Cambridge.
C. Gordon Douglas. M.A. B.M. B.Sc Oxford.
A. Camis [illegible]
[illegible]
Jean Mascart [illegible] Paris
[illegible]

Figure 1.23 Greetings to Prof. Darmstaedter from the members of the International High Altitude Expedition to Tenerife guided by Pannwitz dated March 18, 1910. (Staatsbibliothek Preussischer Kulturbesitz; Darmstaedter collection on Nathan Zuntz)

The shipping management of the Hamburg America Line had provided them with a large cabin for their investigations. Once on board they discovered that the metabolism experiments that Zuntz had developed earlier were too complicated, so during the crossing they were forced to restrict themselves to investigating gas exchange and breathing mechanics while recording pulse frequency and body temperature. Assessment of these experiments revealed that the data obtained at sea corresponded to the control trials in Berlin (Zuntz) and Vienna (Durig). This meant that they were unable to obtain evidence for a particularly stimulating climate at sea.

After the expedition had reached Tenerife, all participants went to the observatory on the Cañadas del Teide (2100 m above sea level) and reached their destination on March 29. They remained there until April 7, and then climbed together to the Alta Vista plateau, located 3260 m above sea level, remaining there until April 10 before returning across the Cañadas to St Cruz on April 13. During the expedition party's stay in the area of the Cañadas and the peak region of the Pico del Teide (3718 m above sea level), the main issue they investigated was "whether the results which Zuntz and Durig had obtained while on Monte Rosa remained applicable even under the altered conditions of the warm climate of Tenerife, or whether the air temperature and light were factors that considerably influenced the previously demonstrated changes in the physiological behavior of man in high altitude climates" (Durig and Zuntz, 1912b, pp. 435–436).

During the investigation they restricted themselves to recording basic physiological parameters, such as pulse frequency, blood pressure and body temperature. Furthermore, they conducted respiratory experiments at various altitudes. A particular focus of their work was the increased radiation of the sun at high altitudes, with its implications for gas exchange and the respiratory mechanism of the organism (Figures 1.24, 1.25).

Zuntz and Durig had already, in 1903, performed similar experiments together on Monte Rosa for this purpose. However, in contrast to the previous series of experiments, the intention at Tenerife was to subject the entire body to radiation from the sun at high altitude. On the southern side of the observatory on Pico del Teide a tent had been erected for this purpose, where the subjects could lie down, and the tent walls could be dismantled quickly in order to take the measurements.

The series of experiments was not without considerable negative repercussions on the subjects' health. Durig's physical state following such a field experiment is described as follows:

> *The effects of the exposure to the sun were intense. Not only did Durig feel a strong burning sensation on his body during the experiment, especially on the chest and upper thighs, but also, as a consequence of the sunning, suffered a quite painful burn on the entire front side of his body, which led to an edema and to blisters forming on his chest, stomach and legs so that even the weight of the blanket at night was extremely painful. The pain was alleviated by the application of anaesthetic ointment. In the evening of the*

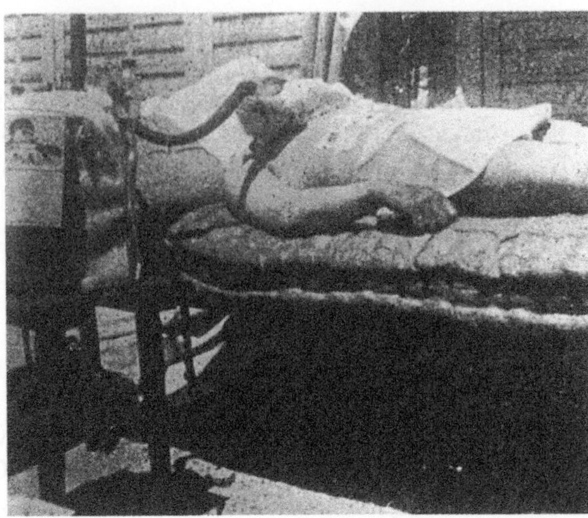

Figure 1.24 Measurements to quantify the gas exchange under resting conditions at high altitude during the international high altitude expedition to Tenerife guided by Pannwitz in the spring of 1910.
(von Schroetter, 1912, p. 11)

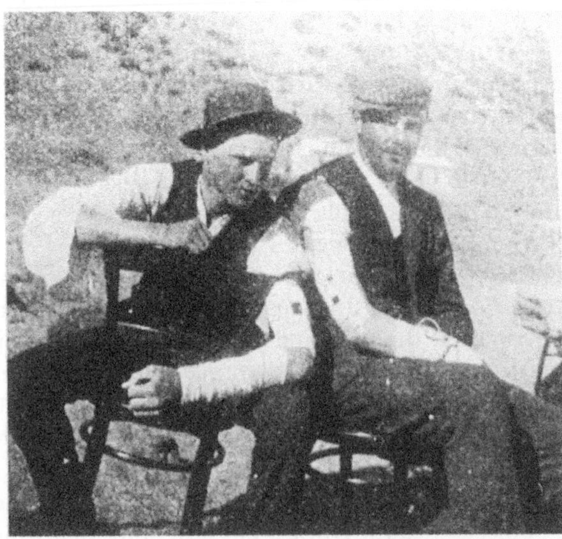

Figure 1.25 Measurements to study the impact of radiation at high altitude during the international high altitude expedition to Tenerife in 1910.
(von Schroetter, 1912, p. 9)

> *day on which he had subjected himself to radiation, the body temperature sank to 35.9°C in the rectum, evidently caused by the loss of heat resulting from the hyperaemic skin.*
>
> *(Durig, von Schroetter and Zuntz, 1912, p. 473)*

When the investigations were over it was apparent that, following such intensive radiation, both gas exchange as well as respiratory mechanics were subject to significant changes, but that these reflected individual fluctuations rather than generally applicable tendencies that could serve as the basis for establishing a role. The most obvious results were obtained by the experiments on alveolar CO_2 tension and pulse frequency, in which the former was proven to decrease while the latter clearly increased under radiation. The degree of alveolar CO_2 concentration decrease was unexpected, since similar values on Monte Rosa could only be recorded in regions located 1000 m higher than on Pico del Teide. As a consequence, Zuntz believed that on the Teide Mountains he might have discovered a factor, characteristic for high altitude climate, which was evoked by the intensive radiation of the sun.

In parallel with the climate physiology studies Zuntz was pursuing in the Canary Islands and on the open ocean, his co-workers Loewy and Müller carried out several expeditions to the North Sea (Sylt) in 1910 and 1911. It was hoped these research trips would shed light on, among other things, the effects of the North Sea climate and the North Sea spas on the human metabolism.

In keeping with Zuntz's own measurements taken during the passage to New York, the experiments on the beach did not produce any results in support of the thesis that metabolic processes might be noticeably altered as a result of a stay at the ocean.

Zuntz also dedicated himself to the issue of in what way, and with what implications, the climate at sea differed from the climate near the coast. Zuntz was well aware that in this context he was dependent on the assistance of scientists from other fields, and in particular on meteorologists. The collaboration with the *Meteorologisches Observatorium Lindenberg* (Lindenberg Meteorology Observatory) (Assmann) and with the meteorology institute at the Royal Agricultural College (Börnstein) had already proved very useful during the Monte Rosa expeditions and the balloon trips he had taken with von Schroetter. Still, Zuntz realized early on that the normal meteorological stations could only provide limited logistical assistance to the research he planned to do on the seaside climate – his queries were too specialized:

> *A great difficulty is emerging for all these experiments as a result of the fact that while the characteristic moments of a climate occur each year, they occur in extremely different degrees. In order to reflect this circumstance appropriately, a series of experiments of this sort must be combined with exact meteorological observations. The customary weather observations as recorded by the meteorological stations do not suffice. Rather, it needs to be remembered that, for example, in the coastal area but at some distance from the beach, the climate might be very different from the climate given where the health resort guests prefer to remain for the greater part of their day.*
>
> *(Zuntz, 1912g, p.514)*

It is easy to imagine an interesting research topic for balneology looming on the horizon here, and indeed Zuntz and his co-workers noted "that up until now, the local climate conditions on the coast, to the extent they are of specific importance for therapeutic purposes, have not attracted the interest they deserve" (Berliner and Müller, 1912, p. 3).

In order to pursue these questions scientifically, Zuntz and Müller developed their own program which looked into the influence that the Baltic Sea climate had on Berlin schoolchildren. The extensive scope of their series of experiments is remarkable: to cite but one example, psychological observations were also to be taken into account, in addition to the physiological attributes. For this reason alone it was ideal to perform the planned study with schoolchildren, since the discipline in school would facilitate an exact implementation of psychological observation.

The planned investigations took place in the summer of 1911 at the Baltic Sea spa of Zinnowitz on the island of Usedom. Since the entire series of experiments was primarily practice-oriented, it is hardly surprising that the board of trustees for the Center for Balneology promised Zuntz and his employees that they would fund this project (Minutes of the meeting of the Board of Trustees of the Center for Balneology on March 2, 1912, Zentralstelle für Balneologie, 1912, p. 15). The project was additionally supported by the *Friedenauer Verein für Ferienkolonien* (Association for Vacation Colonies of the southern (then) town of Friedenau); Müller, Hellwig and Berliner were the scientists responsible for the project.

In correspondence with what the team hoped to discover, the experiments were performed in parallel at two different locations, one directly on the beach and the other at the vacation lodge of the Friedenauer Verein, about 600 m from the coast on the far side of a 450-m wide dune forest.

Already at the beginning of the series of experiments it was becoming clear how extremely the meteorological conditions of the respective locations differed from one another. For example, temperatures on the beach were significantly lower while humidity was much higher than "inland," the term Berliner and Müller used for the site of the vacation lodge. The inland climate conditions reminded both researchers less of a typical ocean climate than of a moderate altitude climate at about 1500 m.

> *But in any case, the dune forest appears to modify the ocean climate to a greater degree than what one tended to suppose previously, and since in Zinnowitz – and in the same way at numerous other Baltic Sea spas – the majority of homes are located in or behind the forest, and accordingly the health resort guests do not spend the greater part of their day at the beach, then it seems ... necessary to determine the climatic difference over several months using exact meteorological methods.*
>
> *(Berliner and Müller, 1912, p. 3)*

For this purpose, both on the beach and inland, measurements were taken or devices installed for determining the temperature, humidity and wind speed. It required considerable effort, both in terms of logistics and personnel, to read and record the corresponding values three times a day at both locations. In order to be able to make the most practice-relevant statements possible, the information was recorded at the times of day at which health resort guests normally were outside. As well thought-out as the experiment was, its complications had been underestimated. The instruments were more sensitive than expected, and the recording staff less exact and reliable than assumed. All in all, the results had to be interpreted with caution. However, it was clear that the differences that people easily sensed between the climatic conditions on the beach and those at the vacation lodge were being confirmed by the measurements, which were even clearer than expected – especially when an ocean wind blew. Thus, even at a marginal distance from the beach, namely at the place "where the settlements belonging to the spas are located, ... the ocean climate could change to a considerable degree and closely resemble the inland climate" (Berliner and Müller, 1912, p. 7). These discoveries of climate physiology had specific, practical implications for the ocean spas on the Baltic Sea, and were to be given more consideration than they had received thus far when building sanatoriums.

Since the study by Berliner and Müller had dealt with one aspect (namely that of the diverse meteorological conditions directly on the beach and on the other side of the dune forest), it was time for physiological and psychological investigations to be performed, which Helwig and Müller proceeded to do. In this project they were able to draw on the numerous concurring experiences of doctors who attributed strong effects on body weight and growth to a stay beside the ocean. Nevertheless, it proved difficult to reconcile these with the concepts and physiological discoveries made thus far. Overall, it turned out to be very complicated to isolate the influence of the climate from other factors, and few definitive discoveries were made.

This study stands out because of the integral approach it took and the diversity of topics it included. For example, the question was posed of "whether every change of habitual abode, this meaning in particular a relinquishment of the unhygienic lifestyle of the metropolitan area" (Helwig and Müller, 1912, p. 2) could not in itself lead to an improved development of the children. Likewise, Helwig and Berliner took into account the social status, as well as the "domestic and personal conditions" (Helwig and Müller, 1912, p. 2), of all the children participating in the study. Thus, alongside physiological and psychological observations, social and environmental aspects were also taken into account.

In the spring of 1912, Zuntz was appointed Director of the newly established Scientific Advisory Council of the Center for Balneology in Berlin. This advisory council had the task of preparing "the decision on the scientific work to be performed and supported" – it is quite understandable that the Board of Trustees felt it was "advisable" that Zuntz assume the directorship of the council (Minutes of the meeting of the Board of Trustees of the Center for Balneology on March 2, 1912, p. 15).

At the same meeting, Zuntz proposed that his investigations at the Baltic Sea now be performed at the North Sea. According to his concept, one group of children from Berlin's Schoeneberg district was to travel to Boldixum on the island of Föhr for the entire summer vacation, while another group was to stay in Berlin's Grunewald forest. The other members of the scientific advisory council raised considerable concerns regarding Zuntz's plans, especially the selection of Boldixum for the planned investigations. They asserted that Boldixum was situated in an area influenced by tideland, and for this reason alone no conclusions could possibly be drawn about the North Sea climate as compared to the Baltic Sea climate. Zuntz acknowledged these objections, and the advisory council decided to first perform the Berlin investigations on children in the vacation camp in Eichkamp while at the same time launching a search for an appropriate location on the North Sea.

In Zuntz's laboratory, however, the researchers did not restrict themselves to moderate European climate conditions, but instead expanded the investigations to study tropical and then, later, desert climates:

> *Thanks to the cooperation with the African researchers Dr. Schelling and Dr. Jaffe, the research that has been carried out for years now in the institute on the physiological effect of various climates was able to be expanded to cover the living conditions of Europeans in the tropics. Dr. Caspari is responsible for processing the plethora of material provided by the gentlemen mentioned, and this research should provide us with practical, meaningful results for the living conditions in our African colonies.*
>
> *(Jahresbericht [Annual Report] the Royal Agricultural College in Berlin, 1908, p. 39)*

Yet even in this research, therapeutic questions played a crucial role. Countries with a primarily desert-like climate, such as Egypt, were becoming destinations for healing and health purposes; providing the scientific foundation for the healing effects attributed to these climates therefore represented a further field of activity for climate physiology research.

In order to gain new knowledge in this area, an expedition left for Egypt in the spring of 1914 under the direction of Zuntz's trusted partner Loewy (Loewy, 1916a, pp. 1–6). The location chosen was the desert site of Bab el Wadi, to the east of Aswan, with climatic conditions that Loewy described as follows:

> *The humidity of the air is down to almost absolute dryness and this for the major part of the day, more or less. Along with the very high air temperature, it can be predicted that this will cause the amounts of water released by the body to be very significant.*
>
> *(Loewy, 1916a, p. 2)*

The main objective of the investigations was to review the generally held opinion that a stay in the desert climate of Egypt had an especially favorable effect on kidney illnesses. This effect was interestingly explained as follows: the body releases large amounts of water mainly through the skin and lungs (*perspiratio sensibilis and insensibilis*), thereby reducing the strain on the kidneys since they no longer have to produce as much urine. Thus, it was suspected that "the dry desert climate altered the metabolism in such a way that the kidney generally worked less, with the skin 'vicariously' stepping in, that is, working more by secreting the water" (Loewy, 1916a, p. 3). This thesis now was to be scientifically reviewed. At the same time, the scientists hoped to draw specific practical conclusions from the results of the expedition that would indicate whether a stay in the desert could be beneficial to people with kidney illnesses or was, rather, unadvisable.

Loewy and his co-workers began by stating that the consideration "that with the restriction of water loss via the kidneys, the concentration of the components dissolved in the urine and thus the concentration work of the kidneys would have to increase" had received little attention thus far. Furthermore, they reported that investigations in their laboratory had provided clear indications that an increased release of water through the skin did not necessarily also mean an increased release of dissolved substances through the skin. In order to support this presumption scientifically, three series of metabolic experiments were performed – one in Berlin and two in Egypt. The objective was to determine to what extent an increase in sweat formation, which is an "increased release of dissolved substances via the skin," went hand in hand with the physical water loss – i.e. the "simple release of steam."

The water loss measured in the desert climate was considerably higher than the comparative values obtained in Berlin, in particular the evaporation via the sweat glands (perspiration sensibilis). According to their results, the increased water loss from the lungs (perspiration insensibilis) played only a minor role. The composition of the sweat was changed very little, if at all, and accordingly hardly any solid components were released through the skin. The experiment proved that the desert climate did not result in a considerable reduction of secretory activity of the kidneys.

The second aspect of the loss theory was also investigated, according to which the kidneys enjoyed a reprieve in the desert climate as a result of reduced urine production. Yet in this regard, too, Loewy and his assistants were unable to provide any scientific confirmation: according to their measurements, the quantity of urine released was not reduced by any means. On the contrary, some of the values measured in Egypt were higher than those which had been obtained during the preliminary experiments in Berlin. The reasoning for this was that "Similarly, no restriction of [the kidney's] fluid loss function occurs under usual eating conditions since any increased water loss through the skin and lung moisture led to a corresponding increased intake of water resulting from a greater feeling of thirst" (Loewy, 1916a, p. 5).

The results obtained by Loewy had consequences regarding the therapeutic use of the desert climate for people with kidney illnesses: "Generally speaking, the kidneys are granted no respite" (Loewy, 1916a, p. 6). While Zuntz was working to intensify his research on the physiology of climate in March of 1912, a decisive, far-reaching development took place in Germany in another area of his work – aviation. In April of that year, under the chairmanship of Prince Heinrich of Prussia, the *Wissenschaftliche Gesellschaft für Flugtechnik e.V.* (Scientific Society for Aviation Technology) was founded in Berlin

(*Jahrbuch der Wiss. Ges. f. Flugtechnik* [Yearbook of the Scientific Aviation Society] 1, 1912/13, pp. 1–38; Harsch, 2001, p. 1). At the same meeting a provisional working committee was established to draw up the statutes and outline the major objectives of the institution (*Jahrbuch der Wiss. Ges. f. Flugtechnik* 1, 1912/13, p. 18). This committee met once again at the beginning of May 1912, and voted to establish a scientific-technical committee. In July 1912, a total of ten subcommittees (a–k) were set up, including the *Ausschuss f für medizinische und psychologische Fragen* (Committee "f" for Medical and Psychological Issues) (*Jahrbuch der Wiss. Ges. f. Flugtechnik* 1, 1912/13, p. 17). As far as reference was made to the committee's members, it was stated that the "*Vereinigung zur wissenschaftlichen Erforschung des Sportes und der Leibesübungen* [Association for the Scientific Investigation of Sport and Physical Exercises]: Professor Nicolai – Berlin; Privy Councilor Professor Zuntz – Berlin" were called in as members by the committee (*Jahrbuch der Wiss. Ges. f. Flugtechnik* 1, 1912/13, p. 17). As stated in Section 3, paragraph 5 of the Scientific Society's founding statutes, one of the Society's objectives was to publish research done in the field of aviation science, which led to the series *Luftfahrt und Wissenschaft* (Aviation and Science), edited by Sticker. One of the first works published in this series that same year was Zuntz's treatise *Zur Physiologie und Hygiene der Luftfahrt* (On the Physiology and Hygiene of Aviation), which today ranks as the earliest and most comprehensive work in the field of aviation medicine in the world.[63]

During the same period, in 1912, Zuntz returned to nutritional issues (von der Heide, Klein and Zuntz, 1913, pp. 765–832) and sports/exercise physiology,[64] and once again made major contributions to the latter field (Mitchell and Saltin, 2003, pp. 258–259; Tittel, 2004, p. 315). Sports and exercise physiology belonged to Zuntz's "long-time favorites" (Zuntz, 1916c, p. 61), especially after his trip to America in the summer of 1908.[65] This journey gave Zuntz many insights into the American lifestyle and left a strong impression, greatly influencing his view and assessment of the European developments and tendencies of his time. This is particularly evident in his work relating to sports physiology, in which he also investigated and addressed athletics and physical activities in various countries, both from societal and political perspectives.

One year after traveling to America, Zuntz composed an article for the journal *Körperkultur und Sport*, an "artistic monthly magazine for hygiene and sports," in which he reported on the "athletic and hygienic impressions gained on a trip to America" (Zuntz, 1910i, p. 9). During his stay Zuntz was lodged at Cornell University in Ithaca (New York State) in a student dormitory, and was thus able to become more familiar with student life than was normally possible for a foreign lecturer. As a result, he particularly noticed that "in America, sports hold an entirely different place in the lives of the students than it does [in Europe]" (Zuntz, 1910i, p. 9), and cited the superior sport amenities at American colleges with their athletic facilities, gyms, etc., as evidence. Zuntz was quite open to objections that might be raised against an over-enthusiastic "cult" of the body, noting that it certainly would not hurt to indicate to the public, "right at the start of any intensive propaganda for physical exercises, that there is the possibility of one-sided exaggeration," and to warn future athletes of too much of a good thing – nonetheless, the athletic movement could rest assured of his support:

> *Indeed, in many instances caring for the body is exaggerated, in contrast to the way we neglect our bodies [in Europe] especially in academic circles. Still, it is certain that the student, when he turns back to his studies after having spent a few hours with physical exercise, will have a different capacity for work than if he had spent the evening drinking beer.*
>
> *(Zuntz, 1910i, p. 10)*

Alongside the "almost complete [alcoholic] abstinence of the majority of the students" (Zuntz, 1910i, p. 10), which Zuntz surely correctly understood to be a consequence of their interest in physical activities and their ambition, the guest lecturer from Germany noticed another positive aspect:

> *In view of these observations, we can certainly hope that the movement for game and sport, which has gained considerable momentum, as a counter-weight to the one-sided mental activity of our youth will not only be beneficial for physical training but rather, under appropriate guidance, may also lend new inspiration to the emotional life of the growing generation.*
>
> *(Zuntz, 1910i, p. 10)*

Seven years later Zuntz expressed very similar views, directly comparing sports in Europe and the USA, as mentioned earlier in this chapter:

To conclude, the issue of scope needs to be regarded systematically, in other words, to what extent people should dedicate themselves to athletic activity in general, and athletic competitions in particular, in their daily lives. Many times we hear the complaint, and it was often expressed to me by Germans who had immigrated to America, that young Americans are so interested in athletic competitions that the entire mental training of the person, his or her interest in philosophical issues, in poetry, paintings and music, suffers as a result.

> *Young Germans, especially those employed in business sectors, find that America lacks the abundance of mental stimulation which they were used to in their homeland, and they cannot muster up enough liking for the things which absorb the entire passionate interest of their American fellow students or co-workers. There can be no doubt that this much too one-sided passion for sports exists is felt here by some individuals, even if they are few. But there can also be no doubt that it currently still represents a healthy alternative to the tendency to turn into a permanent pub resident, and the very superficial discussions associated therewith. In short, we currently do not have occasion to fear the sports movement in Germany. But it is probably not unjustified to warn, at least occasionally, against an exaggerated unilateral emphasis on this activity. Naturally, just as there are specialists for every sort of mental and technical occupation, then in every nation in which physical exercise and sport play the role desired, there will also be a number of people holding a one-sided interest for these things in life. The only protection we have against*

> *these interests forcing other important mental and spiritual valves to a place of secondary importance, without justification, is the organization of our schools, from grammar schools to colleges. Certainly, there are many who would prefer to see sports replace the pubs as a way of relaxing and recuperating from work, and this in particular at our universities.*
>
> *(Zuntz, 1917a, p. 206)*

In addition, his stay in the United States stimulated Zuntz into increasingly addressing issues of nutrition and sport physiology from 1909 onwards (Zuntz, 1901d, pp. 154–174).

Zuntz himself stated that his special new interest in this field was further sparked by the International Hygiene Exhibition, which took place from May to October of 1911 in Dresden. Zuntz had been appointed chairman of the exhibition's division for sports hygiene and sports sciences, and had been commissioned, together with other medical experts and physiologists, to perform sports physiology investigations on participants in competitions. In addition, a forum was to be created in which the methods of sports physiology were to be shown to the public, and "to demonstrate the performance of exercises usual to diverse countries and cultures, in their natural form, not only using pictures, models, devices and other accessory means of all sorts, but also by practical presentation of athletic competitions in the most diverse fields, in order to enable the general public in the broadest sense to be informed" (Zuntz, 1911o, p. III).

Zuntz expressed these ideas in his foreword for the *Bibliography of All Sports*, which was published in the context of the Dresden hygiene exhibition in order to summarize the scattered and complex scientific and popular literature on the topic of sports and physical exercise (Figure 1.26). Zuntz also did not forget the social aspect of athletic exertion that enhanced the "performance of physical exercises," as is evidenced by the following:

> *However, exercise also goes hand in hand with another sort of educational impact of prime importance. On the one hand, it is nowhere as easy as when working together with a large group of young people to practice the spirit of lending mutual support and assistance and to have the right sort of subordination and command ... The right manner of interacting with people develops automatically.*
>
> *(Zuntz, Brahm and Mallwitz, 1911, p. 7)*

The additional tasks of the sports department included, in its scientific division, providing an overview of the typical structure of the organs and how they are influenced by physical exercise. They were illustrated by exhibits such as the life-size muscular system of a human being, including instructions for measuring its proportions, as well as a model of the human gait. In addition, various apparatuses on display measured performance capabilities, metabolism rates and especially breathing, and allowed people to observe their respective influence on the heart's function. In order to study the metabolism during exercise various forms of breathing apparatus were used; these had a dual role as exhibits and tools actually deployed in the sports laboratory. Two sets of

[30]

1913:64

Sonderkatalog

der Abteilung

Sportausstellung

der

Intern. Hygieneausstellung Dresden 1911.

Zusammengestellt

von

Geh. Reg.-Rat Prof. Dr. **N. Zuntz,**

Dr. phil. **C. Brahm**

und Dr. med. **A. Mallwitz,** Abteilungsvorstand.

Verlag der Internationalen Hygiene-Ausstellung
Dresden 1911.

Figure 1.26 The title page of the Sonderkatalog der Abteilung Sportausstellung der Internationalen Hygieneausstellung.
(Zuntz, Brahm and Mallwitz, 1911)

apparatus built upon Zuntz's commission by the company S. Elster in Berlin were available to measure human respiration directly (Figure 1.27).

Other areas of the scientific exhibition were devoted to representations of the exercises in pictures and sculptures, and various aids to be used in exercising; also, indications of possible hazards and damages which could occur as a result of athletic activity. In this context, Caspari, Zuntz's colleague at the Agricultural College in Berlin, was one of a group who presented a graphic depiction of diseased kidneys. In a second, practical part, a comprehensive exhibition presented the most diverse athletic clubs.

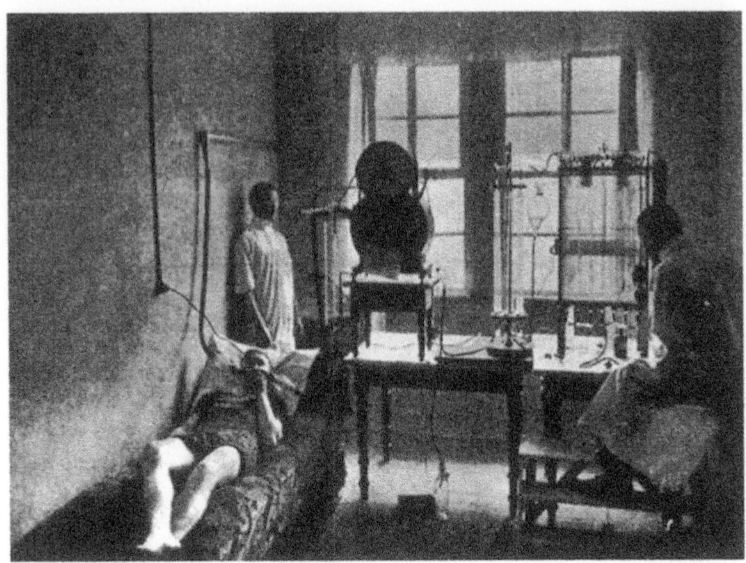

Figure 1.27 A typical installation to measure the oxygen consumption under resting conditions during the International Hygiene Exhibition in Dresden 1911.
(Zuntz, Brahm and Mallwitz, 1911, p. 40)

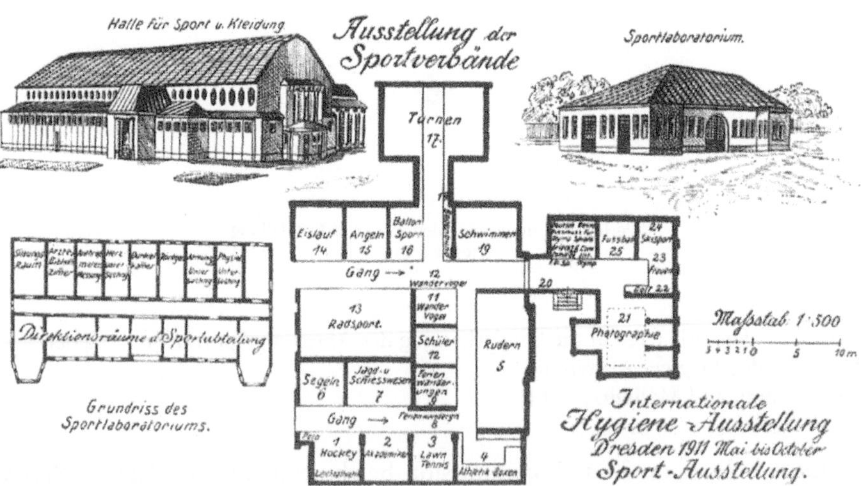

Figure 1.28 Floor plan of the Exercise Physiology Section of the International Hygiene Exhibition Dresden 1911.
(Zuntz, Brahm and Mallwitz, 1911, unpaginated)

Finally, a scientific laboratory, set up solely for this purpose directly next to the *Wettkampfstätte* (competition venue) in Dresden, was available to the scientists; here they were able to analyze the entire physiological behavior of typical representatives of various sports (Figure 1.28). Zuntz and his colleagues hoped to acquire scientifically meaningful results from this series of experiments (Annual Report of the *Königliche Landwirtschaftliche Hochschule* [Royal Agricultural College], Berlin, volume XIX, 1910/11, pp. 58–59). It bears noting, and this was also stressed by Zuntz in the exhibition catalogue, that the equipment in the laboratory had been provided to the researchers without cost by the companies sponsoring the exhibition. Moreover:

> *Showing practical athletic activity is a third unique department of the exhibition made up of the large and exemplary stadium, the swimming and wave pool and, connected thereto, the sports laboratory. This latter is a particularly vivid expression of the purpose of the exhibition: not only to present the existing knowledge and technical means to the public, to the extent this is possible, but also to make a positive impact ourselves by stimulating new research.*
>
> *(Zuntz, Brahm and Mallwitz, 1911, p. 8)*

Zuntz believed that productive scientific work in this field was embodied first and foremost in the sports laboratory:

> *Certainly, ambitious researchers have already, and often, made efforts to subject to a more precise analysis the idiosyncrasies of the body structure and the functions of organs that are particularly active during physical strain such as the heart, lungs, and kidneys. But these efforts were always hindered by the difficulties of creating suitable measurement methods at the place and site [of the sports being performed], and also by the difficulty of subjecting the athletes to any sorts of useful investigations, since for them, any disturbance to the excitement of the competition is unwelcome. These difficulties were not [completely] remedied, of course, but they were significantly minimized by connecting the means for scientifically analyzing the athletes, from a structural perspective, to the sports venue.*
>
> *(Zuntz, Brahm and Mallwitz, 1911, p. 30)*

A comprehensive investigation covered the following measurements that were to be carried out prior to and after the athletic exercise: vital capacities, oxygen consumption, CO_2 content, respiratory quotient (RQ), cardiovascular system (i.e., X-ray of the heart, electrocardiogram, pulse, arterial blood pressure), kidney physiology (microscopic and chemical analysis of urine), and photographic records as a basis for anthropometrical measurements.

In addition, the people participating in gymnastics competitions and other sporting meets were handed the following questionnaire asking them for detailed information on their biography and physical state of being. There were also detailed questions on social habits – to such an extent that Zuntz felt it necessary that he personally guarantee that this questionnaire "will be used for scientific purposes only and otherwise will be kept strictly confidential" (Zuntz, Brahm and Mallwitz, 1911, p. 41):

1. *Name (or first name and first letter of your last name).*
2. *What is your profession and how long have you been working in this profession? Have you changed professions?*
3. *What year were you born?*

4. *In what city were you born?*
5. *What years of your life did you spend in the countryside and which years in a city?*
6. *Did you have any children's diseases and if so, which ones (rickets, diphtheria, scarlet fever, etc.)?*
7. *Have you suffered from joint rheumatism and if so, when?*
8. *Have you had typhus and if so, when?*
9. *Have there been cases of lung tuberculosis in your family and if so, which ones?*
10. *Do you have any physical complaints, for example, racing heart, pains in the side, stomach aches, tendency to have diarrhea or constipation? Do they arise without cause or as a result of what circumstances?*
11. *What sort of schools did you attend and what age were you when you left each of them?*
12. *Did you spend a lot of time outdoors as a child?*
13. *Do you still spend a lot of time outdoors and how long each day?*
14. *Do you drink alcoholic beverages (beer, wine, fruit and berry wine, liquors, brandy, absinthe)? (a) If yes, since when? (b) And approximately how much per day? (c) If no: Did you use to drink alcoholic beverages and how many a day? (d) Since when have you been completely abstinent?*
15. *Do you smoke? Approximately how much do you smoke a day (cigars, cigarettes, long-stemmed pipes, short pipes)?*
16. *Do you consume meat (including poultry and fish)? (a) If so: daily or more than once a day? Can you state how much? (b) If not: Did you use to consume meat and how much or how often per day? (c) Since when have you given up meat entirely?*
17. *Do you regularly consume milk and how much per day?*
18. *If you are a vegetarian: are you (a) lacto-vegetarian (that is, you eat milk, butter, cheese, eggs)? (b) vegan (you consume no animal products)? (c) raw vegetarian (that is, you do not eat bread and other food prepared by cooking)?*
19. *Do you alter the way you live while training and in what manner?*
20. *It would be very helpful to receive some information as to your sex life.*
21. *Have you served in the military? How long and with which part of the armed forces? When did you have your last training exercise for reserve soldiers? If you did not serve in the army, why were you released from this duty?*
22. *For how long have you been exercising (sports, gymnastics)? (a) What exercises did you use to do? (b) What sports do you pursue now? (c) What have you attained (prizes, awards, records)?*
23. *How have you trained up until now?*
24. *Did you practice for the upcoming competition and how?*
25. *What is your approximate height and weight?*
26. *Other comments.*

(Zuntz, Brahm and Mahlwitz, 1911, p. 41)

The results of the studies made in the sports laboratory were to be processed further and then published. Still, the studies in Dresden provided an occasion to establish the *Deutsche Verein zur wissenschaftlichen Erforschung des Sports und der Leibesübungen* (German Association for the Scientific Research of Sport and Exercise) and for the establishment of a corresponding sport-physiology research laboratory on the athletic field of Westend in the town of Charlottenburg (now part of Berlin) shortly before the outbreak of World War I. The first directors of the laboratory were the doctors Nicolai and Michaelis.

As previously discussed, the role of sports at American colleges was a topic which Zuntz had already studied on his trip to the United States in 1908. He reported on his experiences there, and on further work pertaining to sports physiology, on December 9, 1913 at the Urania Lecture Hall in Berlin (Zuntz, 1914b, pp. 439–453). Based on the observation that the manner of competitive and popular sports had changed significantly over the past several decades, attaining an astounding degree of popularity, Zuntz showed, using the example of sport physiology, that applied physiological research benefitted the common good in a multitude of fields.

Zuntz had been inspired in this regard by the experiments he performed on horses and other productive livestock. Starting with the question of the "effect of the animal muscular system as a power [generating] machine," he came to study the "power output under various conditions of human life" (Zuntz, 1914b, p. 439). The practical objectives of exercise physiology, as well as the particular conditions arising during athletic competitions, directed his attention to the relationship of performance to outside conditions, climate effects, attire, and nutrition. At the same time he studied the physiology of exercise to

> *provide clarification not only about the way it takes place, about the way the muscles work together with the most diverse organs of the body, but also in order to answer the question as to what implications these activities have for the development of our body, for its health and capability of carrying out the actual tasks demanded by life. Accordingly, it is to trace the subsequent effect that any intensive athletic exertion will have on the organism, it is to discover the [reasons for the] instances in which physical exertion causes harm, and it is to go beyond the momentary circumstances of the study and educate us as to the later course of life processes, that is, the general capacity of man to perform.*
> *(Zuntz, 1914b, p. 440)*

Zuntz addressed the issue of which nutrition plan was the most effective by first dismissing widely accepted falsities in this area: he believed consuming highly proteinous meats, as propagated and practiced in England, the "fatherland of modern sports" (Zuntz, 1914b, p. 441), to be unsuitable. In the meantime, he stated, an entire series of studies had proven that "though requiring more food than a person at rest, the protein requirement of a person at work is by no means larger than that of the person relaxing" (Zuntz, 1914b, p. 441). Zuntz contended that carbohydrates – that is, sugar and materials providing sugar, starches, etc. – were indispensable for the sudden and intense output of power and energy. Nervous disturbances of athletes, referred to as "over-training" and, in Germany at the time, "beefsteak high," was, in the end, the consequence of a carbohydrate deficiency (Zuntz, 1914b, p. 443).

Zuntz also explored the question of whether a constant limitation of water provision while working out was justified. In his opinion, a reduced intake of fluids could not serve to increase performance, "Yet it seems hardly probable that, given healthy kidneys, anything is achieved by agonizingly limiting water intake; on the contrary, those people who tend towards gout will be threatened by an accumulation of uric acids and other metabolism products in the body's tissues" (Zuntz, 1914b, p. 443).

Just as he had done in other essays, Zuntz again addressed the relationship of physical to mental work alongside the physiological issues:

> *Of course, it is a higher task to train our mental powers, and we cannot afford to damage them while nurturing the physical power and performance of our muscles ... Overcoming weariness, steeling our will for powerful performance, both of which are required in all athletic performance, will certainly also benefit over mental labors. When we have learned to overcome the internal obstacles preventing us from physical work, we will master those that keep us from mental work more easily; endurance and indefatigability increase.*
> *(Zuntz, 1914b, pp. 451–452)*

Although sport physiology was an "old favorite interest" of Zuntz, he was more skeptical about athletic competitions – especially when the only point was to achieve top performance:

> *When I mentioned earlier that war reminds us to think about different fields and aspects of life style, this includes sport activities, especially in view of the very one-sided emphasis on competitions and top performance which occur here. Specifically, wherever such competitions were viewed to be the most essential thing of all, the danger arose that such physical top performance only benefited a small fraction of the population, that the majority of the people pursued their interest in the sport only by watching and providing commentary ... Thus, we are faced with two series of tasks for the further development of exercise, which both demand our particular attention: the harmonic training of the entire person, while avoiding over-emphasizing and unilaterally favoring top performances. In all reports of the last Olympic Games, people bemoaned the fact that Germany was beaten by other countries although its athletes had performed exceptionally well. I cannot really share this regret. If the average performance of our gymnasts and competitors, measured by the number of those who perform satisfactorily, is greater than in those countries [who achieved top performance], then even if we lack such peak performance, I would view that to be an advantage rather than a disadvantage.*
> *(Zuntz, 1916c, p. 62)*

Wherever the sole objective pursued by athletic activity is to achieve maximum performance in competitions, then the "very improper training of specific skills threatens to damage the person as a whole" (Zuntz, Brahm and Mallwitz, 1911, p. 7).

> *Likewise, I would view it as entirely beneficial if physical performance was only regarded as a part of the overall development of the person, in other*

> *words, if what is a theory in America would become a constant reality for us: that a student is only an honorable representative of a college in the field of sports when they have also proven their academic prowess.*
> *(Zuntz, Brahm and Mallwitz, 1911, p. 7)*

It can safely be said that Zuntz used the physiology of sports to illustrate how research in applied physiology could benefit the well-being of the general public in numerous areas.

In the same year, Zuntz was awarded an honorary doctorate as "Dr. agrar. hc." (Annual report of the Royal Agricultural College in Berlin, 1914, p. 9)[66] by the College of Agriculture in Vienna for his extraordinary achievements in the field of nutritional physiology (Bergner, 1968, p. 2, 1986, pp.127–130; Forbes, 1955, pp. 3–15). There can be little doubt that Durig, at that time a professor at the college, played a major role in the realization of this honor for Zuntz.

Shortly before the outbreak of World War I, Zuntz was selected to chair the 6th Meeting of the German Physiological Society in Berlin (Blasius, 1959, pp. 22–23). However, everything suddenly changed dramatically; first, the German Reich declared war on Russia and France in August 1914, which meant that any scientific research now had to be concentrated strongly on research topics that the military conflict dictated; and second, Zuntz suffered a serious accident on Christmas day of 1914. "When he came home late from a meeting, he was run over by a street car. He was seriously crushed, suffering a double rib fracture and an injury of his lung" (Emma (Sarah) Zuntz, undated, p. 55).

The intensity of his department's involvement in research related to the war is illustrated by the following description:

> *In 1915–16 and part of 1917, the director of the institute, Privy Councilor Professor Zuntz, devoted his full enthusiasm to activities in various areas of nutrition and agriculture. The numerous tasks in all sorts of areas arose for the most part from Privy Councilor Zuntz and his colleagues' contact with, and participation in, the most diverse departments and official associations, such as the Central Purchasing Association, War Committee for Fodder Substitutes, Imperial Committee (War Committee) for Vegetable and Animal Oils and Fat, Department of War Sustenance, Imperial Fodder Depot, Imperial Fat Depot, Imperial Wool Depot, Imperial Grain Depot, and so on.*
> *(Report published by the College of Agriculture, 1921, p. 72)*

Over the course of the war, the increasingly limited food supplies prompted Zuntz to conduct nutritional physiological observations on himself (Loewy and Zuntz, 1916, pp. 825–829; Zuntz and Loewy, 1918, pp. 244–264). He published a plethora of articles for the general public on the best possible nourishment under less than favorable conditions,[67] and strongly recommended caffeine drinks in order to enhance the efficiency and well-being of hardworking people (Letter from Zuntz to the *Präsident des Kriegsernährungsamts* [President of the War Nutrition Department] dated March 9, 1917, Bancroft Library, Berkeley).

Since turning 41 in 1888, Zuntz had, together with Loewy, meticulously recorded his metabolism, body weight and body temperature. In May 1916, he determined a "very considerable decrease in the oxidation process" (Zuntz and Loewy, 1918, p. 245) and that he had lost weight, going from 67.5 kg (1912) to 60.6 kg, a result of the war-time food. Shortly after the outbreak of World War I, demonstrating great foresight, Zuntz had already made a strong public appeal that, in view of the threatening shortage, the grain normally used in Germany for feeding pigs be made directly available to the population for food, arguing that this was a more sensible use from a nutritional standpoint. His views ultimately culminated in Zuntz's calling for the official slaughtering of 8–9 million pigs (Eltzbacher, 1914, p. 123; Zuntz, 1915i, p. 23; Letter from Zuntz to Flügge dated February 3, 1915, NL Flügge, No. 172, Archives of the Humboldt University Berlin) – an idea strongly supported by Director Kurzynski of the Department of Statistics in Berlin-Schöneberg (Schulthess, 1915, p. 56). On March 4, 1915, the *Bundesrat* (Federal Council) resolved to arrange for an interim count of pig stock between March 15 and April 15 of that year (Schulthess, 1915, p. 113), and in fact, the pig stock was reduced by 8–9 million animals, exactly as Zuntz had demanded, in the form of a compulsory slaughter in the spring of 1915. This event went down in history as the "Great Pig Slaughter" (Haushofer, 1972, p. 261) or even the "St. Bartholomew's Eve of Swine" (Skalweit, 1927, p. 96), and continued to provoke a great deal of controversy for years, long after the war had ended (Haushofer, 1972, p. 262).

However, nutritional physiological work only represented part of the research conducted in Zuntz's laboratory during the war years. Other major tasks were direct military commissions dealing with protective methods against poison gas, and the development and use of oxygen respiratory equipment. The *Militär-Prüfungs- und Versuchsstelle für Sauerstoffgeräte* (Military Office of Oxygen Equipment Inspection and Experimentation), specially founded for this purpose, was moved to the ground floor of Zuntz's Institute of Veterinary Physiology at the end of 1916 (Annual Report of the Royal Agricultural College, 1921, pp. 72–73). By order of the War Ministry, von der Heide, Zuntz's colleague of many years, took over the reins of this department (Annual Report of the Royal Agricultural College, 1921, p. 72). In 1921, von der Heide summarized the tasks of the military inspection office as follows:

> *The primary responsibility of this military inspection office was to carry out a meticulous inspection of the oxygen equipment, along with the individual attachments and replacement parts which were delivered to the Central Medical Stores from all sorts of factories in the area. For this task the senior medical officer Dr. Ruffing and later Fluegge, a diploma'ed engineer, were assigned to assist the head of the office, as well as 1 infirmary inspector as managing director, 1 sergeant, 3 lance-corporals and 15 regular troops. Additionally, a number of men and women were conscripted for service at a later date.*
>
> *This oxygen equipment – aerophores, army oxygen protection devices, oxygen respiratory equipment, oxygen troop treatment devices, naval respirators,*

Magirus devices, pulmotors and so on – underwent the most thorough of inspections as regards their performance, packaging and immediate serviceability. All told, 484,000 aerophores, 100,000 army oxygen protection devices, 15,000 oxygen treatment devices, 1,000 naval respirators and the same number of Magirus devices, selected at random, underwent a more detailed inspection, not to mention the permanent control of army oxygen supplies as well as the containers, 200,000 steel cylinders with oxygen shut-off valves, some of which had been newly designed. Likewise, 500,000 carbonic acid absorption cartridges for all the oxygen protection devices were checked, then the mouth pieces, the inhalation bags and tubing, and all the continuously deteriorating replacement parts, for which, due to the shortage of raw material substitutes had to be developed and then produced. So it was that, for lack of castor oil, new elastic lacquers were made for the inhalation tubing and bags, just as vulcanized fiber was improved or, instead of rubber stoppers for the oxygen treatment devices and shut-off valves, metal ones were constructed, after countless experiments and analyses, using a special alloy of lead, antimony and tin.

At the same time a number of scientific experiments were performed in such subjects as "The Causes of Oxygen Explosions in Oxygen Tanks," "How Much Work Can a Soldier Perform with the Aid of Sodium Peroxide Cartridges?," "A Comparison of the Efficiency of Aerophores and Army Oxygen Protection Devices with respect to Meterkilograms and Time Units," or "Is Pure Oxygen Dangerous to Inhale?" Moreover, the government had instructed the reporter [von der Heide] to give various lectures on oxygen protection and oxygen inhalation devices at weekly courses held at the Heeresgasschule (Army Gas Training Center) ... To complete the record, it should be mentioned that, in addition to oxygen inhalation equipment for humans, a special oxygen treatment device was constructed for horses stricken by gas poisoning.

(Annual Report of the Royal Agricultural College, 1921, pp. 72–73)

In addition to this, the military inspection and experiment office dealt with issues of aeronautical medicine and psychology, and tested respiratory equipment for pilots in the pneumatic chamber in Zuntz's laboratory. For this purpose, Senior Surgeon-Major Koschel, Surgeon-Major Kun and Senior Physician Kuhn plus personnel were assigned to the laboratory (Annual Report of the Royal Agricultural College, 1921, p. 73). Emma Zuntz noted that Zuntz in addition

visited Aunt Emma in February of 1917 in Hamburg and traveled from there to Kiel, where he wanted to look at the submarines. He did not feel well during the days prior to that – but when did he ever consider his health. He climbed through the icy rooms of the submarines and returned to Berlin with a severe cough ... He fell ill with a serious pulmonary edema during the night. It was a hard struggle between life and death.

(Emma (Sarah) Zuntz, undated, p. 56)

It is very likely that Zuntz was only able to participate in the investigations described above from time to time. In the later years of the war, his health deteriorated rapidly. After a feverish illness (July 1916) with "a typhus temperature curve and a relapse in August" (Zuntz and Loewy, 1918, p. 245), bronchitis (April 1917) and subsequently heart problems and pulmonary edema, Zuntz was only able to recover his strength after spending time in a sanatorium in July of 1917 (Zuntz and Loewy, 1918, p. 245).

In October of 1917, Zuntz celebrated his seventieth birthday. The festivities served as an occasion for many to bestow honors on him, both in Germany and abroad, and to reflect on his person and his life's work. In view of the ongoing war, his contributions to nutritional physiology were a particular focus. As one speaker worded it: "As a result of his generous research, a great service was rendered Germany during the war in the field of nutrition and provisions" (Anonymus, 1917, p. 760). The list reported by Caspari in a laudatio published in the journal *Naturwissenschaften* is impressive:

> *Zuntz is a thinker and a man of science, it is true, but he is also, and in particular measure, sensitive to the fact that practice must also always be enriched by science. On the other hand, science is to serve practice, but practical aspects should not be allowed to be the principles guiding scientific research. This is the reason he gave many lectures to laypeople, and the vivid and understandable manner in which he was able to describe even the most difficult problem made him perfectly suited to the task. He indefatigably strove to advance agriculture, aquaculture, pond fish culture, the sugar industry, the medical profession as exercised by general practitioners, physical exercise, and balneology, and brought his abundant knowledge to bear in each of these fields. He worked diligently on the problems that the World War posed to scientists, and was a pioneer in the field of nutrition for the German population threatened by food shortages, and advanced the work on protection against poison gas.*
>
> *(Caspari, 1917, p. 620)*

According to Loewy (Loewy, 1917, p. 1270) and a report published in a local newspaper in Berlin (*Vossische Zeitung*, October 6, 1917), a bust of Zuntz was presented in the course of the ceremonies and the establishment of a Zuntz Foundation (25 000 *Reichsmark*) for his scientific work was announced. The purpose of the foundation was to prepare a complete edition of Zuntz's publications. Since this never occurred, there is serious doubt whether the foundation ever came into being.

Emma Zuntz reported on the festivities as follows:

> *While [your grandfather] had retired to rest once again, those who wished to congratulate him gathered together. So many people came that it would have seemed oppressive, if not for the same happy anticipation shining in everyone's eyes. We saw so many beloved faces again, people we had missed for so long. Caspari, my old childhood friend, had hurried home from the fields to bring his good wishes to his beloved teacher. Grandfather greeted his guests and sat down on a decorated armchair, and the official ceremony began. You grandchildren gathered all around him, Wolf put his arm around his neck, and when the many honors the speakers listed simply*

> *did not seem to find an end, he softly stroked his face. There is no point in recording the contents of the many speeches. They all lauded his merits in science and his very special personality. Even though I knew a lot about grandfather's work, I was overwhelmed by the abundance of his interests, the diversity of his many fields. The Brigadier General of the Medical Corps* [Generalarzt] *Schulze emphasized that Zuntz's work had defined nutrition as understood by the army, on the battlefield and in the homeland. What made the entire celebration so unique and so moving was the warmth, the true tones of people's hearts speaking that we heard in all of the speeches. Time and again [the guests] expressed deep gratitude in looking back on the past and the cautious hope for the future that this friend, this colleague, the father of his students, would continue to live among them for a long time yet.*
>
> *Neuberg ceremoniously unveiled the bronze bust cast by, which [grandfather's] friends and students had commissioned to show the likeness of the scholar to those succeeding him. The loveliest moments were those when grandfather responded. His words [expressed] the deep gratitude, modesty, and warmth of feeling that only he possessed. He emphasized that each scientist could only contribute small stones to the enormous structure that science was, and that no scientist could ever claim immortality. The only reason he might not be forgotten entirely was, he said, the love that had made his bust be immortalized by Lederer's master hand. The presentation of the [foundation deed of the]* Zuntz-Stiftung, *the earnings of which were to fund the cost of publishing his collected works, made his cheeks blush with joy. The speech by the old privy councilor Thiel was particularly moving; he was the man [grandfather reported to], who had always had a great understanding for him, with whom he had been in close touch since his youth in Bonn, and who had enabled him to establish and build the wonderful institute. Then I would like to remember Caspari's words, who spoke on behalf of the students and employees closest [to your grandfather]. He recalled how grandfather's institute had grown and developed, and then spoke of the spirit that filled this place of work. He found profound words for the rare relationship between students and master. [He said] that for all of them, grandfather was a father, a counselor and helper, not only in the field of science, but [also] in matters pertaining to [their] personal lives. If anyone had suffered need, he was there to bear it with them, and if they rejoiced, the joy became greater by his sharing in it.*
>
> *(Emma (Sarah) Zuntz, undated, pp. 3–4)*

Some months later, the aforementioned bust was apparently placed in the courtyard of the College of Agriculture; the rector in Berlin requested permission from the Minister for Agriculture, State Domains and Forests to do this, in March of 1918, since:

> *At the ceremony on October 6, 1917 in honor of the 70th birthday of Privy Councilor Dr. Zuntz, his colleagues, friends and students presented him with a bust. In accordance with the intention of the donors, it is the wish of the faculty council to have the bust placed in the courtyard of the College of Agriculture as a monument to Privy Councilor Zuntz's many years of distinguished service. I request that the Ministry approve this plan. The sculpture*

> *is bronze and was crafted by Lederer. No expenses will accrue to the State Treasury as a result.*
> *The rector*
>
> *(Letter from the rector of Berlin's College of Agriculture, 1918, fol. 237)*

Official permission was granted on March 14, 1918.[68]

In July of 1918, the Minister for Agriculture, State Domains and Forests filed a petition with Kaiser Wilhelm II that, in consideration of Zuntz's scientific achievements recognized even beyond Germany's boundaries, the fiftieth anniversary of his doctorate award and his advanced age, Zuntz should be awarded the *Königliche Kronenorden zweiter Klasse* (Royal Order of the Crown, 2nd Class) despite the fact that he "had not yet received the *Rote Adlerorden dritter Klasse mit Schleife* (Order of the Red Eagle, 3rd Class with Band)" (Petition of the Minister for Agriculture, 1918, fol. 24, Secret Central Archives). This was followed at the end of July by a notification from the Kaiser that, on the strength of the Minister's report, he had awarded Zuntz the Order of the Crown, 2nd Class. This was to be the last of a total of eight orders Zuntz received in the course of his life:

1. 1870/1871, Kr.d.f.K[69]
2. 1897, *Roter Adlerorden 4. Klasse* (Order of the Red Eagle, 4th Class) (Petition of the Minister for Agriculture, 1918, fol. 24, Secret Central Archives)
3. 1911, *Kronenorden 3. Klasse* (Order of the Crown, 3rd Class) (Petition of the Minister for Agriculture, 1918, fol. 24, Secret Central Archives)
4. 1911/1912, *Ehrenkreuz des Mecklenburgischen Greifenordens* (Medal of the Mecklenburg Order of the Griffon) (Annual Report of the College of Agriculture, 1912, p. 6)
5. 1911/1912, *Ritterkreuz 1. Klasse des Kgl. Sächsischen Albrechtsordends mit Krone* (Knighthood 1st Class of the Royal Saxon Order of Albrecht with Crown) (Annual Report of the College of Agriculture, 1912, p. 6)
6. 1912/1913, Russian Order of St Stanislaus, 2nd Class (Annual Report of the College of Agriculture, 1912, p. 8)
7. 1916, *Eisernes Kreuz 2. Klasse* (Iron Cross, 2nd Class) (Petition of the Minister for Agriculture, 1918, fol. 24, Secret Central Archives)
8. 1918, *Kronenorden 2. Klasse* (Order of the Crown, 2nd Class) (Petition of the Minister for Agriculture, 1918, fol. 24, Secret Central Archives)

Emma Zuntz shared the following reflections on success in life with her children:

> *If we understand success in life, as seen by others, to be the general recognition of a scientist's and scholar's achievements by his colleagues in the field, then grandfather's life was very successful. There was not a student of natural science in the entire world who did not know his name. His institute had a reputation throughout the world; it was visited by scientists from every country. But if we understand success in life, as others often see it, as being granted a chair at a large university, then there can be no doubt that he has been discriminated against. There were times when he would have gladly worked at a university because a more scientific atmosphere prevailed there than at the*

Agricultural College. But [regardless of] how his friends got upset about this injustice, he did not let the pleasure he took in his work or his inner harmony suffer. He accepted it as fate that anti-Semitism opposed his desires. Only once, when Heidenheim himself appointed him as his successor in Breslau [Wroclaw], but someone else was appointed in spite of this, [did] this pain him. Later, after he had established and built his institute, he would not have accepted any appointments. He was indifferent to honors perceived by others, such as honorary doctorates or orders. [Our nanny] Aunt Lisa was all the happier for each recognition [her] friend [received]. That is why we always said, teasingly, we have to go congratulate – Aunt Lisa has been awarded another order again.

(Emma (Sarah) Zuntz, undated, p. 51)

In the latter years of his life there was no end to the scientific honors bestowed on Zuntz. In July of 1918, prior to the conferral by the Kaiser of the Order of the Crown, 2nd Class, Zuntz received an honorary doctorate from the *Tierärztliche Hochschule Hannover* (Hanover School of Veterinary Medicine) (Figure 1.29).[70] Finally, one year later, on August 3, 1919, Zuntz was awarded an honorary doctorate from the Department of Philosophy of the University of Bonn on the occasion of its hundredth anniversary (Figure 1.30).[71] Evidently, this particular honor moved him very deeply. In his letter of thanks to the Department, he wrote:

A greater pleasure could not have been afforded me than the manner in which the honorable Department of Philosophy fulfilled my wish to participate in the hundredth jubilee of my alma mater, and this in a much more splendid way than I had dared to hope … I regard this with gratitude as the culmination

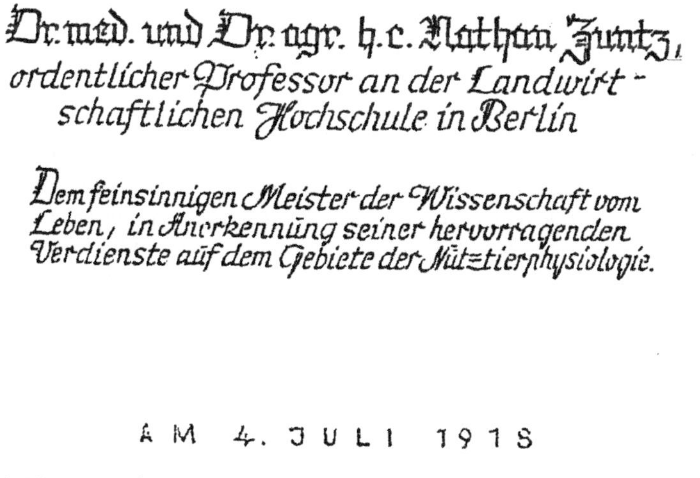

Dr. med. und Dr. agr. h.c. Nathan Zuntz,
ordentlicher Professor an der Landwirtschaftlichen Hochschule in Berlin

Dem feinsinnigen Meister der Wissenschaft vom
Leben, in Anerkennung seiner hervorragenden
Verdienste auf dem Gebiete der Nutztierphysiologie.

AM 4. JULI 1918

Figure 1.29 Honorary doctorate from the Tierärztliche Hochschule Hannover (Hanover School of Veterinary Medicine), dated 4 July 1918.
(Archiv der Tierärztlichen Hochschule Hannover)

DIE PHILOSOPHISCHE FAKULTÄT
DER RHEINISCHEN
FRIEDRICH-WILHELMS-UNIVERSITÄT ZU BONN
VERLEIHT DEM PROFESSOR AN DER LANDWIRTSCHAFTL HOCHSCHULE IN BERLIN
GEH. REGIERUNGSRAT DR. MED. DR. MED. VET. H. C. NATHAN ZUNTZ
DEM SOHNE DER STADT BONN DEM SCHÜLER UNSERER UNIVERSITÄT DEREN LEHRKÖRPER
ER ZEHN JAHRE ANGEHÖRTE FÜR SEINE UNERMÜDLICHEN ERFOLGREICHEN UND VIEL BE-
WUNDERTEN UNTERSUCHUNGEN ÜBER DEN STOFFWECHSEL DER TIERE UND DES MENSCHEN
DIE WÜRDE UND DIE RECHTE EINES EHRENDOKTORS DER PHILOSOPHIE
GEGEBEN AM TAGE DER JAHRHUNDERTFEIER DER UNIVERSITÄT DEM 3 AUGUST 1919
UNTER DEM REKTORAT DES PROFESSORS DER RECHTE DR. ERNST ZITELMANN
UNTER DEM DEKANAT DES PROFESSORS DER BOTANIK DR. JOHANNES FITTING

Figure 1.30 Honorary doctorate from the Department of Philosophy of the University of Bonn on the occasion of its hundredth anniversary, dated August 3, 1919. (Archiv der Rheinischen Friedrich-Wilhelms-Universität, Bonn)

> *of everything the Bonn Department of Philosophy has meant to my education, with all the gifts that I received as a student and later as a lecturer.*[72]

At the end of 1918, Zuntz submitted an application to be allowed to retire for health reasons (Letter from Zuntz to the Minister for Agriculture, 1918, fol. 204, Secret Central Archives). The response of the Minister from January 1919 read as follows:[73]

> *For the reasons you have specified I cannot reject your request to enter retirement on April 1st, 1919, as much as your departure from the professorship which you have administered with the greatest of success over the past four decades will be regretted. I therefore hereby grant my consent to your leaving the civil service on March 31st of next year, while expressing our great gratitude for the services you have rendered for the university and the nation.*
>
> *We reserve the right to determine the amount of your exact retirement pension ... I anticipate receiving the proposals of the faculty for filling the vacant chair for veterinary physiology within two months.*

The faculty of the Royal Agricultural College later chose Scheunert as Zuntz's successor; he had thus far been an *ausserordentlicher Professor* (professor without tenure) at the *Tierärztliche Hochschule* (School for Veterinary Medicine) in Dresden (Annual Report of the Royal Agricultural College, 1921 pp. 24–27 I).

Zuntz's own remarks on the procedures surrounding the appointment to the chair are to be found in a letter from 1920 addressed to an unknown recipient:

> *I would most have preferred to secure Durig, but he had taken on a professorship in physiology at Vienna University half a year earlier. Despite this, he was leaning towards coming, but finally turned the offer down after a lengthy period of hesitation. Amongst my students, my favorite choice as successor would have been Loewy and perhaps also Caspari, but this probably failed mainly on account of the anti-Semitism of my "honored colleagues." In the end, Scheunert in Dresden was appointed, an associate and relative of Ellenberger, and will be taking up the position on April 1st. Personally, I hope that we will get along well together.*
>
> *(Fragment of a letter from Zuntz to an unknown person from February 2, 1920, Staatsbibliothek Preussischer Kulturbesitz (Berlin State Library), Slg Darmstaedter – Nathan Zuntz -, 3d. 1885, fol. 3)*

Emma Zuntz provided a few more details on the relationship between her father and his colleagues and students:

> *As a young man, Goethe once wrote in the fullness of his youth: and everyone near and far are friends. That could have been written as the motto for father's life. He was not at all like most people, who make their friends in youth and became lonelier with age as their own generation dies. He, always young at heart, made new friends again and again. If Mehring was the friend from his youth, then Durig was the friend of his [old] age.*
>
> *(...)*
>
> *As his old friends died before him the young circle of his students had long since gathered around him. Only young Durig from Innsbruck was his equal, as he was inspired by the same wealth of ideas and the same joy he took in his work. How this son of the mountains enjoyed working [with your grandfather] on the Monte Rosa mountain range, how happily he helped his aging mentor when climbing! Nathan's dearest wish was to make Durig his successor at the institute. But the young professor was granted the* Ordinariat *(full professorship) at the University of Vienna – thus this dream disappeared. His son Durig would have made it so easy for him to leave the beloved workplace.*
>
> *(Emma (Sarah) Zuntz, undated, pp. 48–49)*[74]

Fate did not grant Zuntz a long period of retirement, which would perhaps have enabled him to recapitulate his diverse research projects in an edition of his complete works. Nevertheless, the scientific community was quite aware of his remarkable achievements during his career, as demonstrated by the fact that Zuntz was nominated three times for the Nobel Prize[75] – in 1910 by Goldscheider, in 1919 by Loewy, and finally in 1920 by Orth. Interestingly enough, the motivations cited in the nominations differed from one suggestion to the next. While in 1910[76] the extensive work Zuntz had done in the field of physiology (blood, circulation, respiration, thermoregulation) was lauded by Goldscheider, Loewy pointed to his investigations on the metabolism of horses and ruminants. Finally, in 1920, Orth cited his contributions to nutritional research. On the other hand, in 1919 Zuntz himself nominated eight colleagues at the end of his life, these being Barcroft, Benedict, Krogh, Loeb, Neuberg, Oppenheim, Tigerstedt, and von Wassermann. He suggested Barcroft for his work on the oxygen affinity of hemoglobin and metabolism, Benedict for his work on metabolism, and Krogh for his work on respiratory methods, gas exchange, cardiac output, etc. – and in fact Krogh was awarded the Nobel Prize in 1920.[77] Zuntz nominated Loeb, Neuberg and Oppenheim for their contributions to brain physiology and the effects of physical agents on organisms (Loeb); the chemistry and therapy of cancer, the biological effect of light, and the differentiation of carbohydrates and the processes during fermentation (Neuberg); and the study of mental disorders (Oppenheimer). Tigerstedt was nominated by Zuntz for his work on physiology – mainly the invention of new methodological equipment for respiratory and metabolic studies – and von Wassermann, last but not least, for the Wassermann reaction that served to detect a syphilitic infection.[78]

On March 22, 1920, "after many hours of agony" (Zuntz L., 1926, p. 202), he passed away at 1:45 pm in his home at Bleibtreustrasse 38/39 (Sterberegister, 1920, Reg. no. 378). Emma Zuntz closes her description of her father's life with this scene:

> *It was an agreement between us that he would ask me for the last morphine. His eyes were already looking into the unknown – I bent over the suffering father, [and asked,] "What can I do for you?" The will returned to his expression, and he replied, "Morphine, lots of morphine." After the shot, he fell into a light sleep, and woke twice and whispered to me, "Freer and ever more freer."*
> *(Emma (Sarah) Zuntz, undated, p. 61)*

In the death records of the Trinitatis parish, "cardiac weakness" is stated as the cause of death (Sterberegister, 1919–1923, fol. 90). His funeral was held on April 10, 1920, at 12:00 pm at Stahnsdorf's south-west cemetery near Potsdam, with no church representative in attendance.[79]

The estate administration of his scientific correspondence was assumed by the Frankfurt chemist Darmstaedter, who had apparently contacted Caspari with this express wish shortly after Zuntz's death (Staatsbibliothek Preussischer Kulturbesitz, Slg. Darmstaedter "Nathan Zuntz," Letter from L. Zuntz to Darmstaedter, fol. 48–49). Apparently, his friends in the academic community immediately tried to sell Zuntz's large library to support his widow. In this regard, Y. Henderson of the Department of Physiology at Yale University wrote the following note in the journal *Science*:

> *THE LIBRARY OF THE LATE PROFESSOR ZUNTZ*
>
> *TO THE EDITOR OF SCIENCE: A letter received from a friend in Berlin a few days ago brings information of the death of Professore N. Zuntz. The very great services of Professor Zuntz, extending over a long life time, devoted to the advancement of physiology and nutrition, his broad-mindedness and kindly character render his death at this time, when renewal of scientific associations severed by war is so important, peculiarly sad.*

The information comes also that, for the support of his widow who is a hopeless invalid, funds are needed. To this end it is desired to sell the large library which Professor Zuntz had collected. It includes complete sets of practically all of the journals in his field of work. By disposing of the library direct to some purchaser, or purchasers, in this country the advantage of the rate of exchange would accrue to the widow instead of to some book dealer.

I shall be glad to supply the address and further information as I have to any one interested in the purchase of this library.

YANDELL HENDERSON
Department of Physiology,
Yale University

(Henderson, Science, 1920, p. 569)

On January 9, 1921, the medical department of the *Niederrheinische Gesellschaft für Natur- und Heilkunde* (Lower Rhine Society for Natural and Medical Sciences) held a posthumous celebration commemorating Zuntz's death (*Bonner Zeitung*, 1921).

Several days earlier, a memorial plaque had been unveiled at Hundsgasse 26 in Bonn with the inscription "Birthplace of the eminent physiologist Dr. Nathan Zuntz" (*Bonner Zeitung*, 1921). The memorial speech during the ceremony was held by Privy Councilor Verworn, who closed with the affirmation "Physiology will never forget the great services rendered it by Zuntz" (*Bonner Zeitung*, 1921).

Notes

1. Even in the 1847 birth certificate (no. 397, City Archives, Bonn) of Nathan Zuntz, the name of the father was entered as "Lion Zunz," while two years later the birth certificate of Albert, the second son, refers to "Leopold Zuntz" as the father.
2. In a handwritten, undated, only partially paginated report to her children, Emma Zuntz described the life and personality of her father, highlighting a few key events (Miscellaneous Sources).
3. *cf.* letters from the archives of the Rheinische Friedrich-Wilhelms-Universität to the author dated August 10 and 18, 1987.
4. *cf.* information from the archive of the Rheinische Friedrich-Wilhelms-Universität provided to the author on August 10, 1987.
5. Max Johann Sigismund Schultze was born on March 25, 1825, in Freiburg im Breisgau as the son of Karl August Sigismund Schultze (Professor for Anatomy and Physiology in Freiburg im Breisgau, who worked in Greifswald from 1830 onwards). Max Schultze studied medicine at the University of Greifswald, beginning in 1845. In 1849, he graduated as a doctor of medicine; his father had been his counselor. His dissertation was published under the title *De arteriarum notione, structura, constitutione chemica et vita*. In 1850, he was appointed associate professor. From 1850 to 1854 he acted as prosector for his father at the Anatomical Institute of the University of Greifswald. In October 1854, Max Schultze was offered a position at the University of Halle an der Saale and remained there until 1859. Thereafter, he took an appointment in Bonn where he became the director

of the Anatomical Institute. Max Schultze died on January 16, 1874, at the young age of forty-nine. In more than eighty scientific publications, he focused on microscopic-anatomical structures (*cf.* Hirsch, A.: *Biographisches Lexikon der hervorragenden Ärzte aller Zeiten und Völker* (Biographical Encyclopaedia of the Most Outstanding Doctors of all Times and Peoples), 5th volume, Urban & Schwarzenberg 1886, pp. 304 *et seq.*; Bast, T. H.: Max Johann Sigismund Schultze, *Ann. Med. Hist.* 3 (1931), pp. 166–178).

6. *cf.* handwritten *curriculum vitae* by N. Zuntz dated July 6, 1870, Archives of the Rheinische Friedrich-Wilhelm-University Bonn.
7. *cf. handwritten curriculum* vitae by N. Zuntz dated July 6, 1870, Archives of the Rheinische Friedrich-Wilhelm-University Bonn.
8. Eduard Friedrich Wilhelm Pflüger was born on June 7, 1829, in Hanau am Main. His father, Georg Pflüger, was a merchant, politician, journalist, and member of the Vorparlament (preliminary assembly of the first German National Assembly) in the Paulskirche in Frankfurt am Main. Eduard Pflüger studied law from 1848 to 1850 in Marburg, Munich, and Berlin, and graduated as a doctor of law in 1851 in Giessen. In 1855, he took his degree as a doctor of medicine in Berlin; the title of his dissertation was *De nervorum splanchnicorum functione*. Three years later, in Berlin, he wrote his Habilitation thesis on the electrical tonus as part of the formal requirements to become a professor. Pflüger was offered an appointment in 1859 as ordentlicher Professor (full professor) at the University of Bonn in the Physiology Department, and founded the first independent Department of Physiology of the university. Beginning in the year 1864, he focused on the biochemistry of blood, and was the one who sparked Zuntz's interest in this field. Eduard Pflüger became known not only for his numerous scientific publications, but above all as a publisher of *Pflügers Archiv für die gesammte Physiologie des Menschen und der Thiere* (Pflüger's Archive on the Physiology of Humans and Animals), the first volume of which was published in 1868, with the 455th volume being published in October 2008. Zuntz published a large number of his articles in this journal (*cf.* List of works by Nathan Zuntz). He was a member of numerous national and international scientific committees and associations, and was granted a series of prestigious scientific awards and honors during his lifetime. Eduard Pflüger died on March 16, 1910, in Bonn (*cf.* Hirsch, A.: *Biographisches Lexikon der hervorragenden Ärzte aller Zeiten und Völker* (Biographical Encyclopaedia of the Most Outstanding Doctors of all Times and Peoples), 4th volume, Urban & Schwarzenberg 1886, p. 554; von Cyon, E.: Eduard Pflüger, *Pflügers Arch.* 132 (1910), pp. 1–19).
9. *cf.* written notification from Richard Müller to the author dated April 14, 1986. Richard Müller had been the director and full professor of the Department for Animal Nutrition of the *Institut für Tierzucht und Tierfütterung* (Institute for Animal Breeding and Nutrition) at the Agricultural Department until his retirement in October of 1978.
10. *cf.* written notification from Richard Müller to the author dated August 23, 1985.
11. *cf.* Hirsch, A: *Biographisches Lexikon der hervorragenden Ärzte aller Zeiten und Völker* (Biographical Encyclopaedia of the Most Outstanding Doctors of all Times and Peoples), Urban & Schwarzenberg 1886, p. 554; *cf.* Wenig, O.: *Verzeichnis der Professoren und Dozenten der Rheinischen Friedrich-Wilhelms-Universität von 1818–1968* (Register of the Professors and Lecturers of the University of Bonn from 1818–1968), Bonn 1868, pp. 225 *et seq.*

12. *cf.* Zuntz, N.: Über den Einfluss des Partiardrucks der Kohlensäure auf die Vertheilung dieses Gases im Blute (On the influence of partial pressure of carbonic acids on the distribution of this gas in blood), *Zentralblatt M. Wiss.* 5 (1867), pp. 529–533.

    cf. Zuntz, N.: Zur Kenntniss des Stoffwechsels im Blute (Information on the metabolism in blood), *Zentralblatt M. Wiss.* 5 (1867), pp. 801–804.

    cf. Zuntz, N.: *Beiträge zur Physiologie des Blutes* (Contributions on the Physiology of Blood), dissertation (1868).

    cf. Pflüger, E. and Zuntz, N.: Über den Einfluss der Säuren auf die Gase des Blutes (On the influence of acids on the blood gasses), *Pflügers Arch.* 1 (1868), pp. 361–374.

    cf. Zuntz, N.: Über die Bindung der Kohlensäure im Blute (On carbonic acid bonds in blood), *Berl. Klin. Wschr.* 7 (1870), p. 185.
13. Quoted from a letter dated July 7, 1870 (Archives of the Rheinische Friedrich-Wilhelm-University Bonn) from the University Curator to Zuntz, records kept by the office of the Dean of Medicine's office [at the University of] Bonn on the Zuntz "Habilitation" (procedure of fulfilling formal requirements to become a university lecturer), non-paginated sheets.
14. Pflüger's opinion on Zuntz was probably written in July of 1870; records kept by the office of the Dean of Medicine's office [at the University of] Bonn on the Zuntz "Habilitation" (procedure of fulfilling formal requirements to become a university lecturer), non-paginated sheets.
15. *cf.* letter from Zuntz to the dean dated July 9, 1870, Archives of the Rheinische Friedrich-Wilhelm-University Bonn.
16. *cf.* memorial posters for each year of the history of the academy, put up at the Königlich Preussische Landwirtschaftliche Akademie (Royal Prussian Agricultural Academy in Bonn-Poppelsdorf) by Seehaus, –.: Jahrestafeln zur Geschichte der Akademie. *Die Königlich Preussische Landwirtschaftliche Akademie, Bonn-Poppelsdorf, Bonn* (1915), p. 123.
17. *cf.* Schulte, K. H. S.: Bonner Juden und ihre Nachkommen bis um 1930 (Bonn Jews and their Descendants until 1930), *Veröffentlichungen Stadt. Bonn* 16 (1976), p. 547.

    cf. Death Certificate no. 347, record of the Sterbebuch (register of deaths) kept by the personal records office Bonn 1, 1874, City Archives, Bonn.
18. *cf.* written notification from the Stadtarchiv (city archive) in Bonn to the author dated August 13, 1985. In it, the first name of his wife is given as Friederike/ Frieda.
19. *cf.* von der Goltz, T. Freiherr: *Festschrift zur Feier des fünfzigjährigen Bestehens der Königlich Preussischen landwirtschaftlichen Akademie Poppelsdorf* (Publication Celebrating the Fiftieth Anniversary of the Prussian Royal Academy of Agriculture in Poppelsdorf), Bonn 1897. pp. 19 *et seq.*

    cf. letters from Müller to the author dated August 23, 1985 and April 14, 1986.
20. *cf.* Pflüger, E.: *Wesen und Aufgabe der Physiologie* (The Nature and the Task of Physiology), speech held on the inauguration of the new Physiological Institute in Poppelsdorf near Bonn on November 9, 1878, *Pflügers Arch.* 18 (1878), p. 424.
21. *cf.* von Liebig, J.: *Reden und Abhandlungen von Justus von Liebig* (Speeches and Treatises by J. von Liebig), Wissenschaft und Landwirtschaft (academic address, Science and Agriculture) held on March 26, 1861, Leipzig-Heidelberg 1874. pp. 194 *et seq.*, p. 199.

cf. Wittmack, L.: Die Königliche Landwirtschaftliche Hochschule in Berlin, Festschrift zur Feier des 25jährigen Bestehens (The Royal Agricultural College in Berlin, Publication Celebrating its Twenty-Fifth Anniversary), Paul Parey 1906, p. 9.

22. Hugo Thiel was born on June 2, 1839, as the fourth child of the University Secretary and Rechnungsrat (Councilor for Accounting Matters) Thiel, whose family lived in the castle of Poppelsdorf near Bonn. In 1857, Hugo Thiel passed his school-leaving examination. Prior to taking up his studies at the Agricultural Academy in Poppelsdorf, he gained two years of experience as an administrative apprentice in two large estates (Uenglingen near Stendal and Morsbroich near Mühlheim). In 1864, he graduated from the Agricultural Academy Poppelsdorf and was awarded the degree of doctor of agriculture, based on his dissertation on botany and plant physiology, for which Sachs was his counselor. After he had completed the procedure for becoming a university lecturer (Habilitation) in Bonn (1866), he was appointed lecturer, holding the title of Lehrer der Landwirtschaft (teacher of agriculture). From 1867 until 1869, he was director of the experimental field of the Poppelsdorf Academy. Later, he was appointed to a post at the Technische Hochschule Darmstadt (1869) and to the Technische Hochschule of Munich (1872). In 1873, Hugo Thiel accepted positions as Generalsekretär des Landesökonomiekollegiums (Secretary General of the State Economy College) and as Referent für das Landwirtschaftliche Bildungswesen (Expert on Matters of Professional Training in Agriculture), both with the Prussian Ministry for Agriculture. In 1879, Thiel was appointed as Geheimer Regierungsrat (privy higher executive officer) and in 1885 as Geheimer Oberregierungsrat (privy principal) in the Ministerium für Landwirtschaft, Domänen und Forsten (Ministry of Agriculture, State-owned Domains and Forests). From 1880 onwards, Hugo Thiel acted as provisional director of the newly established Landwirtschaftliche Hochschule (Agricultural College) in Berlin and acted, in conjunction with the privy principal Goeppert of the Ministerium für Geistliche Angelegenheiten (Ministry for Matters of Intellectual Affairs) as curator of this institution, for which the minister was directly responsible. In 1897, he was appointed as Wirklicher Geheimer Oberregierungsrat (high privy principal) and as director of Department II within the ministry (being responsible for the state-owned domains). Hugo Thiel supported, in all ways imaginable, the institution and further development of the Agricultural Colleges in Poppelsdorf and later in Berlin. This means that he was one of the most important political contacts that Zuntz could have in the Ministry in Berlin. In addition to these professional activities, Hugo Thiel is also known as the editor of the *Landwirtschaftliche Jahrbücher* (Agricultural Yearbooks), in which Zuntz published many of his essays (*cf.* List of Works by Nathan Zuntz). Hugo Thiel died on January 13, 1918 in Berlin (*cf.* Zuntz, N.: Hugo Thiel, [obituary published in] *Deutsche Landwirtschaftliche Presse* (German Agricultural Journal) 45 (1918a) pp. 161–163; Wenig, O.: *Verzeichnis der Professoren und Dozenten der Rheinischen Friedrich-Wilhelms-Universität von 1818–1968* (Register of the Professors and Lecturers of the University of Bonn from 1818–1968), Bonn 1868, p. 310; Letter from Müller to the author of August 23, 1985).

23. *cf.* von Basch, S.: Über den Einfluss der Athmung von comprimirter und verdünnter Luft auf den Blutdruck des Menschen (On the effects that breathing compressed and diluted air has on human blood pressure), *Med. Jahrbücher* (1877), pp. 489–497.

cf. Lebegott, W.: *Die Ausathmung in verdünnter Luft* (Expiration in Diluted Air), Inaugural Dissertation, Berlin 1882.

cf. Fraenkel, A. and Geppert, J.: *Über die Wirkungen der verdünnten Luft auf den Organismus* (Notes on the Effects of Diluted Air on the Organism), Springer-Verlag 1883.

cf. Dietrich, J.: Die Wirkung comprimirter und verdünnter Luft auf den Blutdruck (The effects of compressed and diluted air on blood pressure), *Arch. Exp. Path. Pharm.* 18 (1884), pp. 242–259.

24. *cf.* the further sources: Winau, R.: *Medizin in Berlin*, De Gruyter 1987, p. 243, and *curriculum vitae* Loewy in the corresponding footnote.
25. *cf.* "Tauf-Nr. 181" dated October 23, 1889, Archives of the Jerusalem und Neue Kirche Parish Berlin, p. 173.
26. Wittmack writes that the Landwirtschaftliche Hochschule (Agricultural College) in Berlin can be traced back to A. Thaer, who in 1806 founded the Königliche akademische Lehranstalt des Ackerbaus (Royal Academic Institute of Agriculture) in Möglin near Wriezen. In 1862, this was moved to Berlin under the name of Landwirtschaftliches Lehrinstitut (Agricultural Institute of Learning). In 1881, the institute was renamed the Landwirtschaftliche Hochschule (Agricultural College) and finally in 1934 was incorporated as a department of the university (*cf.* Wittmack, L.: *Die Königliche Landwirtschaftliche Hochschule in Berlin* (The Berlin Royal College of Agriculture), Paul Parey 1906, pp. 4–15; Asen, J.: *Gesamtverzeichnis des Lehrkörpers der Universität Berlin* (Complete List of Instructors of Berlin University), 1st volume, Harrassowitz 1955, p. III).
27. August Julius Geppert was born on November 7, 1856 in Berlin. He studied medicine in Heidelberg (1875–1877) and Berlin (1877–1879), where he obtained his PhD in 1880. His doctoral thesis was entitled *Die Gase des arteriellen Blutes im Fieber* (The Arterial Blood Gases during Fever). From 1880 until 1885, when he worked with Zuntz, Julius Geppert was second assistant at the Medizinische Klinik (Medical Clinic) in Berlin. In 1886, he went to the Pharmacological Institute of the University of Bonn and in the same year became professor of pharmacology there. In 1893 he was appointed ausserordentlicher Professor (professor without tenure), and in 1899 accepted an appointment as chair of pharmacology in Giessen. His most influential publications addressed the physiology of respiration, anesthetics and hygiene (*cf.* Pagel, J.: *Biographisches Lexikon hervorragender Ärzte des neunzehnten Jahrhunderts* (Biographical Encyclopedia of Outstanding Doctors of the Nineteenth Century), Urban & Schwarzenberg 1901, pp. 592 *et seq.*, Fischer, I.: *Biographisches Lexikon der hervorragenden Ärzte der letzten fünfzig Jahre* (Biographical Encyclopedia of Outstanding Doctors of the Last Fifty Years), Urban & Schwarzenberg 1962b, p. 492).
28. In the history of the University of California it is mentioned that the benefactress Mrs Phoebe Hearst equipped the new laboratories at Parnassus (1899–), among others, with a Zuntz respiration apparatus, and apparently one observer noted, "it is believed that this equipment is scarcely equaled and certainly not excelled in this country." University of California, San Francisco. *A History of the University of California San Francisco* [electronic resource]. University of California Regents, 2001.
29. *cf.* Geppert, J. and Zuntz, N.: Über die Regulation der Athmung (On respiratory regulation), *Pflügers Archiv* 42 (1888) pp. 189 *et seq.* This hypothesis on "Mitinnervation," expressed more than a hundred years ago by Zuntz and

Geppert, is still intensively discussed in scientific literature today (*cf.* Schmidt, R. F., Lang, F. and Thews, G.: *Physiologie des Menschen* (Human Physiology), Springer 2005, pp. 780–781; Dempsey, J .A. and Whipp, B. J.: The respiratory system. In: *Exercise Physiology*, edited by C. M. Tipton, Oxford University Press 2003b, pp. 153, 155; Tipton, C. M.: The autonomic nervous system. In *Exercise Physiology*, edited by C. M. Tipton, Oxford University Press 2003b, pp. 197, 204; Stegemann, J.: *Leistungsphysiologie* (Exercise Physiology), Thieme 1984, pp. 195–197).

30. In this regard, *cf.* Fishman, A. and Richards, D. W.: *Circulation of the Blood – Men and Ideas*, Oxford University Press 1964, pp. 97–98. They stated that Nathan Zuntz "had done pioneering work in the quantitative study of metabolic gaseous exchange," although they felt that in general, "Zuntz and his collaborators were more interested in metabolism than in hemodynamics."
31. *cf.* Zuntz, N. and Magnus-Levy, A.: Beiträge zur Kenntnis der Verdaulichkeit und des Nährwerthes des Brodes (Contributions on the Digestibility and Nutritional Value of Bread), *Pflügers Arch.* 49 (1893), pp. 438–460.

    cf. Lehmann, C., Müller, F., Munk, I., Senator, H. and Zuntz, N.: Untersuchungen an zwei hungernden Menschen (Investigations Made on two Starving People), *Arch. Path. Anat.* 131 (1893), pp. 1–229.

    cf. Lehmann, F., Hagemann, O. and Zuntz, N.: Zur Kenntnis des Stoffwechsels beim Pferde (Notes on the Metabolism of the Horse), *Landwissenschaft Jahrbuch* 23 (1894), pp. 125–165.

    cf. Zuntz, N.: Über den Stoffverbrauch des Hundes bei Muskelarbeit (On the Metabolism of the Dog under Muscular Exertion), *Pflügers Arch.* 68 (1897), pp. 191–211.

    cf. Zuntz, N. and Knauthe, K.: Gesichtspunkte zur Beurteilung praktischer Fütterungsversuche an Fischen (Aspects of Evaluating Practical Fish Feeding Methods), *Fisch. Zg* 1 (1898a), pp. 480–483.

    Zuntz's studies on starving humans must be seen in a broader historical perspective which is beyond the scope of the present publication but nevertheless quite interesting because it deals with the inter-dependencies between science, the public, and the media in addition to covering the physiological aspects (Diezemann, N: *Die Kunst des Hungerns. Anorexie in literarischen and medizinischen Texten um 1900*, Dissertation Universität Hamburg 2005, pp. 103–115).
32. *cf.* Cohnstein, J. and Zuntz, N.: Untersuchungen über das Blut, den Kreislauf und die Athmung beim Säugethier-Fötus (Investigations on Blood, the Circulation, and Respiration of Mammal Fetuses), *Pflügers Arch.* 34 (1884), pp. 173–233.

    The investigations performed in the context of this research on fetal circulation functions certainly only represent a small section of all the research done by Zuntz on circulation physiology. But it is precisely these experiments – conducted once again using brilliantly simple methods – and their results which enjoy high recognition today, especially in the academic literature of English speaking nations (*cf.* Fishman, A. and Richards, D. W., *Circulation of the Blood – Men and Ideas*, Oxford University Press 1964, pp. 750, 765, 791, 802–803).
33. Recently, Rowell recognized the work of Zuntz and Hagemann (1898) on exercising horses and their cardiac output as a remarkable contribution to exercise physiology (Rowell, L. B.: The cardiovascular system. In: *Exercise Physiology*, edited by C. M. Tipton, Oxford University Press 2003b p. 106). A similar opinion was expressed by Hamilton, citing Y. Henderson: "The contributions of these workers [Zuntz and Hagemann] were of such great merit that Yandell Henderson felt that

we should refer to the method of Zuntz and Hagemann rather than the method of Fick." (Hamilton, W. F.: The physiology of cardiac output. *Circulation* 8 (1953), p. 528). A few years after these benchmark experiments by Zuntz and Hagemann, the other close co-workers of Zuntz's laboratory A. Loewy, H. von Schroetter and J. Plesch (Zuntz, N. and Plesch, J.: Methode zur Bestimmung der zirkulierenden Blutmenge beim lebenden Tiere. *Biochemische Zeitschrift* 11 (1909a): pp. 47–60) introduced rebreathing gas mixtures to avoid pulmonary catheterization (Rowell, L. B.: The cardiovascular system. In: *Exercise Physiology*, edited by C. M. Tipton, Oxford University Press 2003b p. 106).

34. Furthermore, based on the publications on the catheterization of the heart in humans and their importance for physiological and medical research by W. Forssmann (Forssmann, 1929) (Germany), A. F. Cournand (United States) (Cournand and Ranges, 1941) and D. W. Richards received the Nobel Prize together with him in 1956 (further details in the Nobel Prize database: http://nobelprize.org/medicine/nomination/nomination.php?-Cournand&action; http://nobelprize.org/medicine/nomination/nomination.php?-Forssmann&action; http://nobelprize.org/medicine/nomination/nomination.php?-Richards&action). The importance of this new method was highlighted by Granger in 1998, when he gave his overview of the history of cardiovascular physiology in the twentieth century (Granger, H. J.: Cardiovascular physiology in the twentieth century: great strides and missed opportunities. *Am J Physiol Heart Circ. Physiol.* (1998) 275: H1925–H1936). Recently, however, Bröer (2002) has raised doubts as to whether it was really proper for Forssmann to have received the Nobel Prize, because Zuntz and others had successfully performed similar experiments in animals earlier (Bröer, R.: Legende oder Realität? Werner Forssmann und die Herzkatheterisierung. *Dtsch. Med. Wochenschr.* 127 (2002), pp. 2151–2154).
35. Quoted according to Bröer, 2002, pp. 2151–2154. *cf.* In this context also: Cournand, A. F.: *From Roots to Late Budding. The Intellectual Adventures of a Medical Scientist*, Gardner Press 1986.
36. *cf.* Herken, H.: Tierexperimentelle Prüfung von Arzneimitteln (Testing Pharmaceuticals using Experiments on Animals), *Dtsch. Ärztebl.* 77 (1980), pp. 2617–2628.

    cf. Wittke, G.: Bedeutung und Begründung des Tierversuchs in der medizinischen Ausbildung (The Significance of Animal Experimentation in Medical Training and the Reasons for its Use), type-written manuscript of the lecture given on the occasion of the inauguration of the Central Animal Laboratory on February 27, 1982 in Berlin.

    cf. Deutsche Forschungsgemeinschaft (German Research Association): Tierexperimentelle Forschung und Tierschutz (Animal Experimentation in Research and Animal Protection), Weinheim 1984.
37. *cf.* Friedlaender, C. and Herter, E.: Über die Wirkung des Sauerstoffmangels auf den thierischen Organismus (On the Effect of Oxygen Deficiency on the Animal Organism), *Z. Physiol. Chem.* 3 (1879), pp. 19–51.

    cf. Fraenkel, A. and Geppert, J. Über die Wirkungen der verdünnten Luft (On the Effects of Thinned Air), Springer Verlag 1883, pp. 1–112.

    cf. Dietrich, J.: Die Wirkung comprimirter und verdünnter Luft auf den Blutdruck (The effect of compressed and thinned air on blood pressure), *Arch. Exp. Path. Pharm.* 18 (1884), pp. 242–259.
38. Adolf Loewy was born on June 29, 1862, in Berlin. He studied medicine in Berlin and Vienna. His doctorate thesis was entitled *Über den Einfluss der Temperatur auf die Filtration von Eiweisslösungen durch tierische Membranen*

(On the Influence of Temperature on the Filtration of Protein Solutions through Animal Membranes), which he submitted in Berlin in 1885. In 1886 he spent a short period in Vienna to study physiology, but then returned to Berlin to work in Zuntz's laboratory. Here, as well as in the pneumatic chamber of the Jewish Hospital in Berlin, he simulated high altitudes to conduct physiological experiments. In 1895, he wrote his Habilitation (second thesis required in the German higher education system to become a professor) and published his important work on respiration and high altitude physiology, *Untersuchungen über die Respiration und Zirkulation bei Änderung des Druckes und des Sauerstoffs der Luft* (Investigations on the Respiration and Circulation when the Air Pressure and its Oxygen Content Change). Zuntz played a significant role in realizing this work and a number of other publications (*cf.* List of Works by Nathan Zuntz). In recognition of this fact, Adolf Loewy dedicated his Habilitation thesis to Nathan Zuntz. In 1900, he became what is referred to as a Titular-Professor (guest professor who is deemed to have the requisite abilities to hold a chair without being granted one), read lectures, gave courses and trained students at the Physiology Institute in Berlin, and additionally worked as a general practitioner. According to Jokl, Loewy refused an appointment in Berlin in 1905 after it was demanded of him that he abandon Judaism and convert "to a respectable religion." In 1909, the *Lehrbuch der Physiologie des Menschen* (Textbook of Human Physiology) was published by Zuntz and Loewy. In 1917 he was appointed ausserordentlicher Professor (professor without tenure) for anatomy and physiology. On January 1, 1921, he was assigned the directorship of the biochemical department of the first medical clinic. Starting in 1922, he assumed directorship for the Schweizerisches Forschungsinstitut für Hochgebirgsklima und Tuberkulose (Swiss Research Institute for High Altitude Climate and Tuberculosis) in Davos. Jokl writes that he was in close contact with Lion Feuchtwanger and Thomas Mann during this time, and further states that during these meetings Mann collected information from Adolf Loewy about tuberculosis and those who fell ill to the disease. Thomas Mann later used this information in his novel *The Magic Mountain*, published in 1924. In 1933, his authorization to teach was revoked. Adolf Loewy died in Davos on December 24 (or 25), 1936, at seventy-six years of age. Today, he should certainly be regarded as one of the definitive pioneers of high altitude physiology research (*cf.* Pagel, J.: *Biographisches Lexikon hervorragender Ärzte des neunzehnten Jahrhunderts* (Biographical Encyclopedia of Outstanding Doctors of the Nineteenth Century), Urban and Schwarzenberg 1901, p. 1040; Fischer, I.: *Biographisches Lexikon der hervorragenden Ärzte der letzten fünfzig Jahre* (Biographical Encyclopedia of Outstanding Doctors of the Last Fifty Years), Urban and Schwarzenberg 1962b, p. 936; Jokl, E.: *Aus der Frühzeit der Deutschen Sportmedizin* (The early years of German sports medicine), German Congress for Sports Physicians, September 27–29, 1984, Berlin, pp. 2–8; Kiell, P. J.: How sports medicine began, *AMAA Newsletter* 2 (1987), p. 10).

39. Ernst Jokl was born in Breslau (Wroclaw) in 1907, attending the Johannes Gymnasium (high school) there, and passed his Abitur (school-leaving examination) in 1925. From 1925 until 1930, he studied medicine in Wroclaw and Berlin. In 1928, he was nominated as a candidate for the 400-m hurdle run in the Olympic Games in Amsterdam. The same year, he completed his exams to be an athletics instructor. In the winter months of 1930–1932, Ernst Jokl received a research scholarship to work with Loewy at the high altitude Physiological

Institute in Davos. In 1931, he became director of the institute for athletic medicine in Wroclaw. In 1933, he emigrated to South Africa. Having arrived there, he received the proposition to assume Loewy's position at the Davos institute. Jokl turned down the appointment and remained in South Africa until 1950. There, he was active in various universities, institutes and state institutions. For two years, from 1951 until 1952, he returned to Germany and gave guest lectures at the Sporthochschule (Athletic College) in Cologne. In 1953 Ernst Jokl emigrated once more, this time to Lexington, Kentucky, in the United States. At the university there he became director and professor of the Exercise Research Laboratories. Jokl died in 1997. His scientific works comprise a multitude of publications, primarily in the area of sports and endurance exercise (*cf.* Wrynn, A. M.: A debt paid off in tears, *Intl. J. Hist. Sport* 23 (2006), p. 1155; Jokl, E. 1975, unpublished autobiography. This includes a *curriculum vitae* Jokl wrote himself, as well as the transcript of the personal conversation of the author of this book with Jokl, which took place on August 24, 1987, in the Physiological Institute of the Freie Universität Berlin).

40. Transcript of the conversation between Jokl and Gunga of August 24, 1987.
41. *cf.*: Written notification from the Jewish Hospital of Berlin to the author dated August 7, 1987.
42. The considerate manner in which Zuntz interacted with others is emphasized in multiple sources:
    cf. Durig, A.: N. Zuntz, *Wiener Klin. Wochensch.* 33 (1920), p. 345.
    cf. von der Heide, R.: N. Zuntz, *Landwirtschaftliche Jahrbuch* 51 (1918), p. 345.
    cf. Loewy, Dem Gedächtnis an N. Zuntz (In memory of N. Zuntz), *Berl. Klin. Wochensch.* 57 (1920), p. 433.
43. Wilhelm Caspari was born in Berlin on February 4, 1872, studied chemistry and medicine in Freiburg and Berlin, and received his doctoral degree at the University of Leipzig in 1895. Afterwards he began work at Zuntz's department of animal physiology at the Königlich Landwirtschaftliche Hochschule in Berlin. For nearly twenty years, together with A. Loewy, he was to be the closest scientific co-worker of Zuntz and a personal friend of Zuntz's family. He worked until 1914 in Zuntz's laboratory and then became a field surgeon on the Western Front in World War I. After the war, he worked at the Department of Cancer in Frankfurt/Main, which he headed until his dismissal by the Nazi regime in 1936. Scientifically, Caspari was deeply interested in research on metabolism and nutrition. He participated in Zuntz's expedition to the Monte Rosa in 1901. He pioneered vitamin and cancer research, especially in relation to the effects of radiation on cell growth and metabolism, and introduced the use of radiation in the treatment of cancer. He died in 1944 in the ghetto of Lodz (Schwartz, E. and Chambers, R.: Wilhelm Caspari: 1872–1944. *Science* 105 (1947), p. 613).
44. *cf.* Letter from Zuntz to Lucius dated April 27, 1888. In this letter, Zuntz requested permission to participate as a shareholder in the Urania. On April 30, 1888, Lucius granted his consent to the same. GStAPK, I. HA Rep 87 B, Geheimes Staatsarchiv Preussischer Kulturbesitz (Secret Central Archives), Files of the Preussisches Ministerium für Landwirtschaft, Domänen und Forsten, B (Ministry for Domestic Affairs, State-owned Domains and Forests), Landwirtschaftsabteilung, No. 20077, fol. 23–25.
45. *cf.* Brooks, G. A., Fahey, T. D. and White, T. P.: *Exercise Physiology*, 2nd edition. *Human Bioenergetics and Its Applications.* Mayfield Publishing Company 1996, p. 14.

46. Angelo Mosso was born in Turino in 1846. Among other places, he studied medicine with Ludwig in Leipzig. Primarily due to this period of studies in Leipzig, Zuntz described Angelo Mosso as having "especially intimate relationships" within the German physiology community. In 1876 he became professor of pharmaceutical instruction in Turin, and some years later he was made chair of physiology there. Starting in 1893, he began to prepare the major, multi-week Monte Rosa expedition which took place in 1894. Angelo Mosso was the founder of the institute for high-mountain research bearing his name, which was located on Monte Rosa. At the same time, under his supervision, the construction of the "Capanna Regina Margherita" was concluded in 1893. This station was located on the Gnifetti peak (4560m above sea level) and was indispensable for the planned long-term studies that were to be performed at high altitude. Mosso named the cabin after the young Queen of Italy (Margherita Teresa Giovanna); she and Mosso had taken several mountain tours in the Alps together. In 1899, Mosso also dedicated his comprehensive book *Man in the High Alps* to the young queen; this book was also published in German (*Der Mensch auf den Hochalpen*). He earned an international reputation for the work he did on exhaustion physiology, for which he used an ergograph he had developed himself. As a high altitude physiologist, he primarily addressed the influence of hypocapnia on the organism. He believed that "mountain sickness" could be traced back to a reduced level of CO_2 in the blood at high altitude. Angelo Mosso died on November 24, 1910, in Turin (*cf.* Pagel, J.: *Biographisches Lexikon hervorragender Ärzte des neunzehnten Jahrhunderts* (Biographical Encyclopedia of Outstanding Doctors of the Nineteenth Century), Urban & Schwarzenberg 1901, pp. 1163 *et seq.*; Rothschuh, K. E.: *Geschichte der Physiologie* (History of Physiology), Springer 1953, pp. 160 *et seq.*; Zuntz, N.: Angelo Mosso, *Dtsch. Med. Wochenschrift* 27 (1911n), p. 31; Zuntz, N.: Angelo Mosso einige Worte des Gedenkens (Angelo Mosso: A Few Words in Remembrance), *Zentralblatt Physiol.* 25 (1911m), pp. 93–94).
47. Hugo Kronecker lived from 1839 until 1914. He was born in Liegnitz, a city near Breslau. He studied in Berlin, Heidelberg, Pisa and completed his doctorate in 1863 in Berlin. In 1868, he went to Leipzig to work with Ludwig. Kronecker completed his Habilitation (second thesis required in the German higher education system to become a professor) in 1872 and was appointed ausserordentlicher Professor (professor without tenure) in 1875. He left Leipzig in 1878 and took up a position as Abteilungsvorsteher (departmental head) at the physiological institute in Berlin, reporting to Du Bois-Reymond. Most likely, he had the opportunity to become acquainted personally with Zuntz over the next two years. In 1885, he left Berlin and assumed the chair of physiology in Berne. During his academic life, Kronecker pursued a variety of physiological issues, focussing on high altitude physiology and in particular on the etiology of "mountain sickness." Together with Mosso, he played a considerable role in the realization and establishment of the international high altitude research station on Monte Rosa (*cf.* Rothschuh, K. E.: *Geschichte der Physiologie* (History of Physiology), Springer 1953, pp. 153 *et seq.*).
48. Simons and Oelz (2001) stated that one could only be astonished about their loyal and respectful relationship because the Zuntz School, with their extremely precise planning, working, and documentation style, was in absolute contrast to the somewhat chaotic but inspired scientific approach of Mosso.

    cf. Simons, E. and Oelz, O.: *Kopfwehberge. Eine Geschichte der Höhenmedizin.* AS Verlag und Buchkonzept 2001, p. 118.

49. Among other, Loewy, Caspari, Müller, Kolmer, and Waldenburg took part in the expedition.
50. In 2006, Wrynn recognized Zuntz's innovative methodological devices: "The construction of instruments to quantify the impact of altitude on the human body was another component of this effort [ballooning]. Much of the research that occurred as the nineteenth century ended came from the German laboratories, particularly the work of Nathan Zuntz. Most importantly, Zuntz and his co-workers developed improved mechanisms for measuring ventilation and oxygen consumption." (Wrynn, 2006, p. 1155).
51. Hermann von Schroetter (Hermann Viktor Anton Thomas Ritter von Schroetter-Kristelli) was born in Vienna on August 5, 1870, as the son of the laryngologist Leopold von Schroetter; he attended the Gymnasium (high school) and the university there. He also completed his doctorates in Vienna, in medicine (May 4, 1894) as well as in philosophy (March 16, 1895). From the summer of 1895 until the end of that year, he worked in various departments of the 1. K. u. K. Garnison-Spital (1st Royal and Imperial Barracks Hospital) in Vienna. From October 1895 until October 1897, he was trained as an "operations pupil" at the university clinic in Vienna. In 1896, together with Mager, who also worked at the clinic of his father, he took his first balloon trip for physiological purposes in Vienna; 1897 saw him take a second trip. Neither of these balloon ascents reached heights sufficient to perform high altitude physiological studies. At this early point in time, von Schroetter had already made contact with Assmann and the Berlin Verein für Luftfahrt (Association for Aviation) through the War Ministry in Berlin (see Chapter 3). In the period thereafter, he dealt primarily with decompression illness. He summarized the results of these findings in cooperation with Heller and Mader in the comprehensive work *Luftdruckerkrankungen*) (Air Pressure Illnesses), published in 1900. In 1902, he participated in the Dritte Versammlung der Internationalen Kommission für wissenschaftliche Luftschiffahrt (Third Conference of the International Commission for Scientific Aviation Journeys) in Berlin. In the course of this event, he went on two balloon expeditions with Zuntz, for physiological purposes. In the following years, von Schroetter devoted himself to the physiological challenges posed by high altitude ascents, focussing on concepts and designs for respiratory devices providing oxygen. As early as 1903, he designed pressure chambers for extremely high balloon travel. In 1906, his publication entitled *Der Sauerstoff in der Prophylaxe und Therapie der Luftdruckerkrankungen* (Oxygen as Prophylaxis and Therapy for Air Pressure Illnesses) was published and was awarded the gold medal at the World Fair in Milan. His paper *Hygiene der Aeronautik and Aviatik* (Hygiene of Aeronautics and Aviation) was published in 1912. Today, this is considered one of the classics of aviation medicine. In order to pursue his climate physiology and tropical medicine studies, he went on several field physiology expeditions, traveling to Sudan, the Canary Islands, and Dalmatia, among other destinations. Another focus of his work was the fight against tuberculosis of the lungs. He was a member of the administrative commission of the International Union against Tuberculosis, and represented Austria as a delegate at the tuberculosis congress in Washington in 1908. During World War I, he was appointed head physician for the aviation forces on account of his many years of research in the field of aviation medicine. After the war, Hermann von Schroetter assumed the directorship of the Alland Sanitarium. He was then appointed to the Sanitätsdepartement des Ministeriums für Volksgesundheit (Sanitary Department of the Ministry of Public Health).

On July 23, 1925, he completed his Habilitation (second thesis required in the German higher education system to become a professor) in internal medicine at the medical department in Vienna. Shortly thereafter, on January 6 (or 8), 1929, Hermann von Schroetter died from complications of severe lung tuberculosis. Without a doubt, he should be considered one of the fathers of aviation medicine (*cf.* the files kept by the Österreichisches Staatsarchiv (Austrian State Archive), Kriegsarchiv (war archive), Haupt-Grundbuchblatt: Hermann Ritter Schroetter von Kristelli; Files on the Habilitation von Schroetter, Österreichisches Staatsarchiv (Austrian State Archive), Verwaltungsarchiv (administration archive), Bestand Unterricht (documents on instruction), non-numbered pages; Standesausweis (official professional identification) H. von Schroetter, Österreichisches Staatsarchiv (Austrian State Archive), Verwaltungsarchiv (administration archive), documents kept by the Soziale Verwaltung (welfare administration), non-numbered pages; staff record sheet H. von Schroetter-Kristelli, Archive of the University of Vienna, non-numbered pages; Die feierliche Inauguration des Rektors der Wiener Universität für das Studienjahr 1928/29 (The Festive Inauguration of the rector of the Vienna University for the Academic Year 1928/29), Vienna 1928, pp. 19 *et seq*.

52. Arnold Durig was born on November 12, 1872, in Innsbruck/Tyrolia. He attended the Gymnasium (high school) there and studied medicine at the local university. He passed all his medical examinations with “excellent” success. After having concluded his studies, he was secondary physician and Volontär (trainee) at numerous clinics in Innsbruck, serving in a wide variety of wards. On January 16, 1898, Durig received his doctorate at the university as a Doktor der gesamten Heilkunde (doctor of general medicine). From May 1, 1894, he held a position as assistant at the Innsbruck Physiological Institute. In 1900 he went to the Physiological Institute in Vienna, and two years later he completed his Habilitation (second thesis required in the German higher education system to become a professor) there. After a period abroad, studying at Oxford, Durig went to Berlin to work in Zuntz’s laboratory. In 1903, he was called to the Hochschule für Bodenkultur (College for Agriculture) in Vienna. Durig and Zuntz undertook their joint expedition to Monte Rosa, continuing the research done during the major expedition of Zuntz, Loewy, Müller and Caspari in 1901. During this time the two men forged a close, lifelong friendship, as documented by the obituary written by Durig. In March 1904, Arnold Durig was appointed ausserordentlicher Professor (professor without tenure) of physiology, and in January 1905 he became ordentlicher Professor (tenured professor). Thereafter, he turned down numerous positions at veterinary medicine schools (Vienna 1900, Innsbruck 1905, Prague and Berlin 1908, Berlin 1911, Berlin 1919). In addition, he is said to have negotiated with a number of other German universities for various appointments. In the summer of 1906, he oversaw another multi-week and carefully planned Monte Rosa expedition, in which Kolmer, Rainer, Reichel and Caspari participated. The results of these studies and the subsequent investigations in the winter of 1906/07 in Vienna were published in 1911 in the *Denkschriften der Kaiserlichen Akademie der Wissenschaften* (Publications of the Imperial Academy of Sciences). Durig received the Lieben Prize from this academy in 1906, and was made a corresponding member in 1911 and a full member in 1915. Durig was certainly significantly involved in Zuntz’s being awarded his honorary doctorate from the department of agriculture at the University of Vienna, presumably in 1913. On December 2, 1918, he was made chair for physiology in the medical department of the University of Vienna, thus becoming Exner’s successor. In total,

his scholarly works comprise approximately 1000 publications, primarily in the fields of metabolism physiology, nutritional physiology, and high altitude physiology. In May 1938, he was forced into retirement for political reasons. Durig was a member and president of numerous scientific institutes, associations and clubs – among others, the Landessanitätsrat (State Sanitary Council) of Vienna, Österreichische Gesellschaft für Volksgesundheit (Austrian Association for Human Health), Österreichisches Zentralkommitee zur Bekämpfung der Tuberkulose (Austrian Central Committee for Combating Tuberculosis), and the Akademie Deutscher Naturforscher (Academy of German Researchers of Nature). He was awarded numerous honors in Germany and Austria. Following his retirement in 1938, he lived at his country house in Vorarlberg until his death on October 18, 1961. On account of his many years of intensive research, in particular on Monte Rosa, Arnold Durig is today considered to be among the pioneers of high altitude physiology (*cf.* staff record sheet of Arnold Durig, files kept by the University of Vienna, non-numbered pages; Die feierliche Inauguration des Rektors der Wiener Universität für das Studienjahr 1962/63 (The Festive Inauguration of the rector of the Vienna University for the Academic Year 1962/63), Vienna 1963, pp. 43 *et seq.*; Strasser, P. and Andreas, R. (eds): *Montafon 1906–2006*, Montafoner Schriftenreihe 2006; Museum des Heimatschutzvereins Montafon, estate of Arnold Durig, unpaginated).

53. Viktor von Podbielski, Staatsminister für Landwirtschaft, Domänen und Forsten (State Minister for Agriculture, State Domains and Forests) from 1901–1906, was born on February 26, 1844, in Frankfurt an der Oder and died on January 20, 1916 in Berlin (*cf. Deutsches Biographisches Jahrbuch* (German Biographical Yearbook), Überleitungsband (transitional volume) 1914–1916, Berlin-Leipzig 1925, p. 366).
54. Bodil Schmidt-Nielsen reported that "maybe the most important outcome of the Heidelberg congress and the visit to Zuntz was that August Krogh met Joseph Barcroft." And he continued, citing from a letter of August Krogh to Lady Barcroft after Barcroft's death in 1947: "I met Barcroft in 1907 in Heidelberg and a few days later again in Zuntz's laboratory in Berlin, and the sympathy between us rapidly became very deep. I have the most vivid memory of those happy, far-off days and the delight of old Zuntz in your husband's micro-method for oxygen in blood: 'Das ist ja für die Kliniker geradezu gefundenes Fressen' (The clinicians will just love it)." (Schmidt-Nielsen B.: *August & Marie Krogh. Lives in Science*, Oxford University Press, American Physiological Society 1995, p. 82).
55. *cf.* in this regard, the following papers:

Zuntz, N.: Zum Ausbau der fischereilichen Wissenschaft (Notes on the Expansion of Fishery Science), *Fisch. Zg* (1904c), pp. 581–583.

Zuntz, N.: Wissenschaftliche und praktische Studien zur Teichwirtschaft (Scientific and Practical Studies on Fisheries), *Fisch. Zg* (1906b) pp. 385–390, 401–406.

Zuntz, N. and Knauthe, K.: Gesichtspunkte zur Beurteilung praktischer Fütterungsversuche an Fischen (Aspects for Evaluating Practical Fish Feeding Methods), *Fisch. Zg* (1898a), pp. 480–483.

Zuntz, N.: Die Aufgaben der Wissenschaft für die Förderung der Teichwirtschaft (The Tasks of Science for Promoting Fisheries), *Fisch. Zg* (1898a), pp. 624–627.

Zuntz, N. and Knauthe, K.: Über die Verdauung und den Stoffwechsel der Fische (On the Digestion and Metabolism of Fish), *Verh. Berl. Physiol. Ges.* (1898b), pp. 149–154.

56. According to a message (email) from the Secretary of the German Physiological Society, R. Koehling, dated May 31, 2007, no material regarding the early history of the Society exists any longer.
57. According to a message from Cornell University to the author, a check of the Cornell "Register" for 1907–08 and 1908–09, which contains course descriptions, schedules, lists of faculty, and often also the names of visiting professors for each college, could not locate any reference to Zuntz in the part describing the summer of 1908. (Letter from Laura Linke, Reference Specialist, Division of Rare and Manuscript Collections, C. A. Kroch Library, Cornell University, Ithaka, NY (USA), dated June 23, 2006.) Furthermore, it would seem that Zuntz was not able to visit Pikes Peak, since there are no documents in the archives of the research station which indicate such a stay (personal communication, J. Reeves (1928–2004) in August 2004 in Xining/China). Regarding Zuntz's opinion about the quality of Armsby's research, W. H. Jordan noted: "When in Berlin in 1913 I asked Dr. N. Zuntz if the United States had any research workers in animal nutrition of equal standing with the best men in Europe, Zuntz answered immediately, "Armsby" (cited after Benedict, F. G.: Biographical Memoir of Henry Prentiss Armsby 1853–1921. *Natl. Acad. Sci. USA, Biographical Memoirs* 19 (1938), p. 278).
58. Jacques Loeb was born on April 7, 1859, in Mayen near Koblenz, Germany. He was a Jewish biologist of German-American origin who is considered to be one of the founders of "general physiology," an experimental field of biological research which started to develop around 1880, although today Loeb's own views on the subject and framework of general physiology are regarded to be "inconsistent and unsystematic" (cited after: Geison, G. L. (ed.): *Physiology in the American Context, 1850–1940*, American Physiological Society 1987, p. 198). Loeb lost his parents at the age of sixteen and started to study philosophy after school. However, he then decided to study medicine in Berlin, Munich, and Strasbourg, where he worked in the laboratory of Friedrich Leopold Goltz. After his graduation (1884) and state examination (1885), he returned to Berlin and worked as an assistant of Nathan Zuntz at the Landwirtschaftliche Hochschule (Agricultural College) (Pace, 1986, p. 10). In 1886, he went to Würzburg to work with the physiologist Adolf Fick, and returned to Strasbourg in 1888. In 1891, Jacques Loeb accepted an invitation to the Bryn Mawr College for Women in Pennsylvania. He started to work as professor of physiology at the University of Chicago in Whitman's "biological physiology" laboratory in 1892, and was put in charge of the physiology course at Marine Laboratory Woods Hole in the same year. According to Geison (1987), it first seemed that Loeb was well suited to Whitman's working style. However, this was not the case. Loeb's passion was to solve problems that would lead to scientific control over organisms. He did not intend to explain the nature of life, and lumped evolutionism together with other aspects of "metaphysics" and Naturphilosophie. So it came that two years later, in 1894, Loeb wrote to Zuntz that scientists succeeded in America "only through 'pull'. In order to have 'pull' one must be a diplomat; but since none of my ancestors was employed in the Foreign Office – Palestinian times excepted, but there the documents are missing – it is not surprising that I have little diplomatic talent" (cited after Geison, G. L. (ed.): *Physiology in the American Context, 1850–1940*, American Physiological Society 1987, p. 197). Loeb was one of the founders of the *Journal of General Physiology*, but by 1902, the leading American physiologists were crowding him out of the professional community. The American Physiological Society Council, led by Chittenden,

removed him from the editorial board of the *American Journal of Physiology* on the grounds that he was no longer affiliated with a major institution (Geison, G. L. (ed.): *Physiology in the American Context, 1850–1940*, American Physiological Society 1987, pp. 198–199). Jacques Loeb died on February 11, 1924, in Hamilton, Bermuda, USA.

Bibliography: Der Heliotropismus der Thiere und seine Übereinstimmung mit dem Heliotropismus der Pflanzen, 1890 (Heliotropism of Animals and its Correspondence to the Heliotropism of Plants); Untersuchungen zur physiologischen Morphologie der Thiere (Investigations on the Physiological Morphology of Animals), 2 volumes (volume I Über Heteromorphose (On Heteromorphosis)), 1891–92, volume II Einleitung in die vergleichende Hirnphysiologie (Introduction to Comparative Brain Physiology), 1899; On the transformation and regeneration of organs, American Journal of Physiology, 4 (1900) pp. 60–68; On the Production and Suppression of Muscular Twitchings, 1906, Vorlesungen über die Dynamik der Lebenserscheinungen (Lectures on the Dynamics of Forms of Living Matter), 1906; The Dynamics of Living Matter, New York 1906; Die chemische Entwicklungserregung der tierischen Eier (künstliche Parthenogenese) (The Chemical Development of Animal Eggs (Artificial Parthenogensis)), 1909; Is species-specificity a Mendelian character? Science 45 (1917), pp. 191–193; Regeneration from a Physico-Chemical Viewpoint, 1924; Die Eiweisskörper (Proteins), Berlin 1924; Fangerau, H. and Müller, I.: National styles? Jacques Loeb's analysis of German and American science around 1900 in his correspondence with Ernst Mach, Centaurus 47 (2005), pp. 207–225.

59. According to Jokl, as an older man Müller supposedly regarded the time spent with Zuntz on Monte Rosa as the best time he ever had in his life (*cf.* Jokl, E. Zur Geschichte der Höhenphysiologie (On the history of high altitude physiology), *Forsch. Fortsch.* 41 (1967), p. 324).
60. Zuntz's successor at the Royal College of Agriculture in Berlin was Börnstein, who was its director from 1908 to 1910.
61. In this context, Zuntz spoke of "Klimatophysiologie" (climato-physiology). Hereinafter, the term commonly employed today, "climate physiology," is used.

cf. in this regard, the following selection of publications:

Zuntz, N.: Künstliches Klima für Versuche am Menschen (Artificial climates for experiments on humans), *Zschr. Baln.* 3 (1911j), pp. 643 *et seq.*

Zuntz, N.: Physiologische und hygienische Wirkungen der Seereisen (Physiological and hygienic effects of ocean travel), *Zschr. Baln.* 4 (1911d), pp. 165–168.

Zuntz, N.: Zur Methodik der Klimaforschung (On the methodology of climate research), *Med. Klin.* (1911h), pp. 854 *et seq.*

Zuntz, N.: Beiträge zur Physiologie der Klimawirkungen (Contributions to the physiology of climate effects), *Zschr. Baln.* 4 (1912g), pp. 523–525.

Durig, A. and Zuntz, N.: Zur physiologischen Wirkung des Seeklimas (On the physiological effect of ocean climate), *Biochem. Zschr.* 39 (1912a), pp. 422–434.

Durig, A. and Zuntz, N.: Beobachtungen über die Wirkung des Höhenklimas auf Teneriffa (Observations on the effect of high altitude climate on Tenerife), *Biochem. Zschr.* 39 (1912b), pp. 435–468.

Durig, A., von Schroetter, H. and Zuntz, N.: Über die Wirkung intensiver Belichtung auf den Gaswechsel und die Atemmechanik (On the effect of intensive lighting on gas exchange and respiratory mechanics), *Biochem. Zschr.* 39 (1912), pp. 469–495.

Loewy, A.: Über den Stoffwechsel im Wüstenklima (On metabolism in desert climates), *Zschr. Baln.* 7/8 (1916), pp. 43–48.

Loewy, A.: Über den Stoffwechsel im Wüstenklima (On metabolism in desert climates), *Veröff. Zentralst. Baln.* 3 (1916a), pp. 1–6.

Durig, A., Neuberg, C. and Zuntz, N.: Ergebnisse der unter Führung von Prof. Pannwitz ausgeführten Teneriffaexpedition (Results of the Tenerife expedition headed by Prof. Pannwitz) 1910. IV. Die Hautausscheidung in dem trockenen Höhenklima (Skin excretion in dry high altitude climates), *Biochem. Zschr.* 72 (1916), pp. 253–284.

62. Among them: Emil Fischer (Nobel Prize 1902, Chemistry), Svante Arrhenius (Nobel Prize 1903, Chemistry), Paul Ehrlich (Nobel Prize 1908, Medicine) and Albrecht Kossel (Nobel Prize 1910, Medicine)
63. For the contents of this publication, please see Chapter 3.4 (Zuntz, N., Zur Physiologie und Hygiene der Luftfahrt (On the physiology and hygiene of aviation), in *Luftfahrt und Wissenschaft* 3 (1912a), pp. 1–67).
64. *cf.* in this regard the following selection of publications:

Zuntz, N.: Zur Physiologie der Spiele und Leibesübungen (On the physiology of games and physical exercise), *Blätter für Volksgesundheitspflege* 11 (1911l), pp. 241–246.

Zuntz, N.: Körperkultur und Sport (Physique and sports), *Salonblatt* (1912e), pp. 1716–1721.

cf. Zuntz, N.: Physiologie des Sportes und der Leibesübungen (The physiology of sports and physical exercise), *Himmel und Erde* 26 (1914b), pp. 439–453.

65. *cf.* in this regard the following selection of publications:

Zuntz, N.: Zur Physiologie der Spiele und Leibesübungen (On the physiology of games and physical exercise), *Blätter für Volksgesundheitspflege* 11 (1911l), pp. 241–246.

Zuntz, N.: Körperkultur und Sport (Physique and sports), *Salonblatt* (1912e), pp. 1716–1721.

cf. Zuntz, N.: Physiologie des Sportes und der Leibesübungen (The physiology of sports and physical exercise), *Himmel und Erde* 26 (1914b), pp. 439–453.

66. According to a written notice to the author of this work, dated September 22, 1987, there are no longer any documents recording this procedure in existence at the Universität für Bodenkultur (University for Agriculture) in Vienna.
67. *cf.* in this regard the following publications:

Zuntz, N.: Meinungsaustausch (An exchange of opinion), *Deutsche Landwirtschaftliche Tierzucht* 37 (1914a), p. 428.

Zuntz, N.: Unsere Ernährung im Krieg (Our nourishment during war), *Die neue Rundschau* 1 (1915f), pp. 405–411.

Zuntz, N.: Einfluss des Kriegs auf Ernährung und Gesundheit des deutschen Volkes (Influence of the war on the nutrition and health of the German people), *Med. Klin.* (1915h), pp. 1–24.

Hehl, H. and Zuntz, N.: Die fettarme Küche (Low fat cooking), *Flugschriften zur Volksernährung* (1915), pp. 1–16.

Zuntz, N.: Tier-Ernährung und Fütterung (Animal nutrition and feeding). In: *Volksernährung im Kriege*, Hobbing 1915a, pp. 100–117.

Zuntz, N.: Einwirkung der Kriegslage auf die Teichwirtschaft (Influence of war on fisheries), *Fisch. Zg.* 18 (1915b), pp. 217 *et seq.*

Zuntz, N.: Zur Einwirkung der Kriegslage auf die Teichwirtschaft (On the influence of war on fisheries), *Fisch. Zg.* 19 (1916b), pp. 127–129.

Zuntz, N.: Die Aufgaben des Arztes beim gegenwärtigen Stande der Ernährungsfrage (The tasks of the physician in the current situation of nutritional issues), *Dtsch. Med. Wschr.* 43 (1917b), pp. 1409–1412.

Zuntz, N.: *Ernährung und Nahrungsmittel* (Nutrition and Food), Teubner 1918a.

Zuntz, N.: Ernährungsfragen (Nutrition questions), *Land und Frau* 2 (1918i), pp. 338 *et seq.*

In the course of these studies Zuntz observed that the growth rate of hair and nails is increased if albuminous substances containing tyrosin and cystin are ingested. It might well be that these observations gave rise to the short note in the the *New York Times* under the title "Offers hope to the bald. German scientist says certain food elements make hair grow" (*New York Times*, 1920, February 11).

68. *cf.* official note of March 14, 1918, made on the letter the rector sent to the Minister on March 6, 1918, Secret Central Archives.

    A visit by the author has shown that today there is no longer a bust of Zuntz at the place cited in the above letter. Thus far, no indications have been found as to its removal, destruction or where it could otherwise be located.

69. *cf.* petition of the Minister for Agriculture, State Domains and Forests to Kaiser Wilhelm II dated July 21, 1918, fol. 24, Secret Central Archives.

    The exact name of the medal received cannot be determined because the file only contains the abbreviation "Kr.d.f.K. 1870/71." According to information provided by Ziems (an expert on matters pertaining to medals, Berlin), this is not an official abbreviation (it could be Kreuz für den Krieg (Cross for the War) 1870/71).

70. *cf.* deed recording the award of this doctorate, Archive of the Tierärztliche Hochschule Hannover (Hanover School of Veterinary Medicine).

    Since the archive of the School of Veterinary Medicine was almost completely destroyed during World War II, no documents are available that record the award ceremony (*cf.* letter from Lochmann (Hanover School of Veterinary Medicine) to the author dated December 21, 1987.

71. *cf.* deed recording the award of this doctorate, Archive of the Rheinische Friedrich-Wilhelms-Universität.

72. Letter from Zuntz to the Dean of the Department of Philosophy dated August 4, 1919, Archive of the Rheinische Friedrich-Wilhelms-Universität, honorary doctorate file for Zuntz.

    It remains remarkable that, in this letter of thanks to the dean, Zuntz does not once mention his doctoral advisor Pflüger when listing his most important instructors at the department and in colloquia. This is further evidence of the deep rift between Zuntz and Pflüger, which Durig alludes to in his obituary (*cf.* Durig, A.: N. Zuntz, *Wiener Klin. Wochensch.* 33 (1920), p. 345).

73. Letter from Fuchs to Zuntz dated January 18, 1919, GStAPK, I. HA Rep.87 B, Geheimes Staatsarchiv Preussischer Kulturbesitz (Secret Central Archives), Files of the Preussisches Ministerium für Landwirtschaft, Domänen und Forsten, B (Ministry for Domestic Affairs, State-owned Domains and Forests), Landwirtschaftsabteilung, no. 20090, fol. 205-205 R.

    The pension amount was later determined at 9874.00 Marks annually (*cf.* pension payment record Nathan Zuntz, GStAPK, I. HA Rep.87 B, Preussisches Ministerium für Landwirtschaft Domänen und Forsten, B Landwirtschaftsabteilung, no. 20090, fol. 215 *et seq.*).

74. Durig was supported not only by Zuntz, who wished that he would come to Berlin, but also by other members of the faculty, who tried over the years to convince him to take a chair in physiology. In this respect, a letter from Abderhalden to Durig on 1911 is remarkable:

"Dear friend!

I do not know what could have broken your ability to work. I can only imagine that you are suffering from neurasthenia. There is only one treatment for this, albeit a radical one – leave the unfavorable conditions! I am convinced that you will quickly and fully recover here. The close proximity to Zuntz, the association with lovely people, the knowledge that you are warmly welcomed by your fellow colleagues: these are all first-rate therapeutic factors.

There can be only one matter that could make you decide against this proposal, namely an illness of one of your organs. The pension terms are worse here than they are in Austria. The law prohibits the Minister from making an exception. I am convinced that you are not suffering from an organic disease. I am sure that it is the unfavorable circumstances you are currently in that are making you feel that your capacities are dwindling. When I visited you in January, I left you and had no concerns about you whatsoever. A sick man does not look as good as you did! I am very excited. What I would like to do most is hurry to Vienna and ask you myself to come. Today I have made it known openly and freely that I know no one else who could be considered [for the position] beside you. If you were to refuse, it will be difficult to fill the chair. Only you can bring the institute [forward]. However, not only the institute and the university need to be considered, but the fact that you personally will be restored by the change. I believe that the prescription [for successfully treating you] is "Berlin"!

The matter is very urgent. [The university] wishes to fill the position as soon as possible. Please make a definite decision and say yes. Permit me to tell you that for a very long time now, I have had an extremely high opinion of you. And in Vienna, I discovered my true affection for you. By nature I am very reserved and may pride myself in never saying anything nice unless it is in fact straight from the heart. More often I curse with my tongue without my heart being truly involved in the angriness. My tongue, however, only speaks warmly if my heart is filled to overflowing. When I tell you that the personal impression [you made on me] in Vienna was such that I took the fondest wish home with me to become your friend and to win your trust, then you will understand what my feelings are about your appointment. I feel with absolute certainty that you will recover here.

I have been treated very well at the veterinary college. I quickly did away with many inconveniences, and permanently so. Everything here is very straightforward. When I wrote to you recently with some bitterness, then that was only because of one person, this being the current rector. He is [the product of] inbreeding. In my view, he does not have the broadness of mind that we both appreciate in Schmaltz. But basically, he is enough of a good-natured fellow. He does not care for me very much because my dog – a splendid core hound – did not show enough respect for his wife!

I would like to share with you, so that you will know how your appointment will come about, my personal list of nominees. Loewy and Cohnheim have already been rejected. You come first, then no one by a long shot, then it is Pregl, Knoop, Trendelenburg, Buerker, Weinland, Tandl. It is a shame about Loewy. I like him very much, but there is no remedy against antipathy. Certainly other suggestions will still be made. I would prefer that we no longer have any reason to think about candidates. You must come. It cannot be any other way. You write that nothing keeps you in Vienna, hence off to healthy circumstances!

If you were here with me, I would not give up for an instant! It would be such a joy if you would accept!

What you think about my prospects in Vienna gives me great pleasure. That would be an extremely great honor for me! I would gladly move to the wonderful city on the banks of the Donau River. However, I am convinced that the whole idea will fail because of Rubner! He is Hans Horst Mayer's most intimate friend! That probably says it all!

May the next mail delivery bring me the reply I long for from you!

With my warm personal regards,

E. Abderhalden"

(Letter from Abderhalden to Durig dated May 24, 1911, Museum of the Heimatschutzverein Montafon, estate of Arnold Durig, unpaginated).

75. for further details, *cf.* Nobel Prize database: http://nobelprize.org/medicine/nomination/nomination.php?-Zuntz&action
76. As early as in 1907, Eduard Buchner (1860–1917), a colleague of Zuntz at the Landwirtschaftliche Hochschule (Agricultural College) in Berlin, had received this award "for his biochemical researches and his discovery of cellfree fermentation."

    For further details, see http://nobelprize.org/nobel_prizes/chemistry/laureates/1907/; http://lem.ch.unito.it/chemistry/nobel_chemistry.html
77. For further details, *cf.* Nobel Prize database: http://nobelprize.org/medicine/nomination/nomination.php?-Krogh&action
78. For further details, *cf.* Nobel Prize database: http://nobelprize.org/medicine/nomination/nomination.php?-Zuntz&action
79. *cf.* Death Register of the Trinitatis parish of the Evangelical Churches in Berlin, folio 90.

    Presumably, Zuntz had waived religious counsel. In view of the relatively long period passing between his death and the burial, it can be assumed that the body was cremated. An inquiry in this respect with the crematories in Berlin was unsuccessful, as archive materials were not sufficient to corroborate this assumption.

    In the meantime, the author of this work has been able to clear the grave of Nathan Zuntz of debris and flora, and has initiated steps to have the grave protected by the authorities.

2 Scientific work

2.1 Definitions of operational terms

For the issues addressed here, the customary scientific terms used in medical physiology do not differentiate the various fields sufficiently. Thus, it seems appropriate to introduce the terms *laboratory physiology*, *field physiology* and *balloon physiology*, and to define them as follows.

In using the term *laboratory physiology*, the author refers to experimental physiological research performed in the laboratory. This term is chosen in order to be able to distinguish the results obtained there from those determined outside of the laboratory environment.[1] By pursuing experimental research in the laboratory, Zuntz was a trailblazer in investigating individual physiological issues in a controlled manner under defined environmental conditions.

In contrast, in using the term *field physiology*, the author refers to studies carried out under conditions which were not artificially created – in other words, studies in which the organism remains in its natural environment during the experiment, and requiring the development of a separate set of light, sturdy and multipurpose physiological instruments. Furthermore, this form of experimentation in particular must be based on a work philosophy characterized by an integrative or "systems biology" (Kitano, 2002, pp. 1662–1664) approach to physiology. A similar approach for this method can be found in earlier scientific literature published in the United States. For example, in 1947, Adolph wrote in his monograph on the physiology of humans in the desert of a "physiology ... studied in the field" (Adolph, 1947 (reprint 1969), p. viii), which he contrasted to the experiments he had performed in the laboratory. Usually, such kind of field studies are characterized by having no preconceived ideas.

When, in 1906, Zuntz referred to his studies as the "Results of Experimental Research in High Mountains and in the Laboratory" (Zuntz *et al.*, 1906) in the subtitle of his work *Höhenklima und Bergwanderungen in ihrer Wirkung auf den Menschen* (The Effect of High altitude Climate and Mountain Hiking on Human Beings), he foreshadowed the concepts of both terms (laboratory physiology and field physiology). This was a sign of how scientific research had begun to leave the confines of the laboratories and later returned to them (Felsch, 2007, pp. 55–80), as well as the starting point of new experimental systems and experimental culture (Rheinberger, 2006, pp. 280–288). It was during the 1930s that Adolph and others further pursued their physiological studies of the influence of hot and cold environments on the physiology of man (Buskirk, 2003, pp. 423–426). Their work was quite similar in structure to the high altitude studies done by Zuntz, Mosso and Bert. In light of the natural conditions under which they were performed, the physiological experiments in hot air balloons should also be categorized under the term *field physiology*. However, in order to differentiate (in conceptual terms) between high altitude physiological studies in a balloon and the experiments done, as part of laboratory physiology, simulating high altitude in a pneumatic chamber, the introduction of a separate term, *balloon physiology*, is justified for this partial discipline of field physiology. Additionally, the particular significance of balloon physiology as a link between high altitude physiology and aviation medicine at the beginning of the twentieth century is expressed using this term.

2.2 Dissertation

In 1868, at the age of twenty-one and after eight semesters of study, Zuntz submitted his dissertation to the department for medical studies in Bonn bearing the title *Beiträge zur Physiologie des Blutes* (Contributions on the Physiology of Blood). The dissertation was forty-two pages long, and was defended with the following theses:[2]

> 1) *The Darwinian theory best explains the relationships between the individual species of the animal and plant kingdom.*
> 2) *The liver is a tubular gland.*
> 3) *Blood contains free carbonic acid.*
> 4) *The heart hypertrophy that accompanies* morbus Brightii *(Bright's disease) cannot be derived solely from the shrinking of the kidneys.*
> 5) *Marriages between blood relatives are a source of many diseases.*
>
> *(Zuntz, 1868, unpaginated appendix to the thesis)*

The study was broken down into an introduction and two chapters, which bore the titles *Die Alkalescenz des Blutes und ihre Veränderungen* (The alkalescence of blood and its changes) and *Ueber das Verhalten des Blutes zu Kohlensäure* (On the reaction of blood to carbonic acid). In his introduction, Zuntz indicated that "upon the instigation of Prof. Pflüger" he had been performing "a series of experiments revolving around the issue of whether free carbonic acid is present in blood or not" for a good two years now (Zuntz, 1868, p. 8). In the following chapters, Zuntz summarized the findings of research on blood that gas physiology had attained. He considered the corresponding investigations by Meyer done in 1857 to be "groundbreaking" (Zuntz, 1868, p. 5): "For the most part, this scientist had already established how oxygen behaves in blood, since he correctly recognized that the oxygen was absorbed by the blood independent of pressure" (Zuntz, 1868, p. 5).

However, a bit further on in his dissertation Zuntz quite justifiably voiced some reservations as to the validity of the results obtained by Meyer. Even Meyer's determinations concerning the oxygen content in blood had been performed using a problematic method, Zuntz asserted, which would inevitably result in even greater mistakes in determining the carbon dioxide content of blood (Zuntz, 1868, p. 5). Zuntz believed the analyses of blood gas performed by other scientists, such as Setchenov, Schöffer, Holmgreen and Preyer, which had been carried out by generating a vacuum, were now invalid, since Pflüger had "made the discovery that not only were certain loose compounds of carbonic acids bound by blood cells, but that this gas was even driven out of the sodium carbonate by the same [blood cells]" (Zuntz, 1868, p. 6). With this statement, Zuntz felt that the issue of whether and to what extent carbon dioxide was present in compound form in blood remained unresolved. Since Zuntz had already intensively addressed similar issues prior to the completion of his doctoral thesis in his first scientific publication (Zuntz, 1867a, pp. 529–533), he wrote in his dissertation:

> *It may be assumed from the experiments described here in excerpts that freshly defibrinated blood, at body temperature and thus [having a] lower partial pressure of carbon dioxide as may occur during the inflation in the air vesicles, which is thus aerated with this gas, only absorbs approximately the same quantity that it does under normal conditions. Since the treatment with the relevant gas mixtures was continued as long as the blood absorbed [such mixtures], there can be no doubt that the same [blood] had chemically absorbed as much carbon dioxide as it could have under the given conditions and that, in addition, it contained a certain quantity of free carbon dioxide corresponding to the partial pressure of this gas. With these experiments the entire issue would already be resolved if one were able to assume that defibrinated blood was identical to living blood. I have discovered this is not the case.*
>
> *(Zuntz, 1868, p. 9)*

Before Zuntz demonstrated how he had obtained this evidence, he first discussed another observation. He had discovered that *in vivo* serum had a clearly alkaline reaction. Zuntz described the structure and sequence of the experiment, which provided him with the necessary evidence, as follows:

> *Using mercury as an agent, I collect canine blood in an airtight cylinder, which at the top has a thin glass tube with a valve. After the colorless serum has separated itself, which occurs when the cylinder is placed in ice, I force the serum out of the cylinder by sinking the cylinder deeper in the surrounding mercury, so that the serum passes through the upper glass tube and a tube with a rubber hose attached to it, whose hook-shaped end is bent upwards, into a test tube placed in the mercury tub and filled with mercury itself. In the dome of the latter, one strip each of red and blue test paper had been attached earlier. The serum, which in this way had not come into contact with air for a single moment, demonstrates an intense alkaline reaction.*
>
> *(Zuntz, 1868, p. 10)*

In order to provide evidence simultaneously that serum also has an alkaline reaction in the presence of free carbonic acid, Zuntz performed the experiment set out below:

> *I collected serum at 0 degrees [Celsius], put it in a balloon-shaped flask and channeled a mixture of air through it that had a concentration of approximately 4–5% carbonic acid, continually shaking the flask until nothing more*

> *could be absorbed. I created said gas mixture simply by inhaling deeply, holding my breath for some time and then exhaling into a gas meter.*
>
> *Then I directed the serum, aerated by this process, which by necessity had to contain free carbonic acids, into a prepared test tube, as described above, and there observed that it evidenced a clear alkaline reaction.*
>
> *(Zuntz, 1868, p. 11)*

In addition, Zuntz established that "following removal from the body, the alkalescence of the blood [undergoes] extremely massive changes, and does so in an extraordinarily short period of time" (Zuntz, 1868, p. 20). In performing his investigations, Zuntz remarked in particular that the alkaline reaction of the blood dropped quickly and that the more alkalescent the blood originally was, the faster these changes occurred (Zuntz, 1868, pp. 20–22). Zuntz assumed that this was caused by acidification within the body, which he connected to the coagulation (Zuntz, 1868, p. 24). It is impressive to read the far-reaching conclusions which Zuntz made based on his observations on the acid–base balance of blood, even if some have since then been proven incorrect:

> *Perhaps I may note the analogous behavior of the muscles, whose coagulation,* rigor mortis, *is likewise accompanied by acidification. The acidification in this case also has to advance to a certain point before it results in myosinuria. As is the case with muscles, in which the acidification that results in coagulation is nothing other than the abnormal escalation of a process which occurs constantly in life, I consider it probable that acids are constantly formed during oxidation processes in living blood, but that they then become the subject of further metabolism at the same rate at which they are created. The fact supporting the argument that acidification is a truly vital process is that it occurs all the more energetically the more one preserves the blood after it has left the veins under the most normal conditions possible. The metabolic disturbance of the blood, which arises after [the blood] has left the body, apparently inhibits the further oxidation of its acids sooner than it does the formation [of the acids.]*
>
> *It is probably quite easy to assume, when one observes how quickly the alkalescence of the blood can change to a significant degree, that these fluctuations are relative to the discharge of carbonic acids in the lungs. I have not yet begun experiments to test this not unlikely assumption.*
>
> *(Zuntz, 1868, p. 24)*

Subsequently, Zuntz pursued the question of "how large a part of the acidification is attributable to the plasma and how large a part to the blood cells" (Zuntz, 1868, p. 25). The experiments he conducted in order to address this topic, using calf's blood made incoagulable, demonstrated "that the acidification is less intensive in the serum than in the blood and that thus, the blood cells play an important role therein in any case" (Zuntz, 1868, p. 26). In addition, Zuntz was able to prove that this reaction was clearly dependent on temperature, inasmuch as temperatures above the physiological standard induced an enhanced acidosis (Zuntz, 1868, p. 26). Conclusions which Zuntz drew from these observations and hypothetical considerations on the development of fevers represent the end of the first chapter of his dissertation. Once again, they show how Zuntz, even as a young scientist, strove to view his physiological research activities in light of their clinical relevance:

> *It appears to me that this observation is apt to throw some light on certain issues quite relevant for pathology. If an increase of the blood's temperature of only a few degrees above the norm suffices to increase the oxidation processes considerably in the same, then this condition must have a strong influence on the course of feverish illnesses. To a certain extent, a fever remains because once it has appeared it is also the cause of its continued existence. The existing rise in temperature causes the level of metabolic activity to increase and this again results in a new rise in temperature. We thus gain the important insight that, in treating a fever, a lot has already been gained in remedying the situation if one can manage to interrupt this constantly self-renewing chain of cause and effect. I see that the withdrawal of warmth has a favorable effect on illnesses with fevers, which treatment has become more and more popular recently. If the drop in temperature does not reduce the metabolism directly in the way which I set out as likely above, it could only have a damaging effect since in the case of healthy persons, every cooling of the skin naturally results in an increased level of metabolic activity in the body, which is, however, a result of those still very unclear mechanisms which enable a consistent body temperature to be maintained.*
>
> *(Zuntz, 1868, p. 27)*

The second part of his dissertation bore the title *Über das Verhalten des Blutes zu Kohlensäure* (On the reaction of blood to carbonic acids). In this section, Zuntz pursued issues of "how living blood behaves with carbonic acids and which compounds the carbonic acid that is normally contained therein forms with the individual components of the blood" (Zuntz, 1868, p. 28).

First, Zuntz defined the absorption coefficient for carbonic acid in blood, which was important for his experiments. At this point, according to Zuntz, only the single set of experiments by Meyer had been published thus far. Zuntz found Meyer's experiment problematic not only due to its having been performed only once, but also because of the method applied (Zuntz, 1868, p. 29). But before Zuntz himself performed a determination of the absorption coefficient for carbon dioxide in the blood, he conducted a control experiment: "In order to check the accuracy of the method, I have made multiple absorption experiments using distilled water and compared the results with Bunsen's figures. They concurred with each other to the greatest extent one could possibly wish for" (Zuntz, 1868, p. 31).

In this connection, Zuntz casually mentioned that he had also succeeded in determining the absorption coefficient in the physiological temperature area of around 39°C, since the aforementioned determinations made by Zuntz and Bunsen had been carried out at 9.5°C (Zuntz, 1868, p. 32). Zuntz felt this absorption coefficient for CO_2 in water to be "noticeably low," namely

> *at 39.0°C = 0.5283*
> *at 39.2°C = 0.5215.*
>
> *(Zuntz, 1868, p. 33)*

In his first publication *Ueber den Einfluss des Partiardrucks der Kohlensäure auf die Vertheilung dieses Gases im Blute* (On the Influence of Partial Pressure of Carbonic Acid on the Distribution of this Gas in the Blood) Zuntz had already determined that in the presence of increased CO_2 partial pressure, more of this gas was absorbed into the blood (Zuntz, 1867a, p. 530). In this regard, Zuntz was surprised to find that:

> *The absorbed quantity of carbonic acid, which increases with increasing pressure ... does not grow in the same proportion as the pressure increases, but rather more slowly. Since the CO_2 physically absorbed increases in proportion to the pressure, one must assume that the carbonic acid additionally absorbed when the partial pressure increases consists on the one hand of a part that increases in the same relation as the pressure does, and on the other hand of a second part that does not increase in the same ratio. The latter share can hardly be regarded as anything other than newly chemically bonded-matter. Thus, if the CO_2 pressure above the blood increases, the blood will show constantly new affinities towards that gas, but in such a way that in the case of high pressure a constant increase of the pressure will increase the quantity of bonded CO_2 to a lesser degree than is the case at a lower pressure. This continual, steady increase of chemically bonded carbonic acid cannot be explained in any other manner than by assuming that an entity (Körper) exists in the blood which bonds itself with carbonic acids according to variable relationships.*
>
> *(Zuntz, 1868, p. 35)*

In order to answer the question in his dissertation as to which blood component is definitively involved in the CO_2 bonding process, Zuntz once again referred to his publication from one year earlier in which he had been able to prove that "cruor absorbs significantly more CO_2 than serum does."[3] At that time, Zuntz was able to conclude in his publication that the still unknown carrier for CO_2 in blood most likely was to be sought in the serum and in the erythrocytes (Zuntz, 1868, p. 533). These advanced experiments done in the context of his dissertation allowed Zuntz to come to the conclusion "that the entity binding the carbonic acids according to variable relationships belongs almost exclusively to the blood cells. That it could be an inorganic salt is hardly believable considering its peculiar behavior" (Zuntz, 1868, pp. 39–40).

Several of Zuntz's insights into blood gas physiology (the role and function of hemoglobin, CO_2) are today considered elementary knowledge in physiology, and were later helpful for other researchers in the field of blood gas physiology, such as Henderson.[4]

Even though some of the clinical conclusions which Zuntz drew from his results cannot be confirmed from today's perspective, the abundance of lasting physiological insights established by this work is nevertheless impressive. The far-reaching significance of Zuntz's dissertation was recognized by Pflüger early on (Pflüger's Report on Zuntz, 1870).

2.3 Broader spectrum of scientific publications

After completing his doctorate in 1868, Zuntz occupied himself for the next fifty-two years of his career and academic life with a large number of topics, some of which have already been described in Chapter 1. The physiology of metabolism, nutrition, respiration and blood gases, under different exercise and environmental conditions, were Zuntz's primary research areas. The multitude of topics that Zuntz addressed in these fields is impressive. At first glance, there is a tendency to fear that his work represents a scholarly polypharmacy; however, proceeding chronologically, this diversity reveals a strong internal connection.

The evolutionary diagram (Figure 2.1) illustrates, as a phylogenetic flowchart, the times at which Zuntz's various research topics appeared while showing their contextual relationships. As can be seen from the figure, Zuntz's entire concept for his thinking and his research is based primarily on his work concerning blood gas, metabolic physiology and circulatory physiology. It is here that one recognizes the strong influence that the Pflüger school of thought had on his work. When he joined the *Landwirtschaftliche Hochschule* (agricultural college) in Berlin, practical physiological issues such as exercise, nutritional and high altitude physiology became his primary focus.

The development of the Zuntz–Geppert respiratory apparatus played a decisive role in the work Zuntz later did during his Berlin period. It enabled him to provide a secure foundation for his metabolic determinations, an essential prerequisite for a large share of the subdisciplines which he later worked on and in which issues of metabolic physiology almost always played a role. This

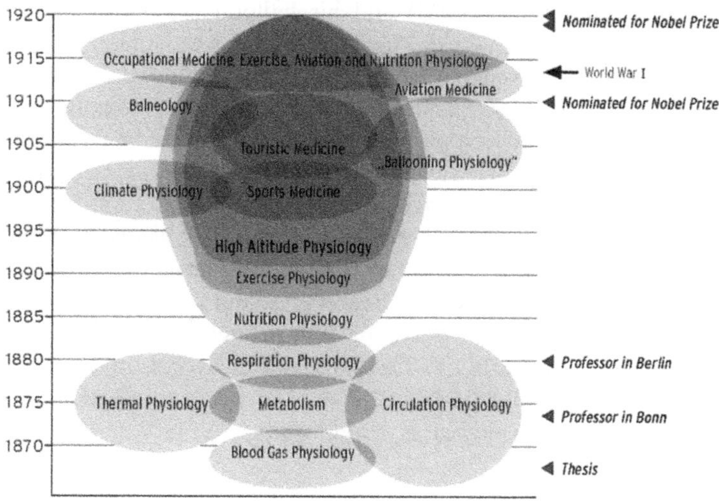

Figure 2.1 Evolutionary diagram to illustrate, as a phylogenetic flowchart, the times at which Zuntz's various research topics appeared and their contextual relationships.

applies in particular to the studies already mentioned, which dealt with exercise and high altitude physiology, which Zuntz embarked on at the beginning of the 1890s. Figure 2.1 shows that subdisciplines of Zuntz's overall work, such as sports physiology, tourism medicine, climate physiology, balneology, balloon physiology and aviation medicine, can be derived from those studies.

The outbreak of World War I and the rapidly decreasing food supply for the German population entailed Zuntz exhaustively occupying himself with issues of nutritional physiology in the last years of his life. In particular, he studied the consequences of undernourishment and prepared guidelines as to how the marginal quantities of foodstuffs were best to be used from a nutritional physiological perspective. In addition, in the *Tierphysiologisches Institut* (Institute for Veterinary Physiology), Zuntz reviewed and developed medical-technical devices for civilian and military facilities.

2.4 Studies on high altitude physiology

The research Zuntz performed on high altitude physiology involved experimental studies in the laboratory as well as field experiments done at high altitude and in balloons.

His key experiments on metabolism were performed in the laboratory with the aid of the Zuntz–Geppert apparatus and a treadmill, as well as in the pneumatic chamber of the Jewish hospital in Berlin. When the Institute for Veterinary Physiology at Chausseestrasse 103 was constructed in 1908, Zuntz received an additional separate large respiratory chamber, the most advanced of its kind worldwide. At this point in time, the high altitude expeditions in the Monte Rosa mountain range (1895–1903) and his balloon ascents in Berlin (1902) were already a thing of the past, and significant key issues of high altitude physiology could already be considered solved. This meant that the respiration chamber served primarily to clear up questions of climate physiology in the context of Zuntz's further field physiological studies. In the course of the Monte Rosa expeditions, it had not been possible to address this set of topics sufficiently.

From the start, Zuntz recognized that, in light of the complexity of high altitude physiology and its considerable methodical problems, he first needed to concentrate on individual phenomena in the laboratory. He believed that in a later phase of research the same phenomena could and should be pursued under field conditions. This consistent two-pronged methodical procedure surely contributed considerably to Zuntz's scientific success in the field of high altitude physiology. The methods and results of his high altitude physiology research will be addressed in greater detail below.

2.4.1 Methodological instruments

After having moved from Bonn to Berlin, Zuntz focused on issues of respiratory physiology. In the course of his work with Geppert concerning the regulatory mechanisms of breathing, the two had developed the Zuntz–Geppert respiration apparatus. This comprised a mouthpiece,[5] a valve mechanism for the separation of expired or inspired gas, a gas meter to quantify the breath volume, and the actual analysis apparatus (Figure 1.9). The latter determined the CO_2 and O_2 volumes given in a sample of expired gas. Due to its size and fragility, this apparatus only allowed metabolic determinations to be made under laboratory conditions. For metabolic measurements in the field, the transportable dry gas meter (Figures 1.17, 2.2, 2.3) was used. Zuntz developed the gas meter specifically for these sorts of studies (Zuntz *et al.*, 1906, p. 163), and presented it following the summer expedition in 1901 at the physiology congress in Turino along with other new instruments (Franklin, 1938, p. 263). Along with the Zuntz–Geppert respiration apparatus, Zuntz's invention was described in handbooks on physiological methods that were widely distributed during this time (Tigerstedt, 1944, p. 129; Müller and Franz, 1952, pp. 499–504). The dry gas meter was a hexagonal box made from sheet steel which weighed only a few kilograms and could be easily carried on the back like a backpack. Inside the box was a bellows. After the subject had closed his nose with a clip (Figure 1.17) and was connected to the gas meter via a tube equipped with a mouthpiece, he would breathe into the gas meter. The bellows inside the box expanded when this happened. This movement was conveyed via an axis on cogwheels to the upper side of the gas meter (Figure 2.2), which in turn was connected to a dial on the back of the gas meter (Figure 2.3). In this manner, the volume entering the gas meter could be determined and read immediately. Figure 2.3 shows the glass tube, approximately 40 cm in length and located on the right side of the gas meter, in which the sample of expelled air intended for analysis collected. At its upper end it was connected by a thick rubber tube to the inside of the gas meter; at the lower end, a longer rubber tube was attached which led to a curved discharge tube. This discharge tube (Figure 2.2) was tied with string to a bobbin on top of the gas meter on a moveable lever, which was likewise equipped with a small cogwheel. Prior to the start of the experiment, the glass collection pipe, tube, and discharge tube were filled with weakly acidic calcium chloride solution. Whenever a sample of the exhaled air was to be taken, the moveable lever on the gas meter was turned so that the cogwheels locked into each other. The bobbin began to rotate, the string unwound, the discharge tube sank, the weakly acidic water flowed out, the water level in the glass pipe fell, and the breath gas from inside the gas meter was sucked into the glass pipe. Since it was ensured that the rotation of the cogwheel axis at the head of the dry gas meter was proportional to the volume exhaled with each breath, the sample collected with each breath exhaled always represented an equal share of respiratory gases. This means that the entire sample of an experiment was an exact average sample of the gas expired (Zuntz *et al.*, 1906, p. 163). The analysis of the samples acquired in this way was carried out in the expedition's accommodations using a simplified version of the Zuntz–Geppert apparatus.

Determining the oxygen absorption of the samples proved difficult. These analyses were performed using phosphor, which meant that the surrounding

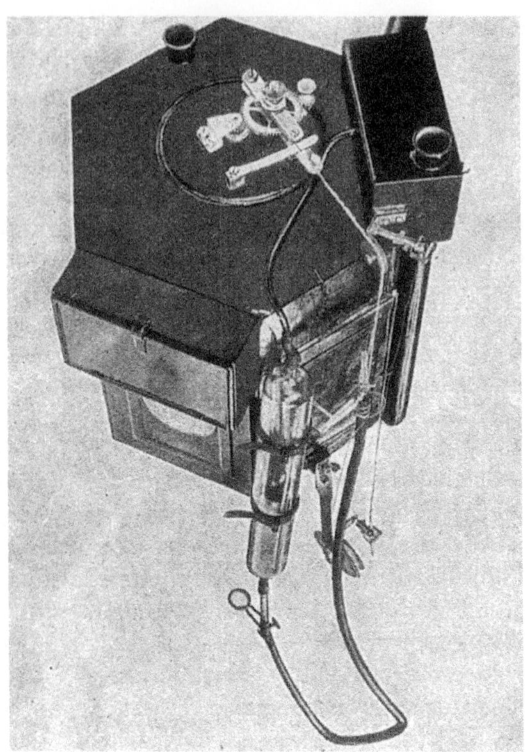

Figure 2.2 The dry gas meter, a hexagonal box made from sheet steel (upper view). (Zuntz *et al.*, 1906, p. 164)

Figure 2.3 The dry gas meter was carried on the back like a backpack (side view). (Zuntz *et al.*, 1906, p. 163)

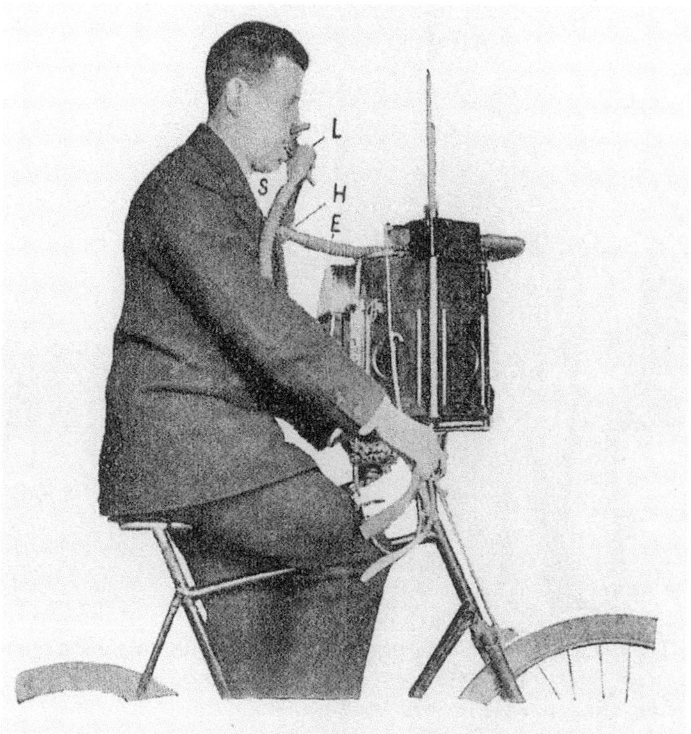

Figure 2.4 Measuring oxygen consumption during bicycling.
(Zuntz L., 1903, p. 7)

temperature could not be below 14°C during a measurement procedure; lower temperatures would have impaired the chemical reactions. This fact forced Zuntz to move his laboratory into the kitchen of the Capanna Regina Margherita (4560 m above sea level), and sometimes into the living and dining room during his major Monte Rosa expedition in 1901 (Zuntz *et al.*, 1906, p. 166). Zuntz and his staff did not only perform metabolic determinations with this transportable gas meter on subjects who were marching, but also on humans while they were bicycling (Zuntz L., 1899, p. 7) (Figure 2.4), typing on a typewriter (von Schroetter, 1925, pp. 323–342) (Figure 2.5), and playing musical instruments (Loewy and von Schroetter, 1926, pp. 1–63) (Figure 2.6). Furthermore, Zuntz and his colleagues used this equipment to measure the energy expended during swimming, which was the first time such an experiment was ever performed (Figure 2.7). Later, Durig used this equipment to determine the effect of different amounts of alcohol comsumption on physical performance during mountain hiking (Figure 2.8). Because Zuntz's respiration apparatus could be used to determine both the carbon dioxide release as well as oxygen consumption, it also became possible to record changes in metabolism quite accurately for relatively short intervals of time (direct determination of the respiratory quotients,

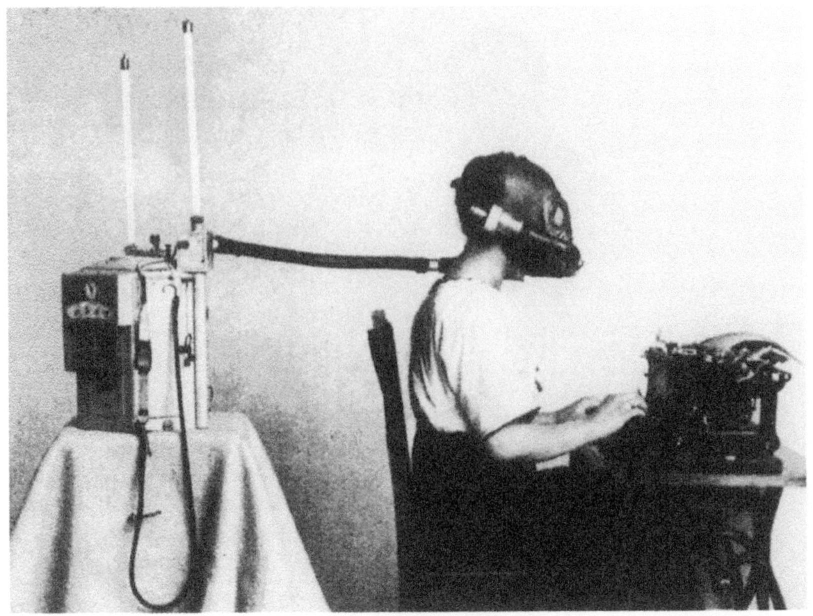

Figure 2.5 The dry gas meter in a metabolic experiment with the subject typing on a typewriter.
(von Schroetter, 1925, p. 324)

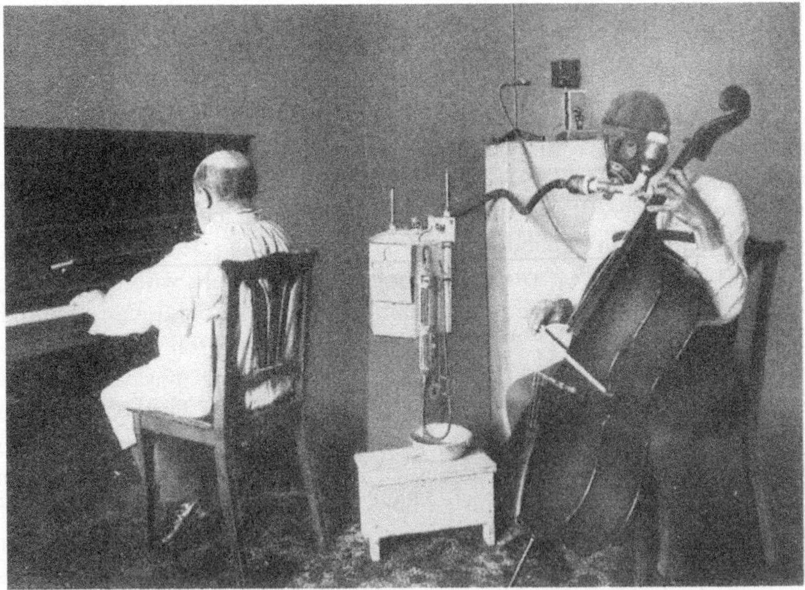

Figure 2.6 The dry gas meter in a metabolic experiment with two subjects performing music.
(Loewy and von Schroetter, 1926, p. 2)

Figure 2.7 Measuring oxygen consumption during swimming. (Zuntz *et al.*, 1906, p. 264)

RQs) (Zuntz *et al.*, 1906, pp. 159–161). This was a decisive advantage as compared to Pettenkofer's respiration chamber procedure, which continued to be widely used at the time[6] and which only permitted a direct measurement of CO_2 release. The Pettenkofer procedure enabled the respiratory quotient to be determined only indirectly by ascertaining the body weight of the subject prior to the start and after the conclusion of the experiment. The difference obtained was added to all bodily excretions (CO_2, water vapor, urine, feces, etc.) and all intakes were deducted from this sum. The amount remaining represented the oxygen intake (Zuntz and Loewy, 1920, p. 656).

A comparison of these methods shows that the Pettenkofer procedure was liable to significant experimental errors in a short series of measurements. However, it proved itself to be comparatively advantageous, as Zuntz also emphasized (Zuntz *et al.*, 1906, p. 158), during long-term experiments lasting hours and days, since this procedure recorded the exchange of gas not only in the lungs but also via the skin. In the introduction to his extensive work on the metabolism of horses, Zuntz described a typical experiment according to this Pettenkofer respiration chamber procedure:

> *The horse stands in a spacious box whose windows are closed and airtight, except for a few openings. Here it can breathe without difficulties, is in similar conditions to when it stays in the stall and can also feed inside. Using an air pump driven by a steam motor, a very large controllable air quantum is constantly sucked through the box and thus the exhalation products of the animal subject are extracted. Between the box and the air pump is a large gas meter which measures the carbon dioxide existing in the box air and in the*

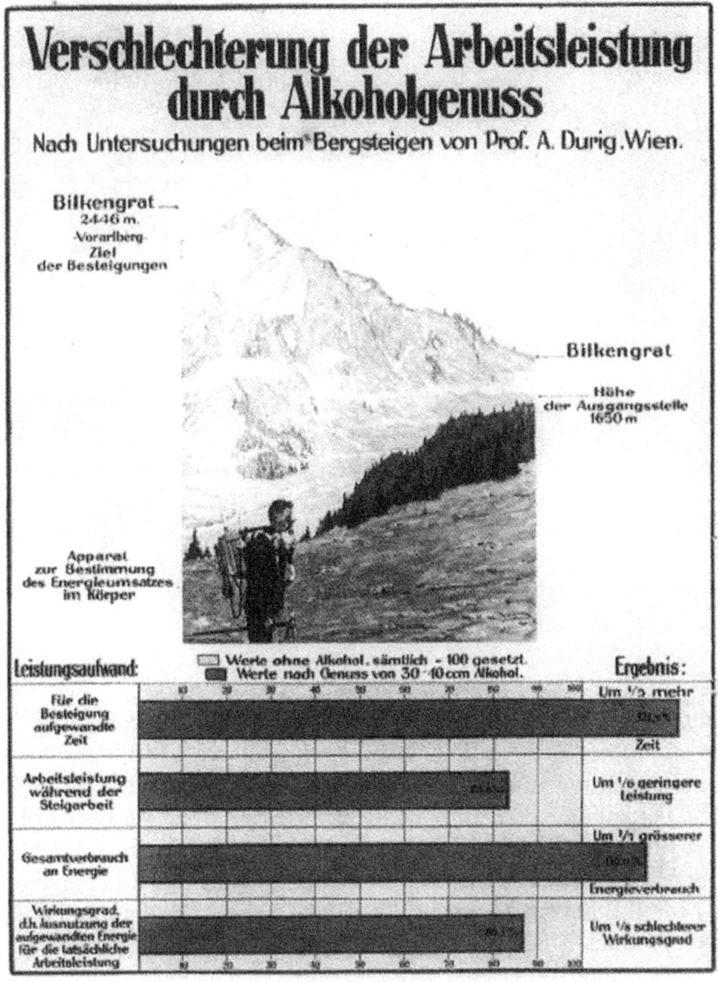

Figure 2.8 The dry gas meter used in a metabolic experiment to determine the influence of alcohol on the ability to exercise at moderate altitude.
(Museum of the Association for the Preservation of the Homeland, Montafon, Estate of Arnold Durig)

> *external air and possibly also the marsh gas. The carbon dioxide production serves as the scale by which to measure the metabolism.*
>
> *(Lehmann, Hagemann and Zuntz, 1894, p. 126)*

The Zuntz–Geppert method of determining metabolism without using the respiration chamber initially fell under sharp criticism from several professional colleagues (Lehmann, Hagemann and Zuntz, 1894, p. 125); its results were even considered untrustworthy and unreliable (Henneberg and Pfeifer, 1890, pp. 258–259). More than anything else, these colleagues objected to Zuntz

Figure 2.9 Schematic side view of the respiration chamber in Zuntz's Institute of Veterinary Physiology in Berlin.
(von der Heide, R., Klein, W. and Zuntz, N., 1913, p. 777)

applying his results, which he had obtained during short-term experiments, to the metabolism of the animal subject for the entire day. Evidently with the intent of reviewing his own methods and so as not to give his critical colleagues any cause for doubt, Zuntz performed control trials in 1891 in Göttingen, at a facility that had been built according to the Pettenkofer principle. After these experiments were concluded, the concerns expressed by Henneberg and Pfeifer as to the Zuntz–Geppert methods were proven to be widely unfounded (Lehmann, Hagemann and Zuntz, 1894, p. 165). Despite this confirmation of his procedure, Zuntz regarded the optimal solution to be a combination of both methods. For this reason he lobbied for years to have a corresponding respiration chamber at his Institute for Veterinary Physiology, but all of his plans in this regard failed because of the lack of space in the old institute.

It was only when the new Institute for Veterinary Physiology in Chausseestrasse 103 was constructed that Zuntz was able to realize his plans and establish his own large respiration chamber (Figures 1.22, 2.9) (Zuntz, 1909b, pp. 473–474). After the facility was completed, the chamber could be operated[7] according to both the Pettenkofer as well as the Regnault–Reiset principle.[8] The chamber had a volume of approximately 80 cbm so that there was enough room inside for a treadmill (Figure 1.15). The human physiological exercise load tests done for the studies on the physiology of marching by Zuntz and Schumburg had been performed on a similar treadmill (Zuntz and Schumburg, 1895a, pp. 54–55). Using the large respiration chamber and the treadmill, which would be called a spiro-ergometric examination facility today, it was possible to address individual physiological issues with the most suitable measurement methods.

In 1911, when the respiration chamber was fully functional and Zuntz increasingly turned his attention to issues of climate physiology and balneology, he determined:

> *It is possible to expose the subject to any temperature, humidity and air movement within the device (by means of internal ventilators) and also to graduate the contents of O_2 and CO_2 in the air at will, thus creating any climate artificially, so to speak.*
>
> *(Zuntz, 1911o, p. 208)*

The respiration chamber enhanced by Zuntz in this way can be viewed as the precursor of today's customary climate chamber, as used in the fields of occupational physiology and environmental physiology.

In 1914, this facility was expanded by an X-ray facility (Figure 2.10). Zuntz used this to determine the size of the heart during exercise load tests. Up until this point in time it had been possible to prepare X-ray records only before and after exercise (Nicolai and Zuntz, 1914, p. 824).

When Zuntz embarked on his high altitude studies, the pneumatic chamber of the Jewish Hospital in Berlin (Figure 1.12) was particularly important

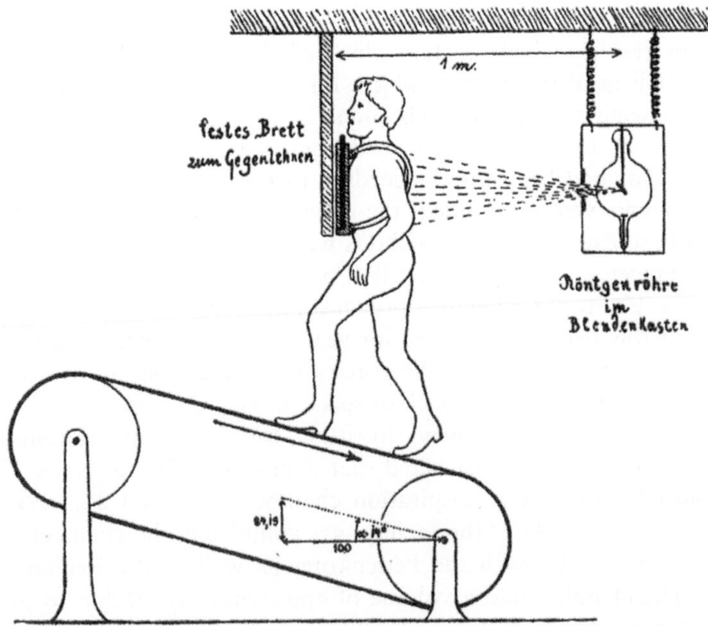

Figure 2.10 X-ray facility in Zuntz's respiration chamber in the Institute of Veterinary Physiology.
(Nicolai and Zuntz, 1914, p. 822)

for him, since it provided him with an experimental laboratory alongside the analysis devices he had available to determine the gas changes. As explained in Chapter 1, pneumatic chambers were increasingly used in the 1860s for therapeutic purposes to treat people with lung diseases. This development began in France (West, 1998, pp. 40–73), and soon thereafter such facilities could be found in numerous cities all over Europe. In 1871, when Bert prepared a list of the pneumatic chambers existing in Europe, no mention was made of such a facility existing in Berlin (Bert, 1878, p. 430). This seems to indicate that the pneumatic chamber in the Jewish Hospital must not have been acquired until the 1870s or 1880s.[9] In any case, Loewy began human physiological studies in the pneumatic chamber in 1891 with the cooperation of Zuntz and Lazarus, and with the financial support of the Gräfin-Bose-Stiftung (Countess Bose Foundation) (Loewy, 1895, p. 3). The objective of the experiments was to simulate sojourns at high altitudes and to study the influence that various air pressures and oxygen concentrations had on an organism.

The respiration and exercise load tests were performed in the pneumatic chamber on the basis of the measurements taken thus far in the institute for veterinary physiology. The load factor during the experiments was, however, not generated by walking on a treadmill – apparently this was not possible due to the restricted space available – but rather by riding a stationary bike (Zuntz, 1906, p. 91).

In 1906, Zuntz summarized the objectives of the experiments he had performed thus far in the Berlin pneumatic chamber in his work *Höhenklima und Bergwanderungen in ihrer Wirkung auf den Menschen* (The Effect of High Altitude Climate and Mountain Hiking on Human Beings) as follows:

> *They form a part of the fundamental preliminary studies with which [our] work in the mountains was then compared. In both cases, the analysis of breathing had the double purpose of providing insights into metabolism, [in other words,] the oxidation processes, while on the other hand enabling conclusions to be drawn from the mechanism of breathing as to the status of the nervous apparatus regulating it, and to thus indirectly [better understand] the entire nervous system as well.*
>
> (*Zuntz* et al., *1906, p. 91*)

Furthermore, the experiments were to help clarify the pathogenesis of mountain sickness – a key issue researched by high altitude physiologists at that time. As is well known, Zuntz assumed that decreased oxygen partial pressure at high altitudes and extreme physical exertion were at the root of this sickness. Since the occurrence of mountain sickness had been observed by various authors, in particular given physical exertion, Zuntz believed that experiments in the pneumatic chamber and balloon ascensions were particularly well suited to study this phenomenon, since in both cases it was possible to study the influence that any decreased O_2 availability had on the organism without having to take the disrupting parameter of physical exertion into account. In

1906, Zuntz gave a vivid description of the physiological effects of his experiments on himself in the pneumatic chamber:

> *If one has the air set to become thinner and thinner within the cabinet, or if one ascends higher and higher in the balloon, while one is at a state of great rest, the only reaction one customarily notices is increasing tiredness and sleepiness which is ultimately so irresistible that one falls asleep and can only be aroused with effort. Normally, no sort of physical discomfort will be felt, with the exception that at times some pressure in the head might become noticeable, a bit of pain in the forehead or temples. Shortness of breath and palpitations of the heart do not normally occur, sometimes deep gasps of breath taken with a bit of a sigh.*
>
> *But this definitely changes as soon as one begins any sort of activity. When sitting up or, to an even greater extent, when bending down, when reading or holding any sorts of instruments or apparatus in one's hand, one feels a noticeable faintness, even dizziness to the extent that one has to steady oneself so as not to fall over. The muscles no longer obey the will. Seeing becomes more difficult, the thought process is impaired. Additionally, both mild and serious headaches occur, a feeling as if the skull is being compressed by a ring, shortness of breath, palpitations of the heart, some people feel nauseous. As soon as one sits or lies down, [all of these symptoms] disappear and only the desire to sleep and, in some cases, a light headache remain.*
>
> *(Zuntz* et al., *1906, p. 449)*

In order to be able to record and describe the observed dysfunctions in greater detail, Zuntz and the others performed detailed psychological tests in the pneumatic chamber. In this context, the subjects[10] were exposed to air pressure which corresponded to being at an altitude of 2500–4500 m. In order to judge the psychological–physical state, they used the following six parameters as benchmarks:

1. *The ongoing mental work was checked by having [the subject] compute the Kraepelin numerical series for 10 minutes. Since the supervisor gave a signal after every minute, the final result could be evaluated both according to the overall performance in the time period of one minute as well as according to the number of errors.*
2. *We checked attentiveness using the Bourdon test by having all the "n's" and "r's" crossed out in a certain printed text. The final result was judged according to the number of missed "n's" and "r's," the number of errors, and finally, according to the time needed to complete the task.*
3. *Manual dexterity was checked by having the subject draw a straight line.*
4. *The memory performance was determined by slowly saying five two-digit numbers to the subject, which they had to repeat immediately. The memory capability was then judged according to the number of incorrect, forgotten, reversed or shifted numbers.*
5. *The speed of simple mental processes, that is, the speed with which a certain activity occurs in reaction to a sensory stimulus, was checked by having the person sit turned away from the apparatus, and then asking them to press a button and to release it the moment they heard the sound of a ball dropping to the ground. The time passing between the heard impression*

> *and the release of the button was measured with a 1/1000 second precision timer (Hipp chronoscope). The experimental psychological laboratory of the university were so kind to provide us with the apparatuses required (Geheimrat, Privy Councilor Stumpf).*
>
> *6. In addition to these tests, the subjects' strength was measured by determining the hand pressure with a dynamometer.*
>
> *(Loewy and Placzek, 1914, p. 1021)*

These sorts of human physiological experiments that scientists performed on themselves were definitely not without danger. This is clearly demonstrated by the fact that, in the course of the above experimental tests, Loewy suffered from an acute ileus and had to undergo emergency surgery. As far as the causal connection between the experiments in the pneumatic chamber and the occurrence of the intestinal obstruction was concerned, Loewy and Placzek wrote:

> *Since in the course of the experiment being addressed here, LOEWY had experienced a particularly strong swelling of the stomach and also discussed this with the other subjects, since furthermore the ileus occurred almost directly after the experiment, there is at least a high probability that as a result of the intestinal movement associated with the rapid expansion of the intestinal gas, an intestinal loop was caught between the existing tracts, or one located between the tracts was bent sharply.*
>
> *(Loewy and Placzek, 1914, p. 1021)*

In order to understand why Zuntz was neither willing nor able to give up the human physiological experiments, it is essential to examine the background for the active high altitude physiological research at the close of the nineteenth century. The following chapter will deal with this and the results Zuntz obtained in his high altitude physiological expeditions.

2.4.2 High altitude physiological expeditions

Towards the end of the nineteenth century, high mountain ranges, especially the Alps, experienced a new stampede of tourists; trips to the Alps had first peaked in the course of the eighteenth century (Raymond, 1986, pp. 1–7). This development meant that an increasing number of mountain railway lines were built and planned in the Alps in order to transport tourists as quickly and easily as possible into the high alpine regions. One of the most famous projects of that time was the plan developed in 1893 by the Swiss industrialist, Guyer-Zeller, to build a train line to the Jungfrau mountain (Gurtner, 1971, pp. 31–35; West, 1998, pp. 97–98).

It is safe to safely assume that a plan of this sort challenged high altitude physiologists to state their position from a medical-physiological perspective. Kronecker, at the time a professor of physiology in Berne, was in charge of the scientists' reactions to the Jungfrau train project (Zuntz *et al.*, 1906). Later, von Schroetter assessed the significance of the project for the further development of high altitude physiology: "It was primarily ... the idea to construct a train line going up the Jungfrau mountain, 4,166 m high, which offered the immediate occasion to perform systematic studies about the effect of thinned air in order to address and clear up health concerns" (Heller, Mager and von Schroetter, 1900, p. 76). These "systematic studies" not only included Zuntz's laboratory physiological research done in the Berlin pneumatic chamber, but also his field physiological studies in the Monte Rosa mountain range and in hot-air balloons.

However, before addressing the methods and results of Zuntz's expeditions to the high mountains, it is important to consider the historical development of high altitude research and its status at the end of the nineteenth and beginning of the twentieth century. Here, the author will limit himself to stating the most significant milestones in the early history of high altitude physiology, since this topic has already been treated extensively by Bert (1878, pp. 22–78), Heller *et al.* (1900, p. 76), Zuntz *et al.* (1906, pp. 1–34), Loewy (1932, pp. 1–11), Jokl (1967, pp. 321–328) and, more recently, West (1998, pp. 1–99).

These sources show that two decisive phases emerge in the development of high altitude physiology, the first stretching back to the sixteenth century. Well-known nature researchers such as Scheuchzer, de Saussure and Humboldt, who at the same time had a lasting influence on the history of other sciences (Hölder, 1960, pp. 39–42, 133, 357), provided the first precise observations of high altitude physiological phenomena in their broader scientific studies. Generally, this meant descriptions of altered physical sensations (nausea, headaches, etc.), which the authors observed in themselves and their companions during their expeditions in the high mountain ranges of the Alps, the Andes, and the Himalayas (Loewy, 1932, pp. 374–375, Heller *et al.*, 1900, p. 62). The second phase of the history of high altitude physiology began in the 1870s in France, marked by the numerous laboratory physiological studies performed by Bert in the pneumatic chamber. This epoch was characterized by increasingly progressive and systematic high altitude research, which was then carried out by Mosso, Kronecker, Zuntz *et al.*, and Durig, to name but a few of the most prominent researchers in this field. Zuntz, while still in Bonn, had carried out his first animal experiments on respiratory physiological issues under various pressure and oxygen conditions at the same time as Bert; however, after Zuntz moved to Berlin to work at the *Landwirtschaftliche Hochschule* (Agricultural College), this focus of his research initially fell behind the methodological issues concerning the analysis of oxygen.

At the start of the 1890s, interest in high altitude physiological research increased once more. Thus, for example, Müllenhoff, among others, held a lecture in 1891 at the *Berliner Physiologische Gesellschaft* (Berlin Physiological Society) summarizing the status of knowledge in this field (Müllenhoff, 1891, pp. 349–350). In this context, Müllenhoff referred extensively to the work done by Bert and the incidental description of physiological observations during balloon ascents prepared by Guy-Lussac and Biot (Guy-Lussac and Biot, 1804, p. 317) and Tissandier *et al.* (Tissandier, 1980, pp. 101–110), among others. From these descriptions, Müllenhoff reached the conclusion that the illnesses observed among mountaineers and balloon passengers (nausea, vomiting, cerebral disorientation) at altitudes of over 3500 m above sea level could primarily be traced back to the unusual physical and mental exertions they had taken upon themselves (Müllenhoff, 1891, pp. 349–350).

Also in 1891, Loewy began his *Untersuchungen über die Respiration und Circulation bei Änderung des Druckes und des Sauerstoffgehaltes der Luft* (Investigation on Respiration and Circulation due to Changes to Air Pressure and its Oxygen Content) in the pneumatic chamber of the Jewish hospital in Berlin. In 1894, Zuntz and Schumburg began their papers on the physiology of marching; in 1895 Loewy completed the essay mentioned above. Both represent the basis for Zuntz's high altitude physiological research, and both form, in their own fashion, the starting point from which he expanded his investigations. While the physiology of marching, the influence of muscle work on metabolism and respiration made up the central focus of Zuntz's work, Loewy's experiments focused primarily on the extent to which the organism tolerated a change of oxygen content and decrease in barometric pressure in the pneumatic cabinet. In terms of their importance for the high altitude research done later, these two different works ideally complemented each other, both methodically and conceptually. This was certainly no accident. Thus, it can be assumed that Zuntz made the decision to become seriously involved in high altitude physiology research at the start of the 1890s.

Zuntz's first high mountain expedition, conducted under the auspices of field physiology, took place with Schumburg and Loewy in August of 1895 in the Monte Rosa mountain range. The fact that Zuntz and the others chose this mountain range at the Italian–Swiss border for their high altitude physiological research will certainly have been based on the relatively favorable orographic and logistic conditions in this area. For example, just two years earlier the Capanna Regina Margherita (Figure 2.11) on the Gnifetti Peak (4560 m above sea level) had been constructed (Mosso, 1899). The station consisted of physiological, zoological, botanical, bacteriological, and physical departments. There was a total of twenty workplaces, which were available to five Italians,

Capanna Regina Margherita, 4560 m ü. M.

Figure 2.11 Schematic drawing of Capanna Regina Margherita (4560 m above sea level). (Zuntz *et al.*, 1906, p. 84)

three Belgians, two each for German, English, French, Austrian, and Swiss nationals, as well as one each for Americans and Dutchmen (*Zbl. Physiol.* 24, 1910, p. 1257). The establishment of this high altitude research station was the result primarily of the ideas and plans of the Italian physiologist Mosso, which he was finally able to realize with the assistance of Queen Margherita of Italy and the Italian Alp Association (Zuntz *et al.*, 1906, p. 30). As can be gathered from the records kept by Zuntz, Schumburg and Loewy, the experiments began on August 19, 1895, in Zermatt and were evidently concluded on August 29, 1895, since the last entries in the experiment records refer to this date (Schumburg and Zuntz, 1896, p. 478). The control trials performed thereafter in the laboratory were carried out in Berlin and, with some interruptions, lasted until spring of 1896 (Schumburg and Zuntz, 1896, pp. 472–475). When Zuntz began planning for the expedition in 1895, he decided to perform the experiments in Berlin (approximately 42 m above sea level), Zermatt (approximately 1600 m above sea level), and on Monte Rosa (between 2800 and 3800 m above sea level). On-site in the high mountains, he chose routes suited for the respiration and metabolism experiments in Zermatt, in the area of the glacier moraine on Monte Rosa and on the glacier itself (Schumburg and Zuntz, 1896, p. 466).

Along with the experiments on the physiology of respiration and of metabolism, which had been the main objective of the expedition in August of 1895, Zuntz was set on pursuing an issue still hotly disputed in scientific circles: the cause of the increased number of erythrocytes during time spent at high altitudes as observed by, to cite but two examples, Miescher (1893, pp. 824–828) and Grawitz (1895, pp. 713–740). At the start of their experiments, Zuntz believed a true erythrocytosis would be impossible; he assumed that a nerve-induced vasodilation was causing the volume fluctuations between the interstitial space and blood, inducing a change to the hematocrit (Schumburg and Zuntz, 1896, p. 491). This assumption was based on the experiments which Zuntz had published together with Cohnstein in 1888 (Cohnstein and Zuntz, 1888, pp. 307–316, 340–341). At that time, Cohnstein and Zuntz drew the following conclusion from their results concerning the shifting of fluid between blood and tissue:

> *The events discussed [nerve-induced vasodilation] suffice to explain the fluctuations in the number of blood cells which we observe in the physiological processes. Thus, all conclusions drawn from these fluctuations as to the generation or loss of these components, as well as the conclusions drawn in this regard in relation to the fever theory are thus accordingly unproven.*
>
> *(Cohnstein and Zuntz, 1888, p. 341)*

Jaquet and Suter quickly disproved Zuntz's assumptions (Jaquet and Suter, 1898, p. 112). It is interesting to note Zuntz's opinion on Miescher's hypothesis. The latter believed the increase in erythrocytes at high altitude was a consequence of a shortage of oxygen (Miescher, 1893, pp. 823–829), and in 1896 Zuntz considered Miescher's hypothesis as "nice" (Schumburg and Zuntz,

1896, p. 490), but stated that his own experiments had completely refuted this theory. On subsequent expeditions, Zuntz and his colleagues pursued this issue multiple times, employing various methods (Zuntz *et al.*, 1906, pp. 172–202). They did not rule out new insights, even if the results called their own views into question or even disproved them. The following statement made in 1906 regarding the scientific point outlined above provides an example of this, and underscores the self-critical attitude towards their own research:

> *Both the results of the experiments in which the behavior of the total hemoglobin quantities circulating in the body was determined when in a high altitude climate, as well as the analysis of the bone marrow have convinced us to acknowledge that haematogenesis increases in high altitudes. This raises the next question: which high altitude factor is to be considered the effective one? This question can also be considered already answered. We have done a large number of experiments that have all shown that it is the thinning of the air that effects haematogenesis.*
>
> *(Zuntz* et al., *1906, p. 200)*

With regard to the metabolism experiments on Monte Rosa, the first expedition came back with the result that, as compared to the measurements in Berlin, the "oxygen consumption ... at high altitude [was] noticeably high during all experiments performed" (Schumburg and Zuntz, 1896, p. 489). Furthermore, they recorded a significant decrease in energy performance (Schumburg and Zuntz, 1896, p. 488). At that point in time, Zuntz and the others ruled out the possibility that this result could be traced back to a relative oxygen deficiency in high mountains. They assumed that "there is something up there which does not have anything to do with an oxygen deficiency, but which manages to lower the degree of working capacity, and furthermore results in the work itself being associated with enormous [amounts] of oxygen consumption" (Schumburg and Zuntz, 1896, p. 489). They assumed that the triggering factors were an "inefficient" (Schumburg and Zuntz, 1896, p. 489) working method of tired muscles, and the impact of abnormal climatic conditions on the organism.

The Zuntz school initially seemed to be faced with unsolvable problems in view of the results obtained in the course of respiration experiments and exercise tests that Loewy had done in the pneumatic chamber, as compared to the data that had been measured in experiments with Loewy at high altitude:

> *The result of the experiments during exertion was more surprising than the result of the experiments made at rest; here, even as low as at the Bétemps hut a considerable increase of oxygen consumption per kilogram meter of exertion took place as compared to the consumption in Berlin – Specifically during exertion, A. Loewy's chamber experiments remained completely constant, even when dilutions were used that exceeded those in the above mentioned experiments by far. Either individual differences were present here or the sojourn at high altitude had a different effect after all on oxygen consumption during exertion than a stay in the pneumatic chamber, perhaps due to the alteration of other meteorological factors.*
>
> *(Loewy, Loewy and Zuntz L., 1897, p. 479)*

In 1895, A. Loewy, J. Loewy and cand. med. L. Zuntz, Nathan Zuntz's son, continued the high altitude physiological research projects in the mountain range of Monte Rosa (Figure 1.16) and in the pneumatic chamber in order to find an answer to this question. After this expedition, however, the authors had to acknowledge once again, in criticism of their own work, that though a multitude of new individual findings were now available that confirmed or supplemented the results of the prior expedition, the knowledge of high altitude physiology as a whole, and specifically concerning its implications for the entire organism, remained perplexing and riddled with holes (Loewy, Loewy and Zuntz L., 1897, p. 538).

This was the state of affairs when Zuntz decided to plan and carry out his most important Monte Rosa expedition in the summer of 1901. He had realized that in order to attain a comprehensive understanding of the multifaceted high altitude processes, it was necessary to carry out an additional, longer-lasting expedition with a larger number of participants. With the financial support of the Gräfin-Bose-Stiftung (Countess Bose Foundation) in Berlin, the medical department, the *Königlich Preussischen Akademie der Wissenschaften* (Royal Prussian Academy of Science) and the German–Austrian Alpine Society, these plans came to fruition (Zuntz *et al.*, 1906, p. XVI).

The results obtained in the course of this expedition will be used as the basis for a thorough analysis of Zuntz's high altitude work in order to present the objectives, contents, and results of his research.

More than any other of Zuntz's treatises, this expedition, which was the core of the work *Höhenklima und Bergwanderungen* (High Altitude Climate and Mountain Hiking) gives considerable insight into the scope of Zuntz's methodological thinking and provides an understanding of the technical and logistical problems he faced. In addition, it conveys a vibrant impression of the physical and mental exertions that the researchers had to endure during the course of the multiweek expedition. It was for a reason that all of the authors wrote their reports such that they were easy to understand by the general public: they believed

> *1) the issues treated [could be] of significance for all those intrigued by the science of life – extending beyond the smaller circle of biologists and doctors (Zuntz* et al., *1906, p. XV).*

and

> *2) the numerous friends of the mountains, who want to enjoy the majesty of creation not only from a visual, aesthetic perspective but also want to be mentally stimulated and wish to understand this phenomenon better. Just as they want to be enlightened about the nature and origins of mountains, valleys, glaciers, they also have the right to become acquainted with the processes which occur inside themselves while drinking from the [metaphoric] "fountain of youth" during their mountain hikes.*
>
> *(Zuntz* et al., *1906, p. XV).*

Zuntz was fully aware of the medical-therapeutic consequences of his research, and in fact had made them his primary objective (Zuntz *et al.*, 1906, p. 2). The research he did on the influence of high altitudes on the organism was to help people "determine in advance in what way a healthy person can best take advantage of health factors and how an ill person can best bring about his or her recovery" (Zuntz *et al.*, 1906, p. 2). At this time, the opinion was generally held in medical circles that "for a not inconsiderable number of visitors, a sojourn in the higher mountain regions in particular did not have any positive effects and was instead even detrimental" (Zuntz *et al.*, 1906, p. 2). According to Zuntz, however, it was particularly people living in urban centers who seriously "needed to partake of fresh air" (Zuntz *et al.* 1906, p. 2). In the high mountains one would not find "the manifold gas and dust-producing products pouring from chimneys, especially factories and technical establishments; primarily coal soot, and larger quantities of carbon oxides, sulfuric and sulfurous acids and carbon dioxides of various sorts" (Zuntz *et al.*, 1906, p. 64). With this comment, Zuntz was ahead of his time in noting the health hazards resulting from the ever-increasing strain on the environment caused by humans themselves.

However, Zuntz believed the mounting strain on the environment to be only one reason for people to flee the city for the mountains. The other, he believed, existed within modern man himself, who "seeks communion with nature, specifically this thrilling impression, in order to find a balance in today's life which, with its flurry, hustle and bustle takes us away from nature too often" (Zuntz *et al.*, 1906, p. 8).

For his experiments, Zuntz subdivided the climate in the Alps into several altitudes:

$>$300–700 m above sea level, mountainous or pre-alpine climate
$>$700 m above sea level, high altitude in the actual sense of the word
  a. $>$700–1200 m above sea level, sub-alpine climate
  b. 1200–1900 m above sea level, alpine climate
  c. $>$1900 m above sea level, high altitude alpine climate.

Zuntz was aware that this breakdown would only be valid in the Alps, and that each of the individual sections was again subject to general physical changes of temperature, air pressure, etc. (Zuntz *et al.*, 1906, pp. 36–39). Zuntz included these environmental parameters in his famous barometric formula (Figure 2.12) (West, 1981, p. 79; West and Wagner, 1998, p. 3), which

For this calculation the following logarithmic formula applies:

$$\log b = \log B - \frac{h}{72\,(256.4 + t)},$$

in which h means the difference in altitude in metres, t the mean temperature of an air column of a height h, B the barometric pressure on the lower, b that on the higher level.

Figure 2.12 The barometric formula established by Zuntz.
(Zuntz *et al.*, 1906, p. 38)

was obviously the outcome of intensive discussions and scientific exchange with the meteorologists Berson and Süring in Berlin (see Chapter 3):

> *However, the climate does not shift with the altitude at the same pace in every zone. The closer we are to the poles, the lower the elevation at which the characteristics of the high altitude climate are already present, while the closer we are to the equator, the higher we must climb from the upper surface of the ocean in order to reach a climate which corresponds to our conceptions of high altitude. In central Europe, an elevation of approximately 1,000 m suffices for this, in Norway this is as low as 500–600 m, in the Bolivian and Peruvian Andes and in the Himalayan mountains, an elevation of over 1,500–2,000 m is required and even at 4,000 m elevation close to the equator, the climate will not evidence all the characteristics of the climate which has already been reached in our Alps at 1,800–2,000 m.*
>
> *(Zuntz* et al., *1906, p. 36)*

It became clear that various climatic circumstances would have different impacts on the organism, the health-enhancing or detrimental repercussions of which Zuntz was hoping to determine.

As participants of the expedition, Zuntz chose his closest staff from the Berlin Institute for Veterinary Physiology (Caspari, Müller and Loewy) as well as his medical students (Waldenburg and Kolmer).[11] The biographical data and the physical state of the participants prior to the expedition in the Monte Rosa region were described as follows:

> 1. *Cand. med. S. Waldenburg, born September 1879, very slender and little accustomed to physical exercise, 168.5 cm tall; was never in the Alps.*
> 2. *Cand. med. W. Kolmer, born July 1879, built with excellent strength and muscularity, very athletic, 168 cm tall.*
> 3. *Dr. med. W. Caspari, born February 1872, lean, very little body fat, not particularly athletic, experienced mountain climber, 173 cm tall.*
> 4. *Dr. rer. nat. et med. Franz Müller, born December 1871, ample body fat, good muscular system, not in shape, experienced mountain climber, 170 cm tall.*
> 5. *Professor A. Loewy, born June 1862, moderate body fat, well-developed leg muscles, not athletically fit, good mountain climber, 152 cm tall.*
> 6. *Professor N. Zuntz, born October 1847, moderate body fat, has not participated in gymnastics for years; undertakes little physical activity, but good mountain climber, 161.5 cm tall.*
>
> *(Zuntz* et al., *1906, pp. 122–123).*

Zuntz intentionally sought out a noticeable variance in anthropometry – i.e., body weight, height, age, athletic, and nutritional state, etc. – of the expedition's participants. The largest possible range of physiological data was to be obtained (Zuntz *et al.*, 1906, p. 122). The preliminary experiments to determine the individual metabolism began for Caspari, Loewy, Müller, and Zuntz on December 14, 1900, in Berlin (Zuntz *et al.*, 1906, pp. 123–124), and were concluded there for Zuntz on December 24, 1900 (Zuntz *et al.*, 1906, appendix table VIII). The fact that the preliminary experiments cannot be termed pleasant has already been described. In the course of the preparations and the preliminary experiments done for the expedition in Berlin, it quickly became apparent that a crucial problem would lie in the adequate nourishment of the expedition participants. On the one hand, tinned foods had to be available for the entire period from August 5 to September 9, 1901, all of which had to have the same composition and to be able to be preserved for weeks on end (Zuntz *et al.*, 1906, p. 124). On the other hand, the expedition participants had "to live off of more or less uniform fare without [demonstrating] dysfunctions of their general well-being or their appetite;" dysfunctions of this sort had to "be avoided at all costs" (Zuntz *et al.*, 1906, p. 123) in order to guarantee that the physiological data from the laboratory and the field could be compared. A cook was to accompany the expedition solely for the purpose of preparing meals in as tasty a manner as was possible under the circumstances:

> *It was her responsibility to prepare the additional main components of our fare: rice and vegetables. Fresh vegetables were of course ruled out because of their varying water content. We had to accept dried preserves – spinach, peas, and carrots. Of these three we had set our hopes on the spinach, but were very disappointed. But once it had been adopted into the program, we were duty bound to eat it every third day.*
>
> *(Zuntz* et al., *1906, p. 125)*

In addition, attempts were made to accommodate the consumption habits of the individual participants as far as possible:

> *Kolmer is a teetotaler. Though Zuntz, Waldenburg, and Müller do not regularly drink alcohol, they do partake of it in moderate quantities now and then at social events. These four abstained from alcohol completely during the length of the experiments and only drank water, sometimes made tastier by adding citric acid and sugar or soda water. By contrast Caspari and Loewy, who were used to consuming approximately 1/2–2 liters of beer a day, at first added to their daily food about approximately 1/2, and later during the marching times, 1 l to 1 1/2 l of a light beer, and on Monte Rosa, the corresponding quantities of red wine. Loewy and Caspari also smoked regularly, Müller now and then, the other participants did not, the consumption of 2–3 light cigars at the most was not considered tobacco abuse.*
>
> *(Zuntz* et al., *1906, pp. 128–129)*

On August 1, 1901, the expedition left Berlin for Brienz (Zuntz *et al.*, 1906, p. 127). The following description of the departure conveys a vivid impression of the abundance of material required in order to successfully perform the investigations:

> *Whoever is used to working in a laboratory furnished with all the modern means of assistance can hardly imagine what needs to be taken along in order to carry out the planned six-week-long experiment on six persons under primitive working conditions. There were 25 kg of chocolate, 42 kg of biscuits and 62 kg of meat to be shipped, not counting the packaging. Two of each were taken of the fragile apparatuses composed of glass tubes for the gas analyses,*

the transportable gas meters plus accessories. Approximately 200 capped bottles and 30 jars for storing urine and feces. Numerous chemicals and preservatives, measuring bottles and measuring cylinders, test tubes and cooking flasks, enamel pots, microscopes, instruments for blood analysis and for measuring electrical phenomena in the atmosphere, thermometers for measuring air and body temperature, devices for recording the pulse, two scales for weighing the daily food – all of this had to be carried. Since, in addition, as discussed in the previous chapter, the preciseness of the investigation on the metabolism depended on the fluctuations of the body weight being determined as exactly as possible, it was of the utmost import that very precise scales were used. We owned one suitably large Garvens jockey scale which measured with an accuracy of 1–2 g even if at loads of 100 kg. In light of its accuracy, we took it with us despite its significant weight and used it in Brienz. But our plan was to divide the expedition into two groups, so we needed a second scale. This one had to be as reliable as possible too, but also needed to be dismantled into individual parts which could each be comfortably carried by one person. After all, it was imperative that the bearers carry them on their backs across the glacier and into the heights. A decimal scale with a 10 g accuracy fulfilled this purpose admirably. After the experiment was completed, the scale was left in the summit hut of Monte Rosa, where it was of great use to Durig and Zuntz in 1903.

Add to all this the necessary mountaineering equipment – glacier ropes, ice picks, winter gear, hiking boots – you can get an impression of the effort and costs required to pack and transport these 30 boxes.

(*Zuntz* et al., *1906, p. 125*)

Zuntz was well aware that after leaving the Berlin laboratory rooms the expedition would be forced to “improvise” (Zuntz *et al.*, 1906, p. 151) in its new quarters. While the accommodations in Brienz (500 m above sea level) were described as “still the most comfortable” (Zuntz *et al.*, 1906, p. 151), the working conditions in this regard at Brienzer Rothorn (2150 m above sea level) already proved to be “more difficult” (Zuntz *et al.*, 1906, p. 151), and conditions were “most uncomfortable” (Zuntz *et al.*, 1906, p. 151) in Capanna Regina Margherita on the Gnifetti summit (4560 m above sea level):

In the unheated rooms the temperature was between 0 to −5 degrees Celsius. Only the kitchen could be heated to something bearable, and when the oven was well-fueled, it could even reach 20 degrees or more. The bedroom had bunk beds and room for 12 people. The guides, the carriers and the workers who were dynamiting rocks for expansion [measures], slept in the top bunks, the four participants of our expedition slept in the bottom bunks wrapped in wool blankets up to their chin, fully clothed and pressed closely to one another. In this cold, it was entirely appropriate that we sleep so close to one another, without this body warmth it would have been impossible to sleep. For the same reason it was completely out of the question to undress in the morning as customary and determine the weight minus clothes. After the undergarments had been weighed separately, they remained on the body for the entire time and the weighing that served to determine the loss of fluids was performed with the undergarments on. Bathing was out of the question since the skin was in a bad state because of the glacial air. The face could only be rubbed with grease. Only our hands were washed with warm water.

(Zuntz et al., *1906, p. 142)*

The expedition reached Brienz on August 4, 1901 (Zuntz *et al.*, 1906, p. 127), and the metabolism experiment began the next day. The perseverance, patience, and concentration with which this research was carried out and the emotions which accompanied these experiments are evidenced by the following section of the report:

On the morning of August 5, the experiment on metabolism began, and with it we were forced to note every movement, every walk, to weigh each drop of water and each gram of food. We were obliged to distance ourselves as much as possible from momentary excitements and moods so that soon, one regarded oneself only as a guinea pig, as an interesting machine. It is also a strange feeling to work for weeks on end without being sure that a result [will be attained]. You stumble about in the dark until after many months, or even years, of tedious analyses, the results begin to materialize. Because of this, one has to comply with every apparently meaningless rule of work and register every fact.

The day was divided as follows: early in the morning, before 6 o'clock, the pulse and body temperature were taken in bed, at exactly 6 o'clock one urinated and then the weight without clothes was determined, on an empty stomach. For breakfast we consumed 200 ccm, that is, a large cup of coffee, prepared the same each day and, depending on our [individual] preferences, various quantities of the rations allotted us for 24 hours of sugar, biscuits with marmalade or butter, some of us also with cheese. Before breakfast the gas exchange experiment was carried out in bed as described earlier using the transportable gas meters, later the breath tests were analyzed, blood analyses and meteorological readings made, urine bottles and feces jars weighed, until around 3 o'clock, after continued, but not too strenuous activity, the midday meal approached.

The food was prepared using simple means in the quite well equipped kitchen and served on the table in six equal portions. The dining table looked quite odd – instead of a flower vase or centerpiece there was a scale and the set of weights! The food was served in the tared pan and this was weighed with its contents on the scale before each [person was given] his sixth of the plate, with the obligation to eat every morsel. Everyone sat at their place, equipped with pencil and notebook in order to take down all of the figures given by the "president" of the table, who also gave each of us our share of food. But we sat down for our daily lunch with the same appetite as for the most elegantly served meal. The rice, prepared with water and butter, was the first course. After the rice pan had disappeared, a second one filled with vegetables was whisked in, which rotated, as mentioned, every three days between spinach, peas, and carrots.

(...)

Along with the vegetables, six steaming meat tins were served and their weight determined first full and then empty. Depending on the preference of the individual, the two beef steaks inside could be eaten at lunch or one could be left over to be consumed in the evening, as – to emphasize this again – was always the case, since the distribution of the quantities so precisely determined for each person was up to the respective individual. A small piece of white bread – weighed of course – was used to wipe the plate clean. We were right to only allow ourselves this one piece for the entire day and otherwise to make do with biscuits. After all, even though bread may have the same weight, this does not mean it will have the same of nutritional since these are differences in the amount of water lost during baking. Incidentally, what is even more important is the requirement to analyze an average sample of each of the respectively purchased quantities of cheese and butter, as is demonstrated by the analyses of the various portions we consumed.

In the afternoon we came together for tea. We had to make do without any milk since this proves to have overly large discrepancies in its composition, and the analysis work would have been so much more complicated. Incidentally, none of us found it difficult to do without [the milk]. In the evening we ate the rest of the daily ration.

(*Zuntz* et al., *1906, pp. 127–128*)

Establishing an exact nitrogen balance of the body required not only that urine and feces be collected completely, but also the perspiration. This meant that after an experimental march, all participants had to undergo a time-consuming and laborious procedure of cleaning the body and washing the clothing:

We proceeded such that we caught the separated perspiration in a special wool undergarment.

Previously this had been repeatedly rinsed out with acid water until the rinsing water contained only traces of nitrogen. After having worn it [during a march], the garment was rinsed out again and the quantity of nitrogen now obtained was determined.

Before donning the perspiration garment, the entire body was washed with warm water; after the experiment another bath was taken and this bathwater was also analyzed for its nitrogen content, as were the bath towels used. The same happened with handkerchiefs used during the march for wiping sweat from the head, face, forehead, and neck. Their nitrogen content was also determined. Thus, we made certain by whatever means we had that no drop of sweat was lost.

(*Zuntz* et al., *1906, p. 170*)

These descriptions and portrayals of the test protocols and procedures sufficiently reveal with what caution, discipline, meticulousness, and personal sacrifice Zuntz pursued his research at high altitudes, using very basic equipment not much more sophisticated than a notebook and pencil.[12]

The following section will explore Zuntz's experiments on metabolism in greater detail. These were carried out while swimming in the lake of Brienz, while hiking, climbing mountains, and in rest periods. The statistics for the metabolism at rest were recorded in the morning before rising from bed (Zuntz *et al.*, 1906, p. 229). In order to obtain results which could be reproduced later for the metabolism experiments done in the field, Zuntz chose three experiment routes on Monte Rosa. The "lower" (Zuntz *et al.*, 1906, p. 129) and the "upper" (Zuntz *et al.*, 1906, p. 132) followed a railway line at about 500 m

Figure 2.13 The exercise testing courses at the top of Monte Rosa. (Zuntz *et al.*, 1906, p. 258)

and 2000–2100 m above sea level, respectively. The third route was close to the Margherita hut on the Monte Rosa glacier (Figure 2.13) (Zuntz *et al.*, 1906, p. 260).

In the course of his research, Zuntz was able to prove that, under the influence of the high altitude climate, the metabolism increased both during physical activity as well as at rest (Zuntz *et al.*, 1906, p. 239). How fast and to what extent the metabolism adjusted to the altered environmental conditions varied dramatically from one individual to the next (Zuntz *et al.*, 1906, p. 267). Additionally, Zuntz observed that the process was reversible – in other words, after a long stay at high altitude, the metabolism readjusted to the normal values, or was even temporarily below normal, once the subject had returned to lower regions (Zuntz *et al.*, 1906, p. 247).

During his experiments analyzing the subjects' performance while mountain climbing, Zuntz came to the conclusion that both the length and the incline of the selected route as well as "the general conditions of the terrain played an extremely dominating role" (Zuntz *et al.*, 1906, p. 267). Zuntz and his colleagues believed the cause for this was to be found in the decreased O_2 partial pressure that is to be found at high altitudes.[13] According to Zuntz, this oxygen deficiency led to increased metabolism in two ways: "for the one part directly as a result of the decreased capacity of the muscles, for the other indirectly as a result of abnormal metabolism products, the formation of which the oxygen deficiency itself had induced" (Zuntz *et al.*, 1906, p. 268). At this point it should be mentioned that, in addition to determining the effects of mountain climbing on the metabolism, Zuntz determined "the energy consumption of the human while swimming, which had not been studied up until now."[14] It turned out that "vigorous swimming was one of the most energetic of physical exercises" (Zuntz *et al.*, 1906, p. 264), and induced unexpectedly high values for the lung ventilation in particular (Zuntz *et al.*, 1906, p. 265).

This led Zuntz to conclude that "while mountain climbing, the capacity of the heart mandates a certain limit; by contrast this limit while swimming is more often set by the respiratory muscles, and often determined by the fatigue of the muscles directly being used" (Zuntz *et al.*, 1906, p. 267). Zuntz's investigations of protein turnover in the high mountains were the basis for dividing the high altitude climate into three levels. While staying at 500 m above sea level, all expedition participants demonstrated a clearly positive protein balance (Zuntz *et al.*, 1906, p. 277), while at levels higher than 1600 m above sea level, those persons in poor physical shape demonstrated a clear protein loss (Zuntz *et al.*, 1906, p. 278). At altitudes of 4500 m above sea level and higher, all participants had a negative nitrogen balance regardless of whether they were physically fit or not (Zuntz *et al.*, 1906, p. 281). From these results, Zuntz and his colleagues drew the conclusion:

> *Thus we see that the mountains exert a very characteristic influence on the existing organism as regards its most important substance and that up to a certain altitude in the mountains – this altitude varying from one individual to the next – an adult will behave similarly to an organism growing under normal conditions. The talk that a journey to the mountains has a fountain of youth effect has thus been given a basis of hard scientific numbers.*
>
> *(Zuntz* et al., *1906, p. 289)*

Blood gas physiology was one of Zuntz's areas of research on Monte Rosa – a topic he had investigated at the start of his scientific career. After the completion of his experiments, it was proven that increased breathing determined the content of CO_2 in the blood – not a direct relationship between CO_2 partial pressure in the lungs and the air pressure, which was assumed by other authors:[15]

> *The more air that is breathed in and out, the lower the carbon dioxide tension in the pulmonary alveoli must be, and compensating for this, the [lower the] carbon dioxide content in the blood.*
>
> *(Zuntz* et al., *1906, p. 300)*

It is in this finding that Zuntz recognized one of the reasons why a stay in high mountains led to a decrease of carbon dioxide in the blood. The other reason, in his opinion, was due to the formation of lactic acids, which increased when the metabolism was working under conditions of oxygen deficiency (Zuntz *et al.*, 1906, pp. 300–302):[16]

> *If such substances [lactic acids] occur in the blood, this is always a sign of abnormal processes happening in the body. This material does not develop in the blood itself, but is instead created under pathological conditions in the body's organs and then discharged by them into the blood. Here they mix*

> *with a portion of the existing alkaline while supplanting the carbon dioxide which previously was connected to this alkaline. The carbon dioxide is set free, travels to the lungs and is released from the body by means of respiration.*
>
> *(Zuntz* et al., *1906, p. 300)*

These findings on the physiology of blood gas induced Zuntz to express "considerable doubts" (Zuntz *et al.*, 1906, p. 302) as to Mosso's acapnia theory. Mosso assumed that the occurrence of "mountain sickness" was instigated by a reduced concentration of CO_2 in the blood, and performed numerous experiments in Capanna Regina Margherita with the intent of proving this assumption. However, in Zuntz's opinion Mosso ultimately did not provide proof of a causal connection between reduced CO_2 concentration in the blood and the occurrence of "mountain sickness" (Zuntz *et al.*, 1906, p. 302).

For Zuntz it was precisely these results, obtained from the blood gas analyses in the high mountains under field physiological conditions, that had obvious and serious impacts on the understanding of general physiological issues. In his view, the results obtained in applied physiological research thus had direct implications on basic research. In this context, Zuntz spoke of significant "expansions" (Zuntz *et al.*, 1906, p. 302) which would benefit physiological research in general as a result of the new knowledge.

In parallel to his investigations done in blood gas physiology, Zuntz and his colleagues recorded the process of respiratory movements with a "respiration belt" (Zuntz *et al.*, 1906, p. 306), which was made of a rubber tube and worn around the chest (Figure 2.14). The movements of the thorax during the breath inhalation and expiration phases were expressed as changes in pressure in the respiration belt, and conveyed via a tube to a membrane. A lightweight lever was situated on this membrane, the tip of which was led along carbon paper stretched across a cylinder. Increased pressure in the tube system (inhalation phase) thus resulted in the membrane being warped, the lever being raised and tracing a corresponding line on the carbon paper. A transportable sphygmograph was used for field research (Figure 2.15). Results showed that at high altitudes, respiration increased in general. As a rule, this increase first occurred at altitudes of more than 1500 m above sea level and became clearly noticeable at altitudes of more than 2000 m if the organism was not athletically trained (Zuntz *et al.*, 1906, pp. 312–315). Additionally, after having concluded his investigations, Zuntz was able to prove that as a result of high altitude training, physical capacity can be enhanced; to this day, this fundamental knowledge continues to be the focus of ongoing research (Zuntz *et al.*, 1906, pp. 314–315; Ward *et al.*, 2000, p. 346).

During their experiments, in which they studied the influence of high altitudes on the cardiovascular system, based on the measurement of blood pressure, pulse frequency and the recording of the pulse pattern (with the assistance of the sphygmographs described above), Zuntz and his colleagues also obtained results very similar to those of the experimental marches (Zuntz

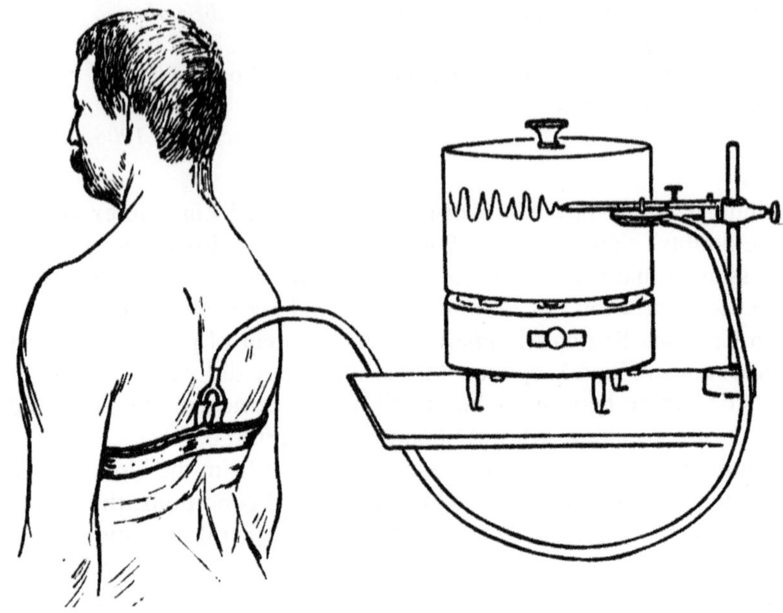

Figure 2.14 Schematic drawing of the apparatus for recording respiratory movements. (Zuntz *et al.*, 1906, p. 306)

et al., 1906, pp. 337–338). Here, too, it was proven that extended sojourns in high altitudes had a clearly fortifying and training effect (Zuntz *et al.*, 1906, pp. 342, 345). Nevertheless, Zuntz warned against regarding the decrease of the pulse frequency resulting after a brief time spent at high altitudes to be an indication of increased capacity (Zuntz *et al.*, 1906, p. 345). Zuntz recognized early on how significant his research would become for modern tourism or travel and wilderness medicine,[17] and emphasized several times and with great urgency the danger people expose themselves to when they spend time in the high mountains:

> *While in high altitude climates and especially while hiking in high mountains, the boundary between healthy and ill behavior in an organ can hardly be more easily overshot and blurred than where the heart is concerned. Nutritional problems and over-exertion of the heart can develop slowly at first, without being perceived. This is precisely where danger lurks, and not a few people have fallen victim to these conditions while hiking in high mountains.*
>
> *(Zuntz* et al., *1906, p. 338)*

> *Likewise a sojourn at high altitude can serve to strengthen the heart muscle and thus improve blood circulation in the body. Nevertheless, caution must also be exercised here. Even the healthy heart of a person not accustomed to physical work will note the increased output expected of it in the mountains.*

Figure 2.15 Zuntz fully equipped for research in the field. (Zuntz *et al.*, 1906, p. 249).

> *A weak heart can easily suffer damages from over-exertion. Thus, it is very sensible to follow the rule according to which weaker persons should [refrain from physical exertion] during the first several days and only slowly allow the stimuli, generated by the muscles' work alongside those generated by the high altitude, to affect the heart. Invalids who do not have sound hearts should not venture into the high mountains at all.*
>
> *(Zuntz* et al., *1906, pp. 473–474)*

High altitude could be applied therapeutically in cases of lung tuberculosis, anemia and rickets (Zuntz *et al.*, 1906, pp. 471–472), and could also "provide a means for toughening persons of delicate constitution" (Zuntz *et al.*, 1906, p. 474).

Zuntz left no doubt that for the healthy and, in certain circumstances, also for the ill, mountain hiking and climbing – for esthetic reasons at the very least – along with rowing and swimming should be included among the sports which best stimulated the entire organism, rather than only individual muscle groups or organs (Zuntz *et al.*, 1906, p. 363). He included weightlifting and "numerous exercises of German apparatus gymnastics" (Zuntz *et al.*, 1906, p. 363) in the latter typical one-sided sports, thus clearly distancing himself from the proposals the Berlin physiologist du Bois-Reymond had published in 1881 (du Bois-Reymond, 1912a, pp. 131–135). Zuntz felt this form of physical activity to be "inexpedient" (Zuntz *et al.*, 1906, p. 363), and for this reason was highly critical of the method of school athletics practiced at the time in Germany:

> *One-sided physical exercises can only be justified when they can be broken down into individual groups, in other words, when those muscle groups of a particular person are subjected to increased activity which are sentenced to relative disuse in the normal life of that person. When this aspect is observed, our German apparatus gymnastics could, in fact, be a very useful instrument. In general, however, such an individualization of gymnastics exercises in our schools is still far from being a reality.*
>
> *(Zuntz* et al., *1906, p. 363)*

Zuntz also expressed the same criticism of military sports, in which the practice of slow marching only partially stimulated the leg and pelvis muscles (Zuntz *et al.*, 1906, p. 363). Similarly, Zuntz was skeptical of (mat) gymnastics; he felt it was "boring" (Zuntz *et al.*, 1906, p. 364), and that this fact would cause people quickly to lose interest in the mandated exercises and to stop exercising altogether after the first few days. Zuntz felt that the decisive motivation for physical activity was sensing "elation in the work" (Zuntz *et al.*, 1906, p. 364) – a feeling that, in his opinion, gymnastics alone was not capable of conveying.

> *Happiness, delight, courage, and energy are to be awoken and promoted by sports. In this regard there is probably no kind of sport which exhilarates, smoothes, and strengthens the human being so harmoniously as Alpinism.*
>
> *(Zuntz* et al., *1906, p. 371)*

In studying *perspiratio sensibilis* (sweat loss via the sweat glands) and *perspiratio insensibilis* (water loss through lungs and skin) at high altitudes, however, their experiments revealed no notably increased water loss (Zuntz *et al.*, 1906, p. 391). Zuntz ascribed this unexpected result to the lower temperatures in the high mountains, which in general reduced the sweat-gland activity (Zuntz *et al.*, 1906, p. 392). He estimated the share of *perspiratio insensibilis* in the total water loss of the body to be insignificant (Zuntz *et al.*, 1906, p. 390). Certainly, Zuntz's views cannot be confirmed from today's standpoint (Hultgren, 1997, pp. 157–164; Ward *et al.*, 2000, p. 31). Nevertheless, it is worth noting that Zuntz was one of the first to warn of the hazards of dehydration during physical activity. In his opinion, the most effective regulation of the body temperature occurred by sweating. As a result, he strongly advocated an increased fluid intake during periods of increased sweating in order to compensate for the loss of water in the body (Zuntz *et al.*, 1906, p. 401). In his chapter "Nutrition of a mountain climber," Zuntz championed the use of coffee or tea in this regard (Zuntz *et al.*, 1906, p. 489). He was very critical of the "unfortunately still so accepted prejudices against drinking liquids when the body is heated" (Zuntz *et al.*, 1906, p. 401) – a position generally acknowledged today.

The case is similar for the hypotheses on "Clothing and hygienic equipment of a mountain climber," which Zuntz addressed in Chapter 16 of *Höhenklima und Bergwanderungen*. Modern clothing physiology has found hardly anything fundamentally innovative to add to this. Clothing for high mountaineering should be chosen so that it protects not only against the cold but also against a rapid change from hot to cold temperatures (Zuntz *et al.*, 1906, p. 408). Protecting the extremities (fingers, toes, etc.) should be given particular attention (Zuntz *et al.*, 1906, p. 409). Finally, maintaining blood circulation is crucial for preventing frostbite:

> *Because of this, every disruption to the blood's circulation, every instance in which the extremities are cut off [from it], every fixed pressure increases the danger of frostbite. Due to this risk, but also in order to keep at bay the unpleasant tingling feeling, the icy coldness and the burning, every tight garter, all pinching elastic gloves and socks, are to be avoided. Gloves should be soft and supple, preferably made of wool. Only the thumbs should be separated, the other fingers should be kept together in one section so that they create collective heat and can move about freely ... Similar principles are to apply for shoes.*
>
> *(Zuntz* et al., *1906, p. 409)*

The clothing material had to meet the following criteria:

> *A combination of porosity with impermeability for dripping water, that is, non-soak ability – those are the characteristics which the outer clothing has to possess in order to meet its double task: maintenance of a not-too-strongly-moving layer of air next to the skin which ensures heat protection and, in the next layer of clothing, to restrict the loss of heat and sufficiently renew the air captured within the clothes for as long as possible so that the perspiration can evaporate at the same degree at which it is formed.*
>
> *(Zuntz* et al., *1906, p. 412)*[18]

Zuntz urgently warned against consuming alcohol with the misguided intent of warming oneself up when the danger of frostbite arose:

> *The biggest danger is posed by the "warming liquor flask." A generous swig generates a cozy feeling of warmth as a result of the increased rush of blood to the skin, but the warmed skin releases more heat, [causing] the body to lose some of its heat reserve.*
>
> *(Zuntz* et al., *1906, p. 410)*

Additionally, tourists venturing into high altitudes should ensure, in order to prevent over-exposure to the sun, that their skin is protected from the increased solar radiation given in the high mountains, by means of applying oil skin cream and by wearing light-colored headwear as protection (Zuntz *et al.*, 1906, p. 418). In addition, mountaineers must find a sensible way of packing and carrying their gear (Zuntz *et al.*, 1906, p. 418). Zuntz believed using a backpack to be optimal for mountain trips, since this was best suited to the carrier's back, and since the body's overall center of gravity would be only slightly shifted towards the rear (Zuntz *et al.*, 1906, p. 419).

The final topics addressed in *Höhenklima und Bergwanderungen* were issues of the effect of oxygen deficiency in high mountains and the etiology of mountain sickness. Zuntz had been particularly interested in both of these topics for years.

In this regard, it must first be said that at the turn of the century it was still heavily disputed whether mountain sickness was a separate form of illness and, if so, which factors were to be held responsible. Once again, Zuntz took a clear position and succinctly reprimanded his contemporary critics who continued to deny the existence of mountain sickness proper: "This skepticism [voiced by critics] who do not want to accept the experiences of others and learn anything [new] is quite unacceptable" (Zuntz *et al.*, 1906, p. 441). According to Zuntz, anyone who climbed beyond a certain altitude would be affected by mountain sickness, although the altitude causing this sickness would vary from individual to individual. Since an evaluation of the etiology of mountain sickness was complicated by general symptoms, in particular states of fatigue and exhaustion, Zuntz came to the conclusion: "The characteristic and specific nature of mountain sickness is first revealed to us under circumstances in which no physical work is performed in the mountains" (Zuntz *et al.*, 1906, p. 442).

These considerations inspired him to draw on the results obtained during balloon trips and in the pneumatic chamber in order to clear up the disputed issues of high altitude physiology, since here "the image [is] the most clear and can easily be subjected to an analysis" (Zuntz *et al.*, 1906, p. 499). When the air gradually became thinner in the pneumatic chamber or in the course of a balloon ascent (Zuntz *et al.*, 1906, p. 418), even though the body was otherwise completely at rest, investigations showed that dysfunctions of the brain (dizziness, darkened vision, etc.) soon resulted. Thereafter the person undergoing the experiment experienced a weakened state, nausea, vomiting, shortness of breath, tachycardia, and headaches. Since these symptoms either disappeared or did not occur at all when artificial oxygen was breathed, Zuntz believed this proved that the cause of mountain sickness was connected to the reduced supply of oxygen at high altitudes (Zuntz *et al.*, 1906, pp. 454, 459, 462, 467).

Zuntz found this opinion confirmed by the experiences related by the balloonists and meteorologists Berson and Süring. During their record ascension in Berlin in 1901, they took oxygen breathing apparatus with them, and thus were able to reach an altitude of almost 11,000 m above sea level. Otherwise, as Zuntz rightly pointed out, the undertaking certainly would have been lethal for both researchers (Zuntz *et al.*, 1906, p. 454).

Zuntz determined that, when the body was at complete physical rest, the described symptoms of mountain sickness appeared at altitudes of over 4000 m above sea level. Under exertion, this altitude could be reduced to 3000 m above sea level, given insufficient amounts of artificial oxygen (Zuntz *et al.*,

1906, p. 454). He believed that the occurrence of mountain sickness depended mainly on three factors that varied from one individual case to the next:

1. the individual's constitution
2. the length of the acclimatization, and
3. the prevailing climate factors in high mountains (Zuntz *et al.*, 1906, pp. 459–467).

According to Zuntz, little was known regarding the influence of the latter. Any therapeutic recommendations Zuntz made for preventing the appearance of mountain sickness basically included guaranteeing the best possible supply of oxygen. During sojourns in the high mountains or balloon trips, either artificial oxygen inhalation apparatuses should be available, he stated, or, in case of danger, there should be immediate descent to lower altitudes (Zuntz *et al.*, 1906, p. 468).

Zuntz was well aware of the significance of his research for the planning, approval, and operation of mountain railways:

> *In view of the mountain railways, some of which are already being built and some of which are still in the project phase, and which are to surmount the highest summits of Europe, the issue of where the limit of danger lies is of great practical importance, in other words, at what altitude a larger percentage of the travelers will presumably be affected by mountain sickness ... We must keep in mind that the Jungfrau train is to travel to altitudes of 4,166 m, the Montblanc train even up to 4,600 m, so we cannot ignore certain concerns. According to these experiences, which completely concur with our theoretical assumptions, we can already foresee now that for many of those reaching the Jungfrau and Montblanc summit with the mountain railway, the time will at the very least not be enjoyable, that they will suffer from more or less considerable discomforts, that some of them will even suffer damages to their health. The speed of the ascent is of utmost importance.*
>
> *(Zuntz* et al., *1906, p. 451)*

From today's perspective, this position taken by Zuntz can only be agreed with.[19]

To conclude this description of the numerous individual results Zuntz obtained in his research for high altitude physiology, the question remains in what overall context Zuntz viewed his research trips in the high mountains. For him, altitude physiology was a way to review the Darwinian theory *On the Origin of Species*, and thus he was well aware of the overall magnitude of his research:

> *In light of these experiences [Zuntz refers to the various adaptations the organisms exhibited to altitude], the question arises whether, in line with the changes which occur before our very eyes while traveling into the mountains, a purposeful adjustment of the constitution has developed in those people living at high altitudes, and what this adjustment could be. With this line of thought, the question ceases to be a physiological one, and becomes anthropological and ethnological. Furthermore, it would need to be researched whether the adjustment to the altitude has led to racial characteristics which*

> *are so permanent that – in contrast to what is exhibited by people living in the lowlands, who rapidly lose the characteristics obtained at high altitude after their return – they would continue to exist if [the person] were to relocate to lower climates ... This is a field opening up many opportunities for research, and its significance will grow in the same degree with which it is transferred by humans to the entire animal world, indeed to the entire living world. At this point, biological research into life at high altitudes can contribute to solving the major problems of the transmutation of forms of creatures and thus the formation of species.*
>
> *(Zuntz* et al., *1906, pp. 493–494)*[20]

In 1902, Zuntz expanded his far-reaching research of high altitude physiology in the mountains and in the pneumatic chamber to include balloon expeditions. In the context of the Third Congress of the International Aeronautic Commission in Berlin in May of 1902, Zuntz and von Schroetter had the opportunity to take their first balloon trip together for physiological purposes; a subsequent flight followed in June 1902 (Schroetter and Zuntz, 1902, pp. 479, 489). During these ascents they were accompanied by the meteorologists Berson and Süring.[21] The objective of the investigations by Zuntz and von Schroetter was to study the conduct of blood and respiratory gas exchange during balloon ascents. For the blood analyses, the prime goal was to review the results obtained by the Zurich physiologist J. Gaule. In the course of two balloon ascents, Gaule had determined a clear increase in the number of erythrocytes and morphologic changes of the erythrocytes (Gaule, 1902, pp. 119–153).

Zuntz and von Schroetter had agreed to take a rabbit with them as a test animal for blood and bone-marrow tests.[22] This animal was killed after the landing by a blow to the head, and dissected twelve hours later. In order to determine the respiratory gas exchange, Zuntz and von Schroetter used the transportable dry gas meter. To ensure the supply of fresh breath gas for the respiration experiments, the inspiration tube of the dry gas meter jutted out of the balloon basket. In addition, the balloon had been filled with hydrogen and, as a result, any contamination of the breath gas during the measurements by carbon oxide and carbonic acids could be excluded (Schroetter and Zuntz, 1902, p. 480).

During their ascents, Zuntz and von Schroetter tested both oxygen breathing apparatuses[23] and oxygen masks for aeronauts (Figure 2.16), which had been developed separately by von Schroetter and by the French physiologist Cailletet and were presented during the Berlin Congress (Record of the Third Assembly of the International Commission for Scientific Aviation, pp. 46–47, 98–129). The demonstration of the oxygen masks was scheduled for the morning of the Eighth Expert Assembly of the Congress on May 24, 1902 (Record of the Third Assembly of the International Commission for Scientific Aviation, p. 5). Zuntz participated in this meeting. The record of the conference reads as follows:

> *Prior to beginning the agenda, Mr. von Schrötter presented another oxygen breathing mask of his own construction and made the following remarks: this mask is distinguished from that designed by Cailletet by several distinctive*

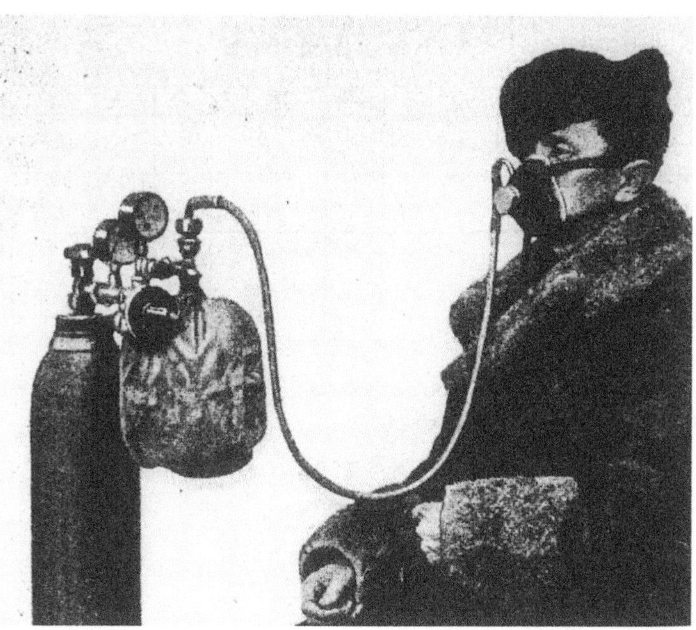

Figure 2.16 Balloonist with oxygen breathing apparatus and special respiratory mask. (von Schroetter, 1912a, p. 30)

> *characteristics. Based on the idea that at an altitude of 8,000–9,000 m, the oxygen to be inhaled is cold, at a temperature of −30 degrees Celsius, he equipped his mask with a preheating device, in which the gas is led through a spiral tube embedded in thermophore material. The mask also has a reduction valve in order to be able to regulate the pressure of the gas to be inhaled to make it comfortable for the lungs. Incidentally, the speaker most strongly objects to the idea that a prolonged inhalation of oxygen is hazardous, [stating] that he subjected himself to the experiment of inhaling pure oxygen for 5 to 6 hours without [experiencing] any sort of disadvantages. The blood only absorbs as much gas as it is capable of ingesting.*
>
> *(Record of the Third Assembly of the International Commission for Scientific Aviation, p. 50)*

On the very same day, at 0:56 pm, Zuntz and von Schroetter took off from the training field of the *Königlich Preussische Luftschifferabteilung* (Royal Prussian Aeronauts Department) for their first physiological balloon expedition. Their second flight took place on June 21, 1902 (Schroetter and Zuntz, 1902, pp. 481, 489). The length of the trip, its progression and the maximum altitudes achieved have been recorded in the barograms of the balloon flights published by Zuntz and von Schroetter (Figures 2.17, 2.18). The balloon expeditions proceeded without major complications; Zuntz and von Schroetter landed safely after their first ascent in Komotau, and after their second in Hennersdorf, close to the city of Neisse.

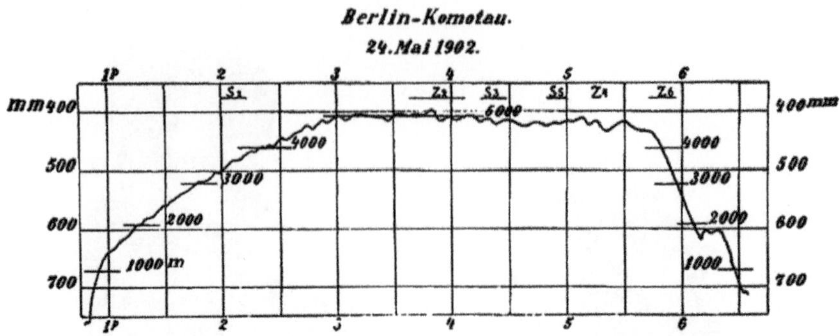

Figure 2.17 Barograms of the balloon expedition of Zuntz dated May 24, 1902. (von Schroetter and Zuntz, 1902, p. 491)

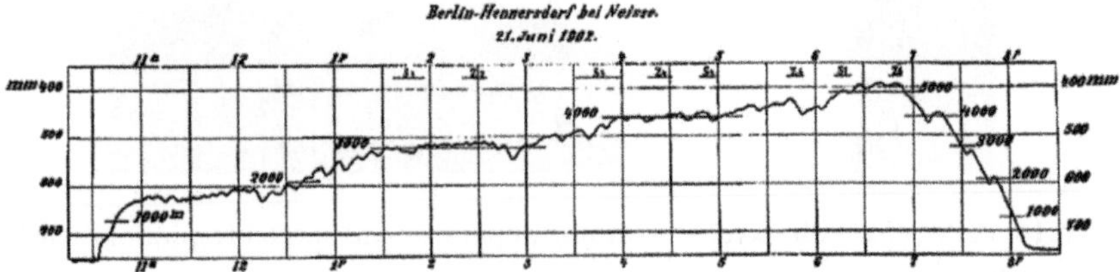

Figure 2.18 Barograms of the balloon expedition of Zuntz dated June 21, 1902. (von Schroetter and Zuntz, 1902, p. 481)

After having evaluated the multiple-hour balloon expedition, the results revealed that the hematological changes as described by Gaule could not be confirmed by Zuntz and von Schroetter. The compounds of bone marrow they obtained from the tibia of the rabbit revealed no indication of an increased formation of young, nucleus-containing erythrocytes (Schroetter and Zuntz, 1902, pp. 485, 499):

> *It should be noted that the bone marrow proved to have reddened and that the animal was an old one. The rib bones also appeared to be redder when held against the light than would have been expected of an animal this age.*
>
> *(...)*
>
> *This finding, of hyperemia of bone marrow, would allow the conclusion to be drawn, provided it could be confirmed in further cases, that a multi-hour stay at an altitude of over 4,000 m perhaps suffices to cause, in accordance with [Friedrich] Miescher's [theories], a hyperemia of the blood-preparing organs, but is certainly much too brief for any verifiable new formation of blood cells to appear.*
>
> *(Schroetter and Zuntz, 1902, p. 486)*

The observations of the behavior of blood pressure and pulse showed no indication of any unusual changes as compared to the measurements which

Zuntz and von Schroetter had taken for control purposes in Zuntz's laboratory in Berlin before and after the ascent (Schroetter and Zuntz, 1902, p. 496). Deviations occurred when oxygen deficiencies appeared (Schroetter and Zuntz, 1902, p. 519). Zuntz and von Schroetter described in detail now they experienced this oxygen deficiency:

> *When the experiments Zuntz performed on von Schroetter came to a close and von Schroetter was to begin reading the gas meter – this took place at 2:27 pm at a pressure of 460 mm and a temperature of −9° – he was clearly not functioning normally. He felt head pressure, experienced a peculiar swaying and a state of confusion of such a nature that it was impossible for him to think correctly and write down the figures he saw. Additionally, a feeling of general weakness overcame him to such an extent that von Schroetter believed it necessary to take up the oxygen tube, and the experiment could not be evaluated due to incorrectly read numbers. Several breaths of oxygen remedied the pathological state at the customary speed; however, during the next experiment, von Schroetter kept the tube in [his] mouth in order to not be caught off guard again by the events described. Had he been at complete rest and inactivity, the use of oxygen would not have been necessary. During the entire trip Zuntz never felt it necessary to breathe oxygen, and he was able to read the meters and take notes without difficulties; his handwriting did not change; only just prior to reaching the highest altitude did he experience a feeling of emptiness in his stomach and slight weakness, which did not, however, differ greatly from similar sensations which Zuntz experienced when skipping normal lunchtime, which was also the case here.*
>
> *(Schroetter and Zuntz, 1902, pp. 483–484)*

Zuntz and von Schroetter were unable to prove an increase of hemoglobin concentration in the blood during their investigations (Schroetter and Zuntz, 1902, pp. 498–499).

During the respiration experiments, the "most important part" (Schroetter and Zuntz, 1902, p. 502) of their investigations, a clear increase of breath frequency became noticeable compared to that of the control trials initially done in the laboratory (Schroetter and Zuntz, 1902, pp. 506, 508). These changes were viewed by Zuntz and von Schroetter to be caused by the influence of particular meteorological conditions in the balloon (cold, significant exposure to the sun) (Schroetter and Zuntz, 1902, p. 512).

Respiration experiments done at an altitude of over 4000 m above sea level then resulted in clear qualitative changes. As compared to the control trials in Berlin, both Zuntz and von Schroetter – despite individual differences – observed an increase of their respiratory quotients (Schroetter and Zuntz, 1902, pp. 506–507). In this context, Zuntz and von Schroetter determined that the subjective sensations caused by oxygen deficiency (see above) did not always concur with the objectively measured changes in the respiratory quotients (Schroetter and Zuntz, 1902, pp. 519–520). They felt that the observed increase of oxygen consumption resulted from increased respiratory work or unaccustomed muscle activities (shivering, uncomfortable sitting) (Schroetter and Zuntz, 1902, p. 505).

In conclusion, it must be said that, from a medical physiological perspective, both balloon expeditions that Zuntz and von Schroetter took in 1902 must be counted as the most successful of their time. Their extensive studies in the balloon have the same systematic structure that Zuntz developed in the course of his high altitude physiological studies on Monte Rosa and in the Berlin laboratory.

But finally – and here there can be no doubt – the successful implementation of the balloon expeditions was dependent on the favorable circumstance that Zuntz and von Schroetter were supported in their research by tried and true logistics and many years of experience gained by other institutions and scientists in the field of aviation in Berlin. This will be discussed in the following chapter.

Notes

1. Kohler thoroughly discussed the history and development of field and laboratory research in biological sciences (*cf.* Kohler, R. E.: *Landscapes and Labscapes. Exploring the Lab–Field Border in Biology*, University of Chicago Press 2002). He made the following remarks which are also valid for the medical/physiological research to be discussed here: "The differences between field and laboratory objectives, places, and practices still shape choices and careers in the border zone, because they are not mere conventions but facts of life. The line between nature and artifice can be blurred but not erased. Natural places cannot be made so lablike that they become unnatural; laboratories cannot be made so natural that they lose the artifice that gives them their power. So, too, with field and laboratory practices. Push quantification or modelling too hard and they become meaningless; take experiments too far afield and they are discredited. Differences between lab and field are a major source of creative innovation. In a patchy cultural landscape, isolation and chance opportunities for mixing – or introgression – make possible novel kinds of practice. The art of border biology will always be to borrow, adapt, and blend – and to know the limits of cultural borrowing. Mixed practices are the bread and butter and the soul of field biology: field practices amplified and lablike, natural history made scientific." (Kohler, 2002, p. 308).
2. Zuntz, N.: *Beiträge zur Physiologie des Blutes* (Contributions on the Physiology of Blood), Dissertation, Bonn (1868) appendix.

   Darwin's evolution theory strongly influenced Zuntz' later work, in particular his studies on high altitude physiology. On the occasion of Darwin's 100th birthday, Zuntz composed two short biographies in which he emphasized the extreme significance of the Darwinian theory for natural sciences and praised the researcher's critical stance even towards his own work (*cf.* Zuntz, A.[N.]: Zu Darwins 100. Geburtstag (On the Occasion of Darwin's 100th Birthday), *Zbl. Physiol.* 23 (1909d), pp. 199 *et seq.*; Zuntz, N.: Charles Darwin, *Med. Klin.* (1909e), p. 298).

   "Morbus Brightii" is a kidney disease that was first described in the last century by Richard Bright (1789–1858) and which involves albuminuria and dropsy (*cf.*: Bleker, J.: *Die Geschichte der Nierenkrankheiten* (The History of Kidney Diseases), Boehringer 1972, pp. 92–107).
3. Zuntz, N.: Über den Einfluss des Partiardrucks der Kohlensäure auf die Vertheilung dieses Gases im Blute (On the influence of partial pressure of carbonic acids on the distribution of this gas in blood), *Pflügers Arch.* 1 (1868), p. 532.

Zuntz believed "cruor [is] defribrinated blood from which, as a result of the displacement of blood cells, a large part of the serum has been removed" (Zuntz, N.: *Beiträge zur Physiologie des Blutes* (Contributions on the Physiology of Blood), Dissertation, Bonn 1868, p. 17).

4. Fishman, A. and Richards, D. W. (eds): *Circulation of the Blood – Men and Ideas*, Oxford University Press 1964, p. 61, and recently West, J. B. (*Respiratory Physiology: People and Ideas*, Oxford University Press 1996, p. 97) have already emphasized the importance of Zuntz's demonstration that the alkali available for the formation of bicarbonate is chiefly contained in the cells, while the bicarbonate thus formed is not found in the cells but in the plasma.
5. In animal experiments, instead of a mouthpiece, normally an airtight locking tracheal cannula was fed into the air pipes of the animal and affixed (*cf.* Figure 1.11).
6. Max von Pettenkofer lived from 1818 to 1901. Together with Voit (1831–1909), he spent decades of his life as a scientist on metabolism research in Munich. Alongside the procedures described above on the determination of respiratory quotients, the development of the calorimeter can be traced back to Max von Pettenkofer; this device can be used to determine heat quantities released by humans (*cf.* Rothschuh, K. E.: *Geschichte der Physiologie* (History of Physiology), Springer 1953, p. 182).
7. An extensive description of the technical structure of this facility can be found in *Landwirtschaften Jahrbuch* 44 (1913), pp. 777–783.
8. The procedure carried out by French physicists Regnault and Reiset permitted, just as the one performed by Zuntz, a simultaneous determination of CO_2 release and O_2 consumption (*cf.* Zuntz, N. and Loewy, A.: *Lehrbuch der Physiologie des Menschen* (Textbook on the Physiology of Man), 1st edition, Vogel 1909, p. 656).
9. According to Jokl, Zuntz and Loewy were to have assumed the construction of the pneumatic chamber in the Jewish Hospital. The chamber was installed in 1889 (*cf.* Jokl, E.: *Zur Geschichte der Höhenphysiologie* (On the history of high altitude physiology), *Forsch. Fortsch.* 41 (1967), pp. 322 *et seq.*). According to information provided to the author on August 7, 1987, the Jewish Hospital in Berlin has no archived documents from this time period.
10. Zuntz participated personally in one of these experiments (*cf.* Loewy, A. and Placzek, S.: Die Wirkung der Höhe auf das Seelenleben des Luftfahrers (The influence of high altitude on the mental constitution of an aeronaut), *Berl. Klin. Wschr.* 51 (1914), p. 1023).
11. Walter Kolmer was born on July 16, 1879, in Vienna. After passing his school-leaving examination, the Abitur, in 1897, he studied medicine in Vienna, Heidelberg, Strasbourg, and Berlin, and completed his doctorate in Vienna. From 1903 to 1914 he was an assistant to Durig, whom he accompanied in 1906 on his Monte Rosa expedition. In 1907, Kolmer completed his Habilitation (second doctorate required to become a professor) in histology at the K.u.K. Hochschule für Bodenkultur (Royal and Imperial College for Agriculture) in Vienna, and in 1913 at the medical department of the same university. In 1914, he was appointed associate professor and followed Durig to the Physiologisches Institut (Physiological Institute) in Vienna in 1919, where he became director of its morphological department. Walter Kolmer died on September 2, 1931.

   (*cf.* Heindel, W.: *Personalbibliographien von Professoren und Dozenten des Histologisch-Embryologischen Institutes der Universität Wien im ungefähren Zeitraum von 1848–1968* (Personal Biographies of Professors and Lecturers of the Histological-Embryological institute of the Vienna University during the

Approximate Period from 1848–1968), Dissertation Erlangen–Nuremberg 1971, pp. 56 *et seq.*).

12. In describing Zuntz' working methods in field physiology, inevitably parallels to the later procedure of Adolph become apparent (*cf.* Adolph, E. F.: *Physiology of Man in the Desert*, Interscience Publishers 1947 (reprint 1969), p. 23).
13. *cf.* Zuntz, N., Loewy, A., Müller, F. and Caspari, W.: *Höhenklima und Bergwanderungen in ihrer Wirkung auf den Menschen*, Deutsches Verlagshaus Bong 1906, pp. 201, 268.

    cf. Schmidt, R. F., Lang F. and Thews, G.: *Physiologie des Menschen* (Physiology of Man), Springer 2005, pp. 789–790.

    The belief that it is primarily the decreased O_2 partial pressure which negatively affects the organism was first advocated by Bert (*cf.* Bert, P. *La pression barométrique* (Barometric pressure), Masson 1878, pp. 1153–1155), which Zuntz expressly referenced (*cf.* Zuntz *et al.*, 1906, p. 173).
14. *cf.* Zuntz *et al.*, 1906, p. 263.

    cf. Nöding, R., Gabler, U., Kipke, L. and Jankowski, R.: Die Anwendung einer modifizierten Anordnung der Douglassackmethode für die Untersuchung der Atmung und des Gasstoffwechsels im Schwimmen (The application of a modified regime of the Douglas bag method for the investigation of breathing and gas metabolism while swimming), *Med. u. Sport* 15 (1975), pp. 238–241.
15. *cf.* in this context today's physiological handbook information (Weil, J. V.: Ventilatory control at high altitude. In *Handbook of Physiology Respiratory System*, edited by N. S. Cherniack, J. G. Widdicombe and A. P. Fischmann, Williams & Wilkins 1986, p. 704).
16. The changes of blood lactate concentrations during high altitude exposure continue to be the subject of intense discussion even today, mainly in view of the so-called "lactate paradox" (Brooks, G. A., Fahey, T. D. and White, T. P.: *Exercise Physiology*, Mayfield Publishing Co. 1996, p. 464).
17. On the contents and objectives of research performed in the field of tourism medicine: Auerbach, P. S.: *Wilderness Medicine*, Elsevier 2007.
18. *cf.* as well: Zuntz, N.: Über die Wärmeregulation bei Muskelarbeit (On thermoregulation during muscle exertion), *Dtsch. Med. Zg.* (1903a), pp. 265–267.

    Zuntz primarily dealt with issues of thermoregulation and clothing physiology in his *Studies on the Physiology of Marching*. In particular, issues of the thermoregulation of the body while under extreme exertion later led Zuntz to study related problems experienced by mine workers, as mentioned earlier. Because of this, Zuntz championed the creation of a "mining hygiene" which today would be a field of labor medicine (*cf.* Letter from Zuntz to the medical coordinator Dr Loebker (Bochum) dated February 20, 1908, Staatsbibliothek Preussischer Kulturbesitz, Slg. Darmstaedter–Nathan Zuntz–, 3d. 1885, folios 88–96).
19. Regarding this issue, *cf.* the following publications:

    Ward, M. P., Milledge, J. S. and West, J. B.: *High Altitude Medicine and Physiology*, Arnold 2000, pp. 48–49.

    Hultgren, H.: *High Altitude Medicine*, Hultgren 1997, pp. 165–178.

    Berghold, F. and Schaffert, W.: *Handbuch der Trekking- und Expeditionsmedizin*, DAV Summit Club 2001, pp. 13–23.
20. This assumption continues to hold true even today. (*cf.* Hochachka, P. W., Gunga, H.-C. and Kirsch, K.: Our ancestral physiological phenotype: an adaptation for hypoxia tolerance and for endurance performance, *Proc. Natl. Acad. Sci.* 95 (1998), pp. 1915–1920.

21. For biographical information on these people, see Chapter 3.
22. A similar experiment had already been carried out in February of 1897 by the Berlin meteorologist Süring at the instigation of von Schroetter (*cf.* Chapter 3).
23. In Germany, primarily the companies Drägerwerke (Lübeck) and Oxygenia-Gesellschaft (Berlin) are to thank for the technical development and construction of these oxygen breathing devices (*cf. Deutsche Luftfahrer-Zeitschrift* (German Magazine for Aeronauts) 17 (1913), pp. 110 *et seq.*; Werner-Bleines, –.: Sauerstoff und Atmungsvorrichtungen für Luftschifffahrt (Oxygen and breathing equipment for air travel), *Illustrierte Aeronautische Mitteilungen* (Illustrated Aeronautics Newsletter) 13 (1909), pp. 1032–1040; von Schroetter, H.: *Hygiene der Aeronautik und Aviatik* (Hygiene of Aeronautics and Aviation), Frankfurt 1909, pp. 29 *et seq.*). According to a written notice from Drägerwerke AG (Berlin branch) of November 4, 1987, to the author, all company records dating from this period have been lost.

3 On the development of aeronautics and aeronautic medical research in Berlin from 1880 to 1918

3.1 Preliminary remark

The investigation of civil and military aeronautics in the context of this study will begin with a brief description of this field's historical development. It will then focus on the extent to which the research institutions of the time created the logistical prerequisites for scientific aeronautic research, and at what point in time, by whom, and in which scope an understanding of the issues pertaining to aeronautic medicine developed in these institutions.

3.2 Civil aeronautics

Berlin can look back on an impressive history in the field of civil aeronautics. This began as early as in September 1788. At that time, the famous French balloonist Blanchard (1753–1809) embarked on his thirty-third balloon flight from the drilling grounds in Berlin's Tiergarten[1] (*Berlinische Nachrichten* (Berlin News), 1788, No. 111 dated September 13, 1788).

Close to 100 years later, on August 31, 1881, the *Deutscher Verein zur Förderung der Luftschiffahrt* (hereafter referred to as the German Society for the Promotion of Airship Flight) was founded in Berlin, the first of its kind in the world (*Journal of the German Association for the Promotion of Airship Flight*, 1882, p. 28; Moedebeck, 1906, p. 329). The inaugural meeting of the association (with seventeen members) took place on September 8, 1881 (*Journal of the German Association for the Promotion of Airship Flight*, 1882, p. 28). As the statutes resolved there determine, the purpose of the society was

> *to generally promote airship flight in every manner as well as to work towards ensuring that all efforts are made to solve the problem of constructing dirigibles airships, and especially to maintain a permanent experimental station in order to test all discoveries made in relation to airship travel and to exploit these discoveries if possible.*
>
> *(Journal of the German Association for the Promotion of Airship Flight, 1882, p. 28)*

By and large, until the end of the 1880s, only planning work took place; there were simply no funds available for realizing the goals set. At the start of the

1890s, a generous donation made by the German Kaiser proved a decisive turning point. Within a few months, the society had the conditions it needed in terms of personnel and logistics, which were unique, the world over. Soon after, Germany was the most experienced nation regarding planning balloon travel and taking balloon flights. In the 1890s, the logistical knowhow and experience of the Berlin society was drawn upon in particular by meteorologists for their research. Of the 200 manned balloon trips taken by the German Society for the Promotion of Airship Travel during the ten-year period from January 30, 1891 until March 30, 1901, forty trips served scientific objectives while the others were purely recreational (*Illustrierte Aeronautische Mitteilungen* (Illustrated Aeronautical News), 1901, p. 95). Moedebeck, a member of the Berlin society and, from 1896 onwards, an elected member of the International Commission for Scientific Aeronautics (Hörnes, 1912, pp. 293–295), was one of the first to speak out in favor of using civil balloon ascents for scientific research (Moedebeck, 1886, pp. 6–11). However, in later years Moedebeck viewed the scientists' ever-increasing use of the balloon society's equipment quite critically:

> *Even today, there is no adequate guidance at all on how to turn balloon flight into something athletic. The associations which are occupied with balloon flight have exclusively, and I would almost like to say, unfortunately, placed themselves in the services of the ancillary sciences of aeronautics and meteorology, and have done so to much too great an extent. In this way, they have transferred their right to exist, which is based on the promotion of airship flight, to the promotion of meteorology. With this statement, I do not wish to accuse the representatives of meteorology as being at fault here, since it is thanks to them, virtually alone, that the balloon has gained such popularity in the past decade in scientific circles. But it is truly time for the other lovers of airship flight to finally summon up their courage and to enforce the large public interest in improving the material used in aeronautic navigation and in promoting it.*
>
> *(...)*
>
> *In order to introduce such useful sporting [events] in airship flight, all airship flight associations should agree on a common procedure. They should appoint representatives charged with the job of creating a program covering all branches of aeronautics and determining the time and place for the competition, and the exhibit to be connected therewith. The associations would furthermore be obligated to ensure the financial security and sufficient advertising of the project, the prizes awarded and the competitive tasks.*
>
> *Only those locations should be considered at which a high participation of the public can be expected, and which have the appropriate facilities to accommodate the events. Berlin, Munich, Vienna and Paris have to take preference before all others.*
>
> *In Germany, Berlin should be considered above others simply for the reason that here the sport, which is just beginning to grow, and the associations who are all working towards the same objective in the interests of promoting airship flight, can take the best and most illustrious development right before the eyes of the Most-High Protector of Aeronautics [the German Kaiser].*
>
> *(Moedebeck, 1897, p. 58)*

In Moedebeck's opinion, the balloon associations should have been focused primarily on ascents in terms of endurance flights, distance flights, speed flights, high altitude flights[2] and flights to specific destinations (Moedebeck, 1897, p. 56). He was completely opposed to manned height ascents in the context of balloon sports competitions, which in turn interested scientists; these, he pronounced, were "dangerous tomfooleries which a number of people have already paid for with their lives" (Moedebeck, 1897, pp. 56–58).

Ever since the tragic end of the height ascent of the *Zenith* in April 1875, when the French balloonists Croce-Spinelli and Sivel lost their lives (Bert, 1878, pp. 1060–1071), balloonist circles had been aware of the danger of height ascents. The French physiologist Bert was the first to champion taking an artificial oxygen supply on such trips, basing his recommendation on physiological considerations. In Germany at the turn of the century, it was primarily Zuntz and von Schroetter who strongly advocated the use of artificial oxygen respiration during such ascents. They did so in lectures such as that given on November 16, 1903 at the 232nd assembly of the *Berliner Verein für Luftschiffahrt* (Berlin Society for Airship Flight) (von Schroetter, 1904, pp. 14–21).

In the following years civil, sport aeronautics underwent a period of tremendous development. In 1912, there were already ninety associations entered in the *Deutscher Luftfahrerverband* (German Association of Aeronauts) (*Jahrbuch des Berliner Vereins für Luftschiffahrt* (Yearbook of the Berlin Society for Airship Flight) 1912, pp. 112–120). The growing popularity of balloon sports (aeronautics) played a decisive role in this development, as did the growing enthusiasm for airplanes (aviation), whose early history is connected to the first glider flights by Lilienthal[3] in Gross Lichterfelde. The emerging fields of aeronautics and aviation consequently led to a dramatic increase in accidents while aloft, which were extensively reported in the press.[4] During this period a number of macabre images were printed, such as that depicting the Grim Reaper dragging a destroyed balloon basket behind him (Figure 3.1).

In the months from May to October 1911, "a disaster occurred at an average of every other day" (von Schroetter, 1912a, p. 127). In the period from 1908 until November 1911, the total number of persons who died in motorized flying accidents alone amounted to 107; one in 1908, three in 1909, 30 in 1910 and 73 in 1911.[5] Just one year later, in October 1912, the total number of accidental deaths had already reached 200, according to Friedländer, the head of the medical-psychological committee of the *Wissenschaftliche Gesellschaft für Flugtechnik e.V.* (Scientific Association for Flight Technology) (Friedländer, 1913, p. 71). The deadly crashes were attributed almost exclusively to technical failure or flight errors, alongside several unexplained circumstances. The accident statistics published in the *Zeitschrift für Luftschifffahrt* (Gazette for Airship Flight) until the end of 1911 do not provide any indication that when the accidents were studied, possible reasons relating to aeronautic medical physiology were taken into account (*cf.* The table "Victims of Flight" published in the *Gazette for Airship Flight*, 1911, pp. 22–25). In this context, the Austrian doctor Weitlaner had remarked, on the occasion of the

Figure 3.1 Contemporary depiction of the hazards of ballooning around 1911 showing the Grim Reaper dragging a destroyed balloon basket.
(Hörnes, 1912, p. 333)

fatal crash of Chavez in 1910,[6] on the significance of the research done on high altitude physiology by Zuntz, Mosso, Bert, Durig and others, not only for aeronautics but also for aviation (Weitlaner, 1910, p. 753). Weitlaner was particularly impressed with the statements of von Schroetter: "Von Schroetter's book '*Hygiene der Aeronautik*' should be most urgently recommended to all aeronauts as mandatory reading" (Weitlaner, 1910, p. 754).[7] Two years later, in 1912, Zuntz used the fatal crash of Chavez to bring the dangers of aviation to the public's attention. One needed to consider, alongside purely technical reasons, the causes for accidents that related to factors of climate, medical physiology and psychology (Zuntz, 1912f, pp. 55–56).

Before this knowledge could gain much ground in German civil aeronautic and aviation circles, flight technology took on an entirely military slant with the outbreak of World War I.

3.3 Military aviation

The first military airship company was founded by the German army command in 1870/71 during the Franco-Prussian War in Strasbourg, but was disbanded shortly thereafter (Mehl, 1911, p. 25). Then, in 1884, the War Ministry founded the airship battalion (Bruce, 1902, pp. 107–108), which reported to the railway brigade until March 31, 1899 *(General Inspektion des Militärwesens* (General Military

Figure 3.2 The aeronaut battalion barracks in Reinickendorf-West in 1902. *(Illustrierte Aeronautische Mitteilungen* 6 (1902), front page)

Inspectorate) *Index PH 9 V, Bundesarchiv (Militärarchiv)* (Military Archive of the Federal Republic of Germany), *Freiburg, p. III*). When the *Inspektion der Verkehrstruppen* (Transport Troops Inspectorate) was founded on April 1, 1899, the airship division thereupon reported to this newly formed authority. The barracks of the *Königliches Preussisches Luftschiffbataillon* (Royal Prussian Airship Battalion) were located in Reinickendorf-West at the time (Figure 3.2).

The awareness of aeronautic medicine within military circles dates back to the turn of the century, when the Third Assembly of the International Commission for Scientific Aeronautics convened in Berlin. However, before the two leading representatives of military aeronautic medicine in Berlin, Koschel and Flemming, are presented in detail, the political significance of the 1902 assembly should be addressed. Beyond purely scientific aspects, the meeting of the International Commission gave the representatives of military, civil, and scientific aeronautics the opportunity to assure each other of their mutually shared strong interest in performing their joint tasks in the realm of airship flight, and of their confidence that they were able to do so. It cannot be stressed enough how strongly this understanding among the various institutions aided the development of airship flight. The daily paper *Berliner Lokal-Anzeiger*, which reported on the course of the convention in Berlin on May 24, 1902, states:

> *The War Ministry yesterday generously permitted a number of those participating in the Third Congress of the International Aeronautics Commission to view the new buildings of the aeronaut barracks at the shooting range in Tegel.*
>
> *(...)*

> *The commander of the aeronaut battalion, Major Klussmann, gave a toast to the Kaiser and also pronounced a cordial toast to the positive cooperation between civil and military aeronautic circles, emphasizing that military aeronautics was providing the building blocks for science. The chairman of the International Commission, Professor Hergesell-Strassburg, thanked the Major for this toast, announcing that military aeronautics not only provided the building blocks, but indeed was to be regarded as a permanent partner. He raised three cheers for the aeronaut battalion.*
>
> *(Berlin Lokal-Anzeiger dated May 24, 1902, No. 237)*[8]

One of the participants of this assembly in Berlin was Koschel, the aeronaut battalion's assistant physician at the time. In the years that followed, Koschel investigated an entire series of topics relevant to aeronautics medicine and physiology with the chief army doctor Flemming, of the Kaiser Wilhelm Institute. Their work had been commissioned by the War Ministry.[9] From the very start, Koschel must have known about the work done by Zuntz and von Schroetter in this field. Records show that, on May 23, 1902, he participated with Zuntz in the Seventh Professional Session of the Assembly of International Commission, at which, among others, von Schroetter gave his lecture *Zur Physiologie der Hochfahrten* (On the physiology of high altitude flight) (Record of the Third Assembly of the International Commission for Scientific Aeronautics, 1902, pp. 46–48). We may assume that Koschel was inspired by the statements made by Zuntz and von Schroetter at this assembly. Only five months later, on October 10, 1902, the aeronaut battalion submitted a petition to the *Königliche Inspektion der Verkehrstruppen* (Royal Transport Troops Inspectorate) which requested that assistant doctor Koschel be allowed to participate in medical studies during balloon ascents. The petition was approved by the War Ministry on November 13, 1902 (Letter of the War Ministry No. 550/10.02.A6 dated November 13, 1902, files of the *Bundesarchiv* (German Federal Archive) in Freiburg, PH 9V/25, p. 137), which indicates that the army command must have become aware of the significance of medical and physiological research in aeronautics, and the urgent need for this. This is demonstrated by further postal correspondence between the aeronaut battalion, the Transport Troops Inspectorate and the War Ministry at the end of 1903 and beginning of 1904. The letter from the aeronaut battalion dated December 13, 1903 and addressed to the Transport Troops Inspectorate is printed a facsimile in Figure 3.3.

In the War Ministry's letter of response dated January 9, 1904, addressed to the Transport Troops Inspectorate, four balloon ascents were approved for head physician Flemming "and since these are officially commissioned studies ... they shall be remunerated at the same conditions applicable for official trips by officers of the battalion" (Letter of the War Ministry to the *Inspektion der Verkehrstruppen* (Inspectorate of the Transport Troops) dated January 9, 1904, files of the *Bundesarchiv* (German Federal Archive) (*Militärarchiv*, Military Archive) Freiburg, PH 9V/25, p. 166). The research done on the selection of flying personnel according to medical criteria – an issue of great import for

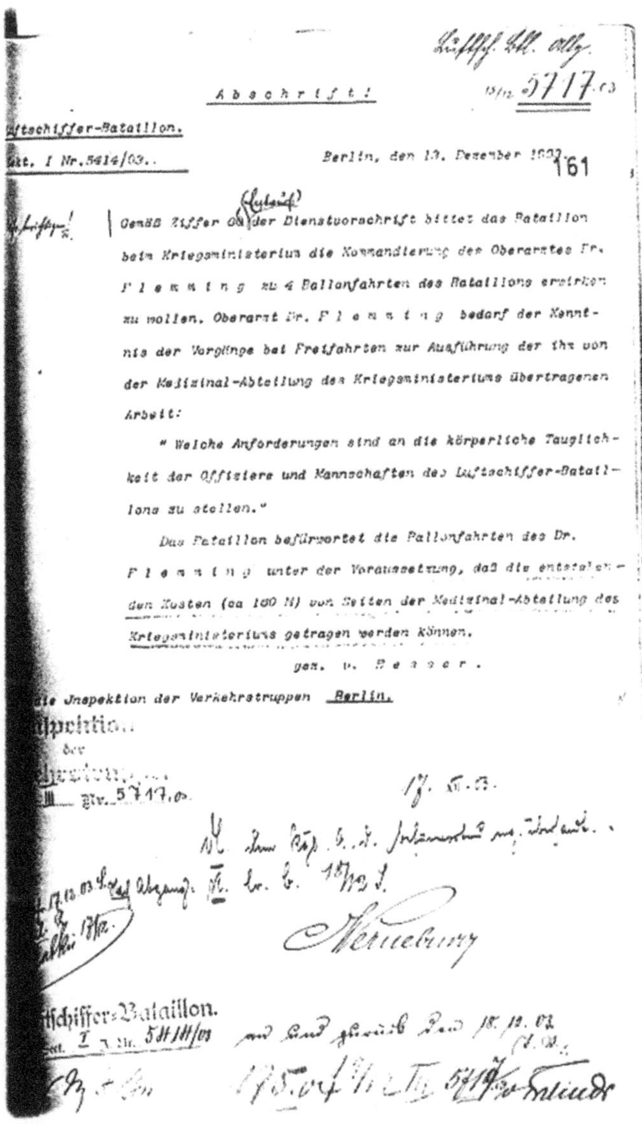

Abschrift!

[Luf]tschiffer-Bataillon.
[Abt]. I Nr.5414/03..

Berlin, den 13. Dezember 1903.

Gemäß Ziffer 0 der Dienstvorschrift bittet das Bataillon beim Kriegsministerium die Kommandierung des Oberarztes Dr. F l e m m i n g zu 4 Ballonfahrten des Bataillons erwirken zu wollen. Oberarzt Dr. F l e m m i n g bedarf der Kenntnis der Vorgänge bei Freifahrten zur Ausführung der ihm von der Medizinal-Abteilung des Kriegsministeriums übertragenen Arbeit:

" Welche Anforderungen sind an die körperliche Tauglichkeit der Offiziere und Mannschaften des Luftschiffer-Bataillons zu stellen."

Das Bataillon befürwortet die Ballonfahrten des Dr. F l e m m i n g unter der Voraussetzung, daß die entstehenden Kosten (ca 100 M) von Seiten der Medizinal-Abteilung des Kriegsministeriums getragen werden können.

gez. v. B e s s e r .

[An d]ie Inspektion der Verkehrstruppen Berlin.

Figure 3.3 Petition submitted by the airship battalion to the War Ministry for consent to four balloon ascents for senior physician Flemming.
(German Federal Archives, Military Archive, Freiburg/i. Breisgau)

the army command – later became a focus of Koschel's work. In addition, in 1910 Koschel conducted multiple balloon ascent experiments on blood pressure, blood count, and the influence of reduced artificial oxygen respiration at great altitudes. He was well aware that Zuntz had already published authoritative works on these questions. In this context, Koschel referred to the treatise

Zur Physiologie und Hygiene der Luftfahrt (On the Physiology and Hygiene of Aeronautics) by Zuntz an "excellent work" (Koschel, 1912, p. 116).[10]

Koschel performed several of the studies cited above together with Flemming, as is set out in the report on the scientific high altitude flight of the balloon *Harburg III* on May 26, 1911 (Schubert, 1911, p. 16, dated 25 July). They were joined on this ascent by Krusius,[11] a *Privatdozent* (junior professor) of ophthalmology at the Charité hospital, who "performed experiments on the curative aspects of sun exposure at great heights on eye tuberculosis (animal experimentation) and made other observations on the eyes of the fellow travelers" (Schubert, 1911, p. 16). Flemming also used these high altitude flights, which reached heights of 6000 m above sea level, to test his newly-constructed oxygen masks, which can be viewed as prototypes for the systems that later became customary for aeronauts:

> *By means of an attached T-tube and using 1½ m-long rubber tubes each, every passenger in the basket was provided with an oxygen supply. The supply tube was attached to a mask especially created according to Dr. Flemming's instructions, which in turn could be affixed to a cap equipped with a neck covering. This system has the great advantage that one breathed through the nose, not the mouth, and that, in the event of unconsciousness, the tube remained in place.*
>
> *(Schubert, 1911, p. 16)*

The cap equipped with a neck covering (Figure 3.4) evidently resulted from Flemming's experiences two years earlier during another balloon ascent for scientific purposes:

> *We had given particular attention to the impact of sunlight. As is well-known, exposed body parts sustain serious inflammations from radiation, also during glacier expeditions (glacial sunburn) and sporting balloon ascents. In order to become familiar with the nature [of this condition] on our own bodies, we left our heads and necks exposed. The results of this six-hour, unimpeded sunlight radiation at an altitude of 4,000 to 8,000 m was astounding, at least for the one among us who moved about in direct sunlight for his experiments more than the others. An enormous swelling of the skin occurred, which, except for the fact that there was no fever (missing proteinuria), completely resembled the formation of erysipelas. The inflammation did not climax until 48 hours after the ascent. It impacted the skin of the head with hair, the neck and the entire face, especially the eyelids whose swelling, on the morning after the second night, kept the eyeballs completely covered. The alteration of the physiognomy of the face was one of such extremity that even close acquaintances did not recognize the subject at first. Not until the third night did these objective symptoms abate, as did the unpleasant burning and tight feeling and over the next eight days, the skin peeled off twice, in long strips.*
>
> *(Flemming, 1909b, p. 1022)*

The research done by Koschel and Flemming in the field of aeronautics for more than a decade presumably played an integral role in the adoption of these two scientists as official regular members of the "subcommittee 'f' for

Figure 3.4 Flemming's construction of an oxygen mask incorporated into a cap with neck protection.
(Mehl, 1911, p. 231)

issues of medicalpsychology" in 1912, when the *Wissenschaftliche Gesellschaft für Flugtechnik e.V.* (WGF, Scientific Association for Flight Technology) was established (*Yearbook of the Scientific Association for Flight Technology*, 1913, pp. 9, 11, 17; Harsch, 2001, p. 1). As this association soon expanded its work, it was renamed the *Wissenschaftliche Gesellschaft für Luftfahrt e.V.* (Scientific Association for Aeronautics) in 1914.

With the founding of this subcommittee, Flemming, Koschel, Halben[12] and Zuntz, who had also been elected to this subcommittee (*Yearbook of the Scientific Association for Flight Technology*, 1913, p. 17), now found themselves united in one group. This concentration of leading aeronautic physicians in Germany at one institute led to visible success as early as 1913, such as when Koschel reported on the development of guidelines for the selection of flying personnel on June 5, 1913, at the annual general meeting of the *Wissenschaftliche Gesellschaft für Luftfahrt*. Koschel characterized the general practice of pilot selection in the associations as follows:

> *Up until a few years ago, no physical examination was required for the pilots of aircraft. The associations appointed nearly anyone as pilots and assigned them companions without the assurance that, alongside being knowledgeable in piloting, which was indeed tested, the pilots also met certain health standards. After surgeon major Dr. Flemming had already drawn attention to this situation, it [still] wasn't until the suggestion of a series of doctors called together by* Privatdozent *(junior professor) Dr. Halben that the* Deutscher Luftschiffertag *(German Aeronauts Congress) in Breslau mandated that aspiring pilots undergo a doctor's physical examination.*
>
> *(Koschel, 1914, p. 144)*

Towards the end of 1912 and at the beginning of 1913, Koschel was commissioned by the *Deutscher Luftfahrerverband* (German Association of Aeronauts) to assist in developing a corresponding questionnaire (Koschel, 1914, p. 144). At the end of April 1913, Koschel called for "a survey on health dysfunctions in hot-air balloons, airships or planes" in the *Deutsche Luftfahrerzeitschrift* (German Aeronauts Journal). The issues raised in Koschel's questionnaire reveal that he was fully aware of the problems facing aeronautic medicine, physiology, and psychology. Nonetheless, there can be little doubt that the items particularly addressing physiology can be traced back to considerations and research previously carried out by Zuntz.

Finally, in December of 1913, preparatory talks were held with the German Association of Aeronauts, at the instigation of the *Wissenschaftliche Gesellschaft für Flugtechnik* (WGF, Scientific Association for Flight Technology) and the military authorities, "for the purpose of obtaining consent that physicians, who are yet to be appointed by the W.G.F., receive general permission to examine the aeronauts and to ask the questions set out in the questionnaire" (*Yearbook of the Scientific Association for Flight Technology*, 1914, p. 218). At the same session, on December 21, 1913, the WGF recommended the following doctors for this task: Zuntz, Flemming, Koschel and Halben (Yearbook of the Scientific Association for Flight Technology, 1914, p. 218).

In January 1914, the nominated doctors were granted general permission from the German Association of Aeronauts to access the airfields operated by this association, and to perform their investigations.[13] In this context, chief physicians Flemming and Koschel were primarily responsible for the survey at military airfields (*Yearbook of the Scientific Association for Flight Technology*, 1914, p. 218).

With the outbreak of World War I in 1914, the work of the scientific committees of the WGF was put on hold until 1918 (*Yearbook of the Scientific Association for Flight Technology*, 1920, pp. 20–21). During the war, Zuntz and Koschel crossed paths once more at the end of 1916. When the newly founded *Militär-Prüfungs- und Versuchsstelle für Sauerstoffgeräte* (Military Testing and Experimental Department for Oxygen Devices) was moved to the ground floor of the *Tierphysiologisches Institut* (Institute for Veterinary Physiology), it was Koschel who was ordered to pursue his research there (Report of the *Landwirtschaftliche Hochschule* (Agricultural College) in Berlin 24–27, 1921, pp. 72–73). In the pneumatic chamber of the institute, Koschel tested a variety of respiratory apparatuses and carried out physiological-psychological studies on aeronauts and pilots (Report of the *Landwirtschaftliche Hochschule* (Agricultural College) in Berlin 24–27, 1921, p. 73). These studies were presumably connected to the development of a "military medical certificate" for pilots (see below), which, following an order issued in May 1916, was to be obligatory for pilots from then onwards.

Military Medical Certificate
Pursuant to the order issued the head of field health division N. 10544.16 dated May 23rd, 1916.
Department ordering the examination:
Surname and first name of the examinee:
Rank:
Troop division:
Purpose of examination:
Place and date of examination:

A) Preliminary history
I. The answers to questions 1–22 are to be completed by the examinee himself to the extent possible and signed by him.

Questions 1–7 relate to the period before service was entered.

1. *Date of birth?*
2. *Have the parents, siblings or close relatives of the undersigned suffered from any mental or nervous illnesses?*
3. *Has the examinee suffered in the past from serious physical illnesses, nervous illnesses or symptoms of illness related there to, such as cramps, nocturnal unrest, etc.? When and for how long?*
4. *Did he participate in gymnastics and other sports as a boy?*
5. *Was he ever involved in an accident during the activities listed above, or upon another occasion? What physical damages, in particular concussions or bone breakages, did he experience as a result? Following the accident, did he experience any nervous symptoms? What were they and how long did they continue?*
6. *Before, during or shortly after examinations in school or later in his career, or on any other stressful occasions, did he suffer from nervous symptoms (headaches, inner nervousness, inability to collect one's thoughts, increased weariness, sleeplessness, low spirits, etc.), and for how long?*
7. *What profession did he have prior to piloting activities or prior to entering the service?*
8. *Had he already served on the battlefront prior to the start of piloting activities and if so, in what rank?*

9. *During this period (when and how long), were there any physical symptoms of illness, especially of the heart, the respiratory or digestive organs, or signs of nervous exhaustion, and what were they traced back to? Was he wounded? When? What sort of wound was it? How long did the treatment last? When did he return to the front?*
10. *What caused him to switch to piloting activities?*
11. *When and where did the training as an airplane pilot or observer take place? (Please underline which title is applicable.) Please also see No. 13.*
12. *Did he experience any particular difficulties during training, for example, unsuccessful landings (breakages), slipping or the like; did any physical injuries result and did he have any nervous symptoms subsequently? In particular, was he forced to suspend flying activities for a time? Did his inner peace and sense of safety decrease? How long did these feelings last?*

12a. *What illnesses did he have during the training period? When and for how long?*

13. *How did his further piloting career develop? After how long a period or after how many flights and how many solo-flights did he perform the required examinations? When did he arrive at a* Flugpark *[training depot and point of distribution of aviation personnel to front line squadrons], when at a field aviation division or a fighting squadron, etc.?*
14. *How often has he flown over the enemy or how many orders were completed?*

    Please specifically note the following preliminary comments to Nos. 15–18!

    When answering the following questions 15–18, the impression these incidents (please list individually) made upon his nervous system is to be emphasized in particular. This may be, for example, that he felt uneasy immediately thereafter, had less self-confidence, had to suspend flying activity for a time, etc. Please ask how long these disturbances lasted.

    Incomplete answers to questions 15–18, in particular any answers that show the preliminary comment was not taken into account, will mean that the questionnaire will have to be returned, thereby delaying the vacation procedure and/or the transfer to the J. d. Flieg.!
15. *Was he shot at often and accurately, and how did this affect his nervous system? (Please note the preliminary comment to Nos. 15–18!)*
16. *Was his airplane, his pilot or observer or he himself hit during these flights or did he see another airplane crash close by or experience a similar [incident]? Please list these events individually with their dates and times. How did these events affect his nervous system? (Please note the preliminary comment to Nos. 15–18!)*
17. *During his active service at the front, did he have emergency landings, break-downs or the like, or did he have to overcome hazardous situations which turned out fortunately, such as the airplane slipping and being caught again, carburetor fire which went out prior to landing, etc.? Please list these events individually and provide details of when they occurred. Were there injuries? How did these events affect his nervous system? (Please note the preliminary comment to Nos. 15–18!)*

18. *While flying above the enemy, did he encounter any difficulties as a result of engine failure or through strong headwinds with low fuel quantities or any other incidents? How did these events affect his nervous system? (Please note the preliminary comment to Nos. 15–18!)*
19. *When did he first notice physical or nervous symptoms and how did these appear?*
    a) *Symptoms prior to the flight (nervousness, fear, racing heartbeat, etc.).*
    b) *Symptoms during or directly after the flight. During rapid ascent (ear pressure, anxiety, etc.). During rapid descent (ear pressure, head pressure, pressure in the temples or forehead, sense of lightheadedness, seeing spots or stars, difficulty breathing, etc.). During steep spiraling flights (dizziness, etc.). While at high altitudes: please describe exactly at what altitudes the symptoms began and when they disappeared and whether the high altitude flights also became more difficult after other nervous symptoms had appeared outside of the flights (anxiety, lightheadedness, sleepiness, feelings of fear, obsessive thoughts, heart problems, vomiting, etc.). As a result of exhaust gases, petrol emissions etc. (headaches, nausea, vomiting, etc.). With which type of airplane? Because of engine noises (hard of hearing, etc.). As a result of drafts and the cold (freezing, actinic conjunctivitis, catarrh of the upper respiratory tract, etc., rheumatic symptoms). As a result of strong radiation (inflammation of the eyes, visual impairments, etc.). As a result of physical exertion while operating the steering system, the machine guns, etc. (muscle pain, muscle cramps, weariness of certain limbs, general feelings of tiredness, etc.).*
    c) *Symptoms during the rest of the time.*
    d) *What illnesses did he have during the time after training was concluded? When and for how long?*
    e) *When and for how long has he already been on leave or gone through health rehabilitation programs supervised by a doctor, and with what success?*
20. *Did the symptoms increase gradually or did anything cause them to increase more rapidly?*
21. *What events trigger the current symptoms or cause them to increase in such a manner that the examinee is forced to temporarily or permanently suspend his activity, and to what does the examinee attribute the triggering of the current symptoms or their increase to the current degree?*
22. *Please list symptoms which have not yet been stated.*

Signature

II. Any comments and additions by the certificate issuer on the prior history and development of the current problem.

B) Examination findings (pursuant to D.A.Mdf.Z. 183)

In all cases in which an organ is afflicted for which a specialist physician is available, the examination is to be carried out by this physician wherever possible. In any case, the heart must be examined, if at all possible, by a physician specializing in internal diseases and, if possible, an orthodiagraphic determination of the heart's size should be prepared.

C) **Evaluation**
1. *What is the examinee suffering from?*
2. *Is the examinee only temporarily or permanently unable to continue service as an airplane pilot or observer?*
3. *Can it be anticipated that he will be able to fly again? How will this be achieved?*
4. *What is the timeline for the above?*

Name, rank, troop division of the certificate issuer

(Koschel, 1922, pp. 14–16)

Most of the questions listed in this certificate clearly deal with issues of the psychological resilience of the pilot, an area which Koschel had been focusing on for years and which was later designated "psychophysiology" (Ruff and Strughold, 1944, pp. 5–6). The aspects of physiological medicine addressed in question 19b had already been addressed in 1912 by Zuntz in his paper *Zur Physiologie und Hygiene der Luftfahrt* (On the Physiology and the Hygiene of Aeronautics) (Zuntz, 1912f, pp. 13–54).

3.4 Scientific aeronautics

3.4.1 Meteorology

The history of scientific aviation in Berlin stretches back to the end of the nineteenth century, when the meteorologist and physician Assmann (Assmann *et al.*, 1899, pp. 120–121) began to plan and implement scientific balloon ascents for purposes of meteorological research. The first of this sort of high altitude flight took place in 1888; five others followed in 1891 (Assmann *et al.*, 1899, pp. 108–109).

At the start of the 1890s, Assmann succeeded in convincing scientists in Berlin, such as Helmholtz, von Siemens and Förster, to support his plan that the German Society for the Promotion of Airship Flight should perform fifty flights the purpose of which would be the scientific investigation of the atmosphere. The required capital, amounting to 50,000 Marks, was to be procured by submitting a direct appeal (*Immediateingabe*) to the Kaiser. However, after the *Akademie der Wissenschaften* (Academy of Sciences) had forwarded the project with its consenting statement in 1892, the authorities only approved half of the amount requested from the reserve fund of the General Staff account. According to Assmann, this cut in funding would have had the consequence that "the planned program would not only have been called into question, but also certainly ruled out" (Assmann *et al.*, 1899, pp. 108–109) if the Kaiser had not himself granted the funds still required. Lieutenant Gross of the *Luftschifferbataillon* (aeronaut battalion) assigned the task of designing and constructing the balloons; he was also responsible for training the young meteorologists Berson[14] and Süring[15] to become balloon pilots (Assmann *et al.*, 1899, pp. 139–140).

In the years that followed, leading up to 1899, the German Society for the Promotion of Airship Flight conducted a total of seventy-five manned balloon ascents. The majority of these ascents had scientific objectives, while a small number pursued military and sporting goals (Assmann *et al.*, 1899, pp. 139–140). Flight records are available for all ascents, since they were recorded in *Wissenschaftliche Luftfahrten* (Scientific Flights) (Assmann *et al.*, 1899, pp. 1–142). Berson was the first person within Germany to report on several medical physiological observations made on the occasion of his high altitude flight with the balloon *Phönix* on May 11, 1894 (Assmann *et al.*, 1899, pp. 68–70). One year later, Berson provided an extensive description of his observations in the journal *Das Wetter (The Weather)*:

> *I had achieved and surpassed altitudes of 6,000 m four times before and recalled that given similar air pressure and temperature, I had always felt relatively good. However, three of these ascents exceeded this boundary only insignificantly; during the one mentioned earlier, on May 11th, which not only exceeded 7,000 m but even attained 8,000 at the highest altitudes, there were moments when I was in a sleepy state with my vision clouding, and I was forced to pull myself together – meanwhile my companion, Lieutenant Gross, suffered more from heart palpitations and shortness of breath ... However, I constantly breathe oxygen – and then only perceive a small feeling of dizziness in my head, accompanied by moderately strong palpitations of the heart and otherwise am completely in a position to observe, consider, write ... However, as soon as I – even for just a few seconds – am forced to remove the mouthpiece of the tube due to work in the basket, or intentionally for purposes of physiological determination, I am afflicted by very violent palpitations of the heart, then I come close to lurching about and quickly reach for the life-giving gas-tube. Once I was astonished to notice how easily my eyes were falling shut despite everything: I shook myself, berating myself harshly, fully aware that much is at stake here.*
>
> *(Berson, 1895, pp. 4–5)*

H. von Schroetter, who was researching the causes of caisson sickness in Vienna at this time, had evidently questioned Assmann in Berlin in 1896 regarding physiological observations of this kind, and received the following message, which has been preserved and is available as a copy:

> *Copy ad III 25104* *Grünau in der Mark, July 13, 1896*
>
> *The experiences as regards the physiological impact of thin air, collected during balloon ascents undertaken by the German Society for the Promotion of Airship Flight, have not yet been published in a summarized form and we have forgone extensively publishing parts of the results of our scientific ascents as they are still in the preparation process. Nevertheless, several interesting individual reports have been published.*
>
> *In the 8th issue of the* Zeitschrift für Luftschiffahrt und Physik der Atmosphäre *[Journal for Airship Flight and the Physics of the Atmosphere] (published in Berlin by Mayer & Müller), 1894, an essay has been published by Captain Gross on the height ascent of "Phönix" in May 1894, which*

contains some information that may be important for your purposes. In the 12th issue of the same journal, 1894, an article by Berson on the height ascent of December 4th, 1894, introduces some extremely significant material on this topic, especially since it reports on the highest altitude achieved by a human thus far: 9,150 m.

Furthermore, an essay by Berson entitled Eine Reise in das Reich der Cirren *[A Journey into the Empire of the Cirrus Clouds] appearing in the 1st issue of the meteorological monthly journal* Das Wetter *in 1895 (published in Braunschweig by O. Salle) [contains] interesting details on this topic.*

Recently the Archiv für die gesammte Physiologie *[Archive for Cumulative Physiology] volume LXIII (published in Bonn in 1896 by Emil Strauss) published an important article by Schumburg and Zuntz entitled* Zur Kenntnis der Einwirkungen des Hochgebirges auf den menschlichen Organismus *[Notes on the Effects of High Mountains on the Human Organism], which likewise provides extremely valuable information on this subject.*

Signed

(Heller et al., *1900, p. 1186)*

It cannot be ruled out that the "important article" (Schumburg and Zuntz, 1896, pp. 461–494) was the recommendation that led to von Schroetter contacting Zuntz.

This shows that, up until this point in time, medical physiological observations in balloons were performed almost exclusively by meteorologists. However, from now on, physiologists and physicians developed a growing interest in the systematic research of these phenomena. In the subsequent period, the field of physiology medicine developed into an independent branch of research in aeronautics that quickly became a major field of interest.

3.4.2 Physiology medicine

Three months after von Schroetter had questioned Assmann in Berlin regarding medical physiological observations made during balloon ascents, the former went on his first balloon physiology expedition in Vienna with his colleague Mager; the details of the ascent were reported in the *Neues Wiener Tagblatt* newspaper on October 11, 1896:[16]

Vienna Physicians in a Balloon

A scholarly ascent

A few days ago, two local physicians, the gentlemen Dr. von Schroetter jun. and Dr. Mager, assistant at the clinic of Professor von Schroetter in the Allgemeines Krankenhaus *hospital, performed an interesting experiment. They boarded the gondola of the balloon "Vater Radetzky" and ascended into the air with this balloon for medical purposes. During this journey, the two physicians took a large number of instruments and apparatuses with them, instruments for examining the changes experienced by the human organism when passing from one layer of air to the other; they wanted to measure precisely the changes their bodies underwent in the process. The experiment has an even more important purpose since it is connected to studies on a relatively young disease about which little is known, in terms of treatment as well as general knowledge, this being the so-called "caisson disease," which lock workers suffer from and which is accompanied by complicated symptoms.*

(...)

In order to now probe more deeply into the nature of this illness, the physicians, who have been investigating caisson sickness for quite some time now, have performed an experiment with the balloon during which one also rises from thicker layers of air into thinner ones. During their journey, they observed the accompanying physical phenomena such as altered blood circulation, pressure on the pulse and respiratory organs, disturbances of the hearing and of the nervous system. Since one must remain still in the gondola in order to not rock the balloon excessively, the two physicians were unable to collect any great quantities of material during this first ascent. They ascended to 3,000 m altitude, that is, not high enough to discover true parallels in the atmospheric differences that people experience when exiting the caisson, which is filled with compressed air, into normal air layers. Still, this first scholarly ascent resulted in interesting details that will be put to use in comparison work. As reported to us, the ascents are to be continued systematically. The experiences and knowledge gathered in this regard will be recorded by the physicians responsible for the experiments.

The physicians' balloon ascent, presumably unique in its kind, was performed under the direction of Lieutenant Pruckmüller from the military-aeronautics institute. It lasted three and a half hours and ended smoothly at a slow, normal speed near Bisenz.

At approximately the same time, in autumn of 1896, von Schroetter must have contacted Berson with the request that he take a rabbit with him in the balloon for physiological experiments. This can be gathered from the response letter written by Berson to von Schroetter dated February 1897.

Editorial offices of the Zeitschrift für Luftschifffahrt und Physik der Atmosphäre *(Journal for Airship Flight and the Physics of the Atmosphere)*
Berlin, February 5th, 1897

Most honored Sir!

It proved to be absolutely impossible for me to fulfill your request and take a rabbit with us on the balloon ascent of November 13th/14th. I wasn't given the telegram until I arrived at the airship division at 1:00 a.m. at night where it had been forwarded from the meteorological institute; I even had inquiries made with all the non-commissioned officers of the division but no-one had [a rabbit] and the division is too far outside the city for it to have been possible to acquire one in the middle of the night prior to the ascent, which was scheduled to take place at 2:30 am. So now I certainly won't forget to take an experiment animal on the 18th, when I will be making another ascent. Incidentally, in addition to our unmanned "probe balloon" and the one in which I will be conducting a solo ascent, it is highly probably that a second manned balloon with one officer and the meteorologist Dr. Süring will be making an ascent. I want to propose that these gentlemen also take an animal with them since the maximum altitude of the two balloons will probably be different, this will presumably not be without interest.

For this purpose, I would like to ask that you kindly instruct me by briefly describing all key aspects that I should consider. Perhaps even with regard to the selection of the animal; age? gender? etc. Please forgive me if I'm asking ridiculous questions, but if the thing is going to be done, then I want to make sure I haven't missed anything so that we will obtain useful results ... To whom should the animal/s be given for dissection afterwards? Would you prefer a certain physiologist? Or should I just contact the physiology institute? Perhaps you could arrange something in this regard in advance. As stated, I am happy to do whatever it takes so that the results obtained correspond to what you have in mind.

The ascent will again be an international one (Paris, Berlin, Petersburg, Munich, possible also Warsaw and Strasburg) and will take place in the presence of the French and Russian ambassadors; it is very probable that His Majesty, the Kaiser, will also be in attendance.

Since I still haven't fulfilled my promise made last autumn, I actually needn't possess the "sorry courage" to remind you of my less than reliable personality. However, I still can't foresee how I will find the time to compose a more comprehensive report, as I had planned. If instead it would be sufficient to briefly answer your questions, then I would do this immediately after the ascent on the 18th – if you, most honored sir, still have use for this.

May I now request a prompt answer to my questions, especially as regards the rabbits to be experimented on? A kind hastening of the same would be very welcome in light of the required preparations.

I remain respectfully yours,

Your very devoted

(Quoted according to a letter from Berson to von Schroetter dated February 5, 1897)

Berson's balloon ascents took place as planned on February 18th. However, only his colleague Süring conducted the experiments that von Schroetter had requested (Assmann *et al.*, 1899, pp. 113–118). Süring did so on the basis of detailed instructions given him by Berson, which the latter had most likely received from von Schroetter:

Rabbit I.
To be killed at maximum altitude by means of a rabbit punch (hold on to the animal's hind legs with the left hand and strike the hanging head sharply with the narrow side of the palm of your hand between the throat and the ears, repeat if necessary.)

Rabbit II.
After landing, kill it in the same manner and dissect it just like Rabbit I.
1) Skin the animal, submerge it in water and open, under water, the chest and stomach cavities, carefully observing:

1. *Whether gas bubbles can be observed in the blood vessels, especially in the blood vessels of the mesentery (that is, the thin membrane to which the intestines are attached) and in the blood vessels which run across the heart;*

2. *Whether an extravasation of blood can be seen on the heart;*
3. *Whether the lungs are filled with blood.*

Rabbit III.
After landing, it is to be decapitated using a knife or axe. The head should be placed, as is, in a glass container and liquid should be poured over it which is to be created by mixing about ½ l water with the contents of the small flask labeled 'formalin' until the head is completely submerged in liquid, close the container with a well-sealed cork and send it to Mr. von Schroetter in Vienna, IX/2, Mariannengasse 3.

1) If a doctor is available it would be sensible to request him to perform the dissection.

(Quoted according to a letter from Berson to von Schroetter dated May 5, 1897).

The dissection report of this first, systematically planned and methodically thought-out animal experiment for aeronautic physiological purposes still exists. The dissection was performed by the Colmar county veterinarian and the slaughterhouse inspector on February 19, 1897. They sent the following report to von Schroetter and Berson:

Colmar in Prussia, March 4th, 1897

Most honored Doctor!

Attached I have taken the liberty of enclosing photogenic drawings of the three dissected rabbits. Those points at which gas bubbles were found in the vessels are marked blue. The thymus glands which exhibited numerous red spots have been shaded yellow. The photogram also reveals that the lungs of the first rabbit are darker (containing more blood) than those of the other two. On the lungs of the third rabbit, red stripes are also clearly visible (probably as a result of the chest cage being crushed).

Most respectfully,

The rabbit given to us today, February 19th, 1897, by Dr. Süring was killed at 12:45 pm as instructed and then dissected. The stomach and chest cavities were opened under water.

Dissection report on three rabbits taken on a balloon ascent for the purpose of a physiological experiment. It was to be determined:

1. *Whether gas bubbles were present in the vessels, especially in the vessels of the mesentery,*
2. *whether the lungs exhibited a concentration of blood,*
3. *whether blood has hemorrhaged on the heart.*

Rabbit I, killed at an altitude of close to 4,000 m using a rabbit punch.

There are no air bubbles in the post cava and mesentery vessels, the lungs exhibit a concentration of blood especially in the rear portions; in the retraction state. No hemorrhaging under the epicardium, sub-epicardial hemorrhaging. Thymus – one gland situated in front of the lungs – is permeated with numerous blood spots the size of pinheads. There are no abnormal quantities of air in the intestines.

Rabbit II, about 19 hours after landing, 12:45 pm, killed using the rabbit punch and dissected immediately thereafter.

In the post cava there were several larger air bubbles, none in the mesentery vessels. Upper part of the colon from the ileum to the further parts of the colon were swollen by gas and strongly tympanitic; the haustra – pocket-like colon sacculations – were swollen up like balloons. Lungs were noticeably pale; substantial alveolar emphysema; pulmonary alveoli expanded, filled full with air. In the lung artery leading from the right heart chamber, there were numerous large and many small gas bubbles which confluenced into one large bubble after some time and then came to rest in the lower ventricle portion. Heart systolic, contracted, coronary vessels almost empty, no hemorrhaging beneath the epicardium. When the right heart chamber was pressed, so much air came out that the lung arteries were visibly stuffed with air and expanded to 1.5 cm length.

Rabbit III, about 20 hours after landing, killed by decapitation.

Lungs dotted in red, pulmonary alveoli expanded; basic color of lungs is pale, mostly with red stripes up to 2 cm long. The same are on the interior of both lungs in the direction of the larger vessels and on the rib surface, especially the right lung running parallel to the ribs (could the thorax have been crushed?). In the coronary vessels of the heart there were numerous air bubbles; five point-shaped hemorrhages beneath the serous epicardium close to the front coronary channel. Coronary vessels, namely those of the left half of the heart, are strongly filled. Numerous large air bubbles are in the antechamber and the large vascular stems, aortic arch and the post cava.

When pressing down on the right heart chamber, a 1.5 cm visible air bubble emerged which completely filled the lumen of the lung arteries. In the lung veins there are small air bubbles; none in the mesentery vessels:
Colmar in Prussia, March 3rd, 1897.

signed
County veterinarian – Slaughterhouse inspector

(Quoted according to: Sektionsbericht des Kreisthierarztes und Schlachthausinspektors *(Dissection Report of the County Veterinary and Slaughterhouse Inspector) dated March 3, 1897)*

In the time following these experiments, the cooperation between meteorologists and physicians during balloon ascents, which had begun with great promise at the end of the nineteenth century, intensified even further. Without doubt, the young von Schroetter played a decisive role in this development. In July 1901, he took a high altitude flight with Berson and Süring in Berlin at 7500 m above sea level[17] and conducted experiments with the two meteorologists in Lazarus' pneumatic chamber at the Jewish Hospital (von Schroetter, 1902, p. 90). In the pneumatic chamber and in the balloon, he examined physiological symptoms such as respiratory frequency, blood pressure and the pulse as well as the influence of decreased oxygen intake on the organism. Early on, von Schroetter reached the conclusion that "an abundant program for physiological observations will be available for the future" (von Schroetter, 1902, p. 94). Evidently, over the years a close friendship developed between von Schroetter and Süring, extending beyond the scientific objectives they shared.

In May and June of 1902, in the context of the Berlin convention of the International Commission for Scientific Airship Flight, Zuntz and von Schroetter undertook their balloon ascents for physiological purposes; their objectives, methods, and results have already been described. These two ascents marked the start of systematic balloon-physiological research in Germany. At this point in time, brief reports of similar studies had already been published by French scientists at the international level, the most prominent of these being works published in 1901 by Henocque (1901, pp. 1003–1006), Hallion and Tissot (1901a, 1901b),[18] Calugareanu and Henri (1901, pp. 1037–1039), Bonnier (1901, pp. 1034–1037), and Jolly (1901, pp. 1039–1040). The French scientists had performed experiments, some of them supported by the Aéronautique Club de France, on changes to the blood, respiration, arterial blood pressure, and the impact on the vestibular organ during balloon ascents on May 12, June 24, and November 8, 1900 (Henocque, 1901, p. 1003), as well as on November 20 and 21, 1901 (Calugareanu and Henri, 1901, p. 1037; Jolly, 1901, p. 1039). In this context, Hallion and Tissot emphasized that the precision of experiments in a balloon was still plagued by considerable technical difficulties (Hallion and Tissot, 1901a, p. 1030). In 1907, Soubies' comprehensive dissertation *Physiologie de l'aéronaut* (Physiology of Aeronauts) provided a summary of the experiences of primarily the French scientists over the course of the years. Acknowledging the international development of balloon and high altitude physiological research, Soubies referred to the works of Zuntz, Loewy and von Schroetter on this topic in several instances (Soubies, 1907, pp. 9, 38, 52, 62).

In Germany, in the years between 1902 and 1912, it was primarily Zuntz's student, von Schroetter, who addressed the "hygiene of aeronautics" in detail.

An article of the same name by von Schroetter in the *Denkschrift zur ersten internationalen Luftschiffahrtsausstellung* (Commemorative Work on the First International Airship Flight Exhibition) in 1909 in Frankfurt am Main was published with the motto chosen by von Schroetter himself: anyone wishing to explore the physiological (hygiene) effects of ballooning must commit himself to high altitude ballooning (Figure 3.5).

His charts published in this treatise – I. *Gewöhnliche Atmung im Luftballon* (Normal Respiration in the Balloon) and II. *Sauerstoffatmung in grossen Höhen* (Oxygen Respiration at High Altitudes) (Figure 3.6) – show how notably knowledge had increased since 1902 and had become more detailed; there can be no doubt that Zuntz contributed considerably to this increase with his research. In the column "Comments" in charts I and II, von Schroetter indicated the necessity of an enclosed basket for high altitude flights of over 10,000 m above sea level on the basis of physiological considerations (von Schroetter, 1909, p. 214) (Figure 3.6). On November 16, 1903, at the 232nd Assembly of the Berlin Society for Airship Flight, he himself presented the corresponding draft for a "basket of the future,"[19] with Zuntz in attendance (Figure 3.7).

Von Schroetter's ideas and plans for developing a "basket of the future" represented the first serious attempt on the part of science to employ pressure

Motto: Um die hygienischen Verhältnisse
im Korbe des Luftballons richtig
zu würdigen, ist es notwendig, daß
der ärztliche Beobachter selbst
„in das Reich der Cirren" emporsteige.
H. v. S.

Figure 3.5 The goddess Hygienia breathing oxygen, and the von Schroetter motto of his research in the balloon ("In order to truly understand the hygienic situation given in the basket of a balloon, the physician must himself rise up into the kingdom of the cirrus clouds").
(von Schroetter, 1912a, p. 1)

I. Gewöhnliche Atmung im Luftballon.

Atm.	Höhe	Luftdruck	Sauerstoffspannung der äußeren Luft		Sauerstoffspannung der Lungen- bzw. Alveolarluft		Kohlensäuregehalt der Exspirationsluft	Blutgase Gehalt an Sauerstoff in Vol.-Proz.	Anmerkung
		mm	mm	%	mm	%			
1	0	760	160	21	120—110 aber auch 95	18—17 aber auch 13 u. darunt.	Die Alveolarspannung d. Kohlensäure anfangs (bis zu einer Sauerstofftension v. ca. 60 mm) konstant, dann sinkt dieselbe.	20	Die Alveolarspannung des Sauerstoffes sinkt rascher als der äußere Druck. Je nach der Atmung (Tiefe derselben) Differenzen von 1/3—1/2 gegenüber der Tension der äußeren Atmosphäre. } Die Alveolarspannung des Sauerstoffes liegt infolge von Wasserdampftension, schädlichem Raume, CO_2-Gehalt der Lungenluft tiefer als die O_2-Tension der äußeren Atmosphäre. Erhöhte Lungenventilation und Herzarbeit sowie bessere Ausnutzung des Sauerstoffes seitens der Gewebe kompensieren ganz oder teilweise die Dichtigkeitsabnahme des Sauerstoffes der Atmosphäre; gesteigerte Ventilation kann Differenzen bis zu 1/3 oder 30 mm der Sauerstoffspannung ausgleichen, kompensieren usw.
	2500	550	115	16	Bei zweckmäßiger Muskelarbeit höher.				
	3000	520	110	15	Bei guter Ventilation noch genügend.				
	3500	480	104	14	90	12,5			
	4000	450	95	12,5	70 — Der Sauerstoffmangel beginnt sich zu äußern, der Sauerstoff reicht z. Deckung des Gesamtblutes nicht mehr aus.	10	Die Kohlensäureausscheidung wird durch gesteigerte Ventilation sowie größeren Säuregehalt des Blutes vermehrt. — Erhöhung des respiratorischen Quotienten weist auf qualitative Veränderung des Stoffwechsels.		
	4500	421	90		Auf d. M. Rosa ca. 40—50 mm beobachtet.			Können schon um 10% und mehr in ihrer Menge herabgesetzt sein.	Bei Mehranforderungen können schon Erscheinungen als Folge von, bes. lokalem, Sauerstoffmangel auftreten. Schon bei 50 mm Alveolarspannung beginnt merkbare Dissoziation des Oxyhämoglobins.
	5000	410	85	11	Relativer Sauerstoffmangel. Im Ballon ca. 60 mm beobachtet.				
	5500	380	80	10,5	Hier Abnahme auf 35—30 mm oder 6% schon möglich.				Daher: Um vollkommen normales Befinden zu erreichen, wird Sauerstoffatmung notwendig.
1/2	6000	350	75	9,5	35—30	6	(Kohlensäure ist ein wichtiges Anregungsmittel für die Atmung.)	Arterieller Sauerstoffgehalt gleich jenem der Venen unter normalen Verhältnissen. Grenze des Lebens.	Atmung in diesen Höhen im Ballon flach; daher können leicht Tensionen von 35 mm = 5 % in der Lungenalveole bestehen. Diese Spannung für die Hämoglobinversorgung des Blutes auch bei Ruhe nicht mehr ausreichend.
	7000	310	66	8,8	Beginn des absoluten Sauerstoffmangels, da die Alveolarspannung bloß 25 mm oder 4% sein kann.				
1/3	8000	270	59	8	25	4			Noch höher tritt starke Dissoziation des Oxyhämoglobins ein.
	8500	265			Genügende Alveolartensionen können überhaupt nicht erreicht werden; Atemmuskeln und Herz beanspruchen an sich hohen Bedarf.				Die Sauerstoffrespiration hat mittelst einer Maske zu erfolgen.
	9000	240	50	7					Geschlossener Korb mit erhöhter Sauerstoff-Tension wird zur Notwendigkeit.
	10000	217	46	5	†				
	11000	190							
1/4	12000	170	30	4					

II. Sauerstoffatmung in großen Höhen.

Atm.	Höhe	Sauerstoffdruck der Atmungsluft (bei künstlicher Respiration)		Alveolartension nach obigem	Physiologisches	Anmerkung
		mm	%			
1	0	760	100			Druckbedarf. Als Minimum für die Hämoglobinversorgung werden 35—40 mm O_2-Druck beansprucht. Die Alveolarluft muß enthalten an mm: 46 entsprechend Wasserdampfspannung; 32 Kohlensäure-Partialdruck; 40 Hgb.-Versorgung (gegen Dissoziation); 44 Druckverlust durch Sauerstoffverbrauch seitens der Gewebe. Summa 162 mm entsprechend einer Höhe von 12 500 m; da aber unter diesem Grenzwerte das Leben ein pathologisches ist, da überdies der Sauerstoff nicht rein sein, gesteigerte Anforderungen eintreten können (bei flach bleibender Atmung), Kältewirkung usw., so soll der Grenzwert, bis zu welchem die, kontinuierliche, Sauerstoffatmung mittelst Zylinders und Maske durchgeführt werden darf, mit 11 000 bzw. 10 000 m = ca. 220 mm Barometerdruck praktisch angesetzt werden; dann wird der geschlossene Korb (s. oben) zur Notwendigkeit.
1/3	8000	250	30	35 mm oder 6% und weniger (theoret.)		
1/4	11000	190	25			
	12500	160	17,6		Bei 160 mm Bestand der Sauerstoffrespiration bei Ruhe noch theoretisch möglich.	
1/5	13000	141				
	14500	122	16		Der Wert 116 entspricht nicht mehr dem wegen Wasserdampfspannung, Kohlensäuregehalt, Hämoglobinversorgung nötigen Druckanteile.	
1/6	15000	116				
1/8	17000	90	12			

Figure 3.6 Graph by von Schroetter setting out the physiological and pathophysiological effects of high altitudes on the organism and the limits of prophylactic use of oxygen.
(von Schroetter, 1912a, p. 16)

chambers in aeronautics. Ultimately, this means that von Schroetter's draft can be considered the predecessor of pressure chambers in aeronautics and space flight.

The year 1912 marked a turning point in the history of aeronautic medicine in Germany for multiple reasons: von Schroetter published his *Hygiene der*

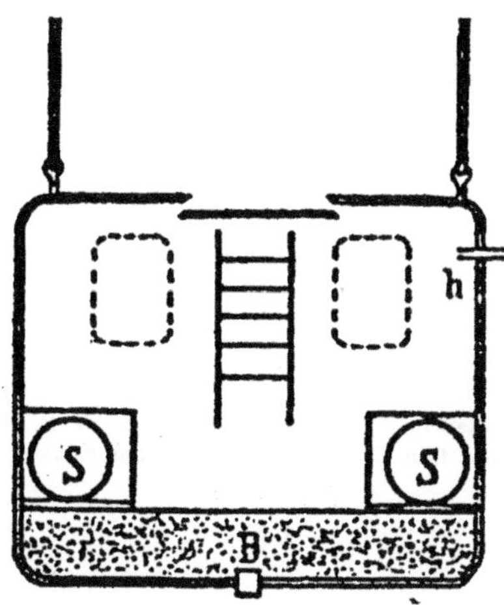

Figure 3.7 The Hermann von Schroetter design of a "basket of the future" (S: *Sauerstoffbehälter* (oxygen container); B: ballast; h: not explained in more detail). (von Schroetter, 1904, p. 18)

Aeronautik und Aviatik (Hygiene of Aeronautics and Aviation) and Zuntz published his paper *Zur Physiologie und Hygiene der Luftfahrt* (On the Physiology and Hygiene of Aeronautics) a short time later; also, the *Wissenschaftliche Gesellschaft für Flugtechnik* (Scientific Association for Flight Technology) was founded in Berlin, with a subcommittee for medical psychological issues.

Since Zuntz's activities in the context of the Scientific Association for Flight Technology have already been addressed, the contents of his treatise *Zur Physiologie und Hygiene der Luftfahrt* (On the Physiology and Hygiene of Aeronautics) will now be discussed.

Zuntz's publication was made possible by funding granted by the Magdeburg *Verein der Luftschiffahrt* (Society for Airship Flight), the *Deutsche Luftschiffahrt-Gesellschaft* (German Airship Flight Company) and the personal assistance of Sticker, the publisher of the journal *Luftfahrt und Wissenschaft* (Aeronautics and Science) (Zuntz, 1912f, p. 7) (Figure 3.8).

Several months prior to the publication, towards the end of October 1911, Zuntz undertook a further balloon ascent for medical physiological purposes with the zeppelin *Schwaben*, and in 1912 flew several laps around the Johannisthal airfield in Berlin in an airplane (Zuntz, 1912f, p. 7). These two research flights presumably provided direct occasion for Zuntz to summarize his entire aeronautic medical and physiological experiences in this work.

The publication *Zur Physiologie und Hygiene der Luftfahrt* is broken down into eight chapters. In the first chapter he refuted, by his own calculations of the labor expenditure for flying, the speculative views held by other authors that man could perhaps fly using his own power (Zuntz, 1912f, p. 2). The second chapter described the accompanying circumstances which led to the

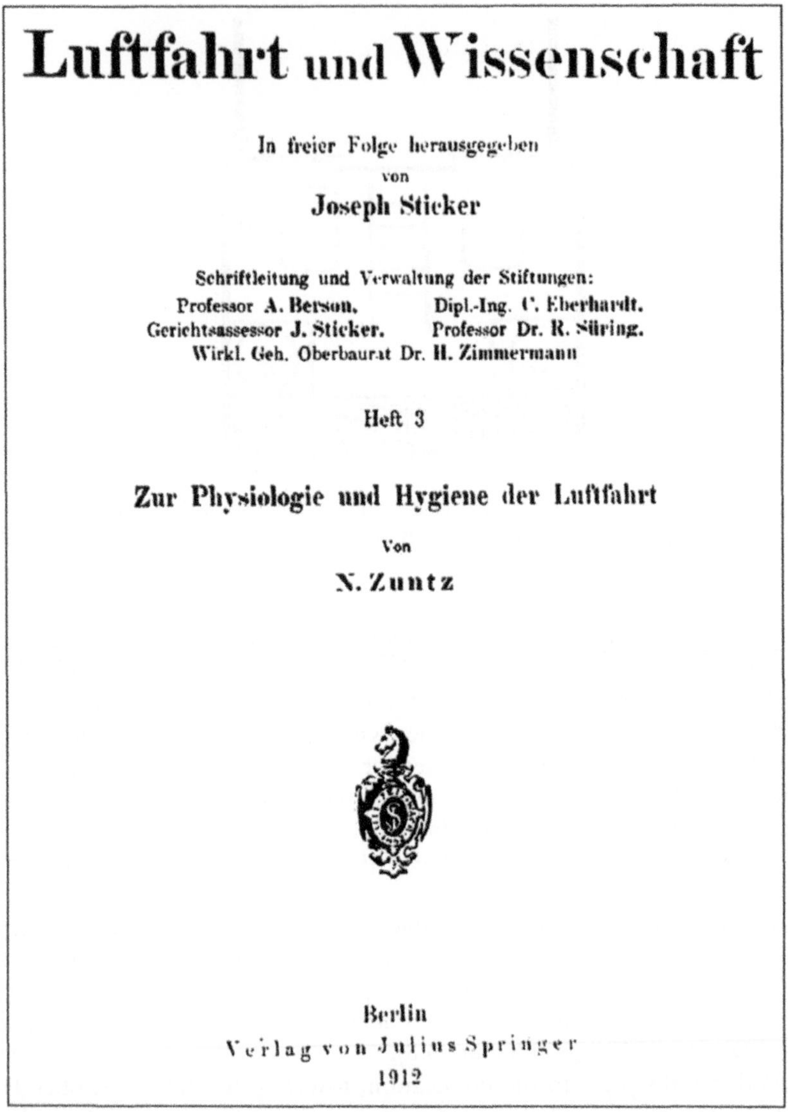

Luftfahrt und Wissenschaft

In freier Folge herausgegeben
von
Joseph Sticker

Schriftleitung und Verwaltung der Stiftungen:
Professor **A. Berson**, Dipl.-Ing. **C. Eberhardt.**
Gerichtsassessor **J. Sticker.** Professor Dr. **R. Süring.**
Wirkl. Geh. Oberbaurat Dr. **H. Zimmermann**

Heft 3

Zur Physiologie und Hygiene der Luftfahrt

Von
N. Zuntz

Berlin
Verlag von Julius Springer
1912

Figure 3.8 Title page of *Luftfahrt und Wissenschaft – Zur Physiologie und Hygiene der Luftfahrt* (On the Physiology and Hygiene of Aviation) published in 1912.

completion of the work and briefly outlined the actual task at hand. Two topics can be seen as the focus of the remaining chapters, which Zuntz himself characterized as follows:

> *Sensory organs and response to their agitation. Breathing at great altitudes and given rapid changes in altitude.*

> *The first chapter deals in particular with airplane operation, since here we use our senses under very unusual conditions and need to respond logically to their impressions.*
>
> *Respiration plays a role primarily due to its indirect impact on brain activities during those high altitude flights which exceed 2,000 m; but the ability of our respiratory apparatus to perform is decisive for the altitudes which man can reach in a balloon.*
>
> *(Zuntz, 1912f, pp. 5–6)*

The fact that Zuntz placed particular value on the physiology of the senses may at first appear surprising, since his field of scientific activity was, after all, determined essentially by his work on metabolism, nutrition and high altitude physiology. However, a glance at the titles of his lectures shows that Zuntz had been intensively involved with this topic for years. During the 1904 summer semester, for example, Zuntz gave a lecture on the senses "Taste and Smell." Zuntz considered somato-visceral sensitivity (sense of touch, sensitivity to depth, especially sense of strength and positioning), the sense of sight, and the sense of equilibrium to be the senses that are of crucial importance for aviation:

> *For all these apparatuses [the sensory organs], the customary impulses, and especially the success of the customary response to these impulses, are delayed when we are in the air instead of standing on solid ground with our feet.*
>
> *These things are of the highest practical significance for the pilot who has to react to the finest adjustments of the state of his airplane at every moment of flight, and in some instances must react so quickly that there is no time for conscious thought. In this context, it can be fatal for him when innate reactions or those muscular reactions, acquired in early youth, which have become reflexive as a result of constant repetition but are adjusted to the mechanical conditions on solid ground, interfere negatively with the actions which are required for steering and balancing the airplane. For this reason, physiology can provide crucial aspects for the construction of airplanes or their steering equipment by means of precise studies performed on the unconscious movements that occur when the vestibular system is disturbed.*
>
> *(...)*
>
> *Everyone knows to what extent our physical bearing is influenced by sudden, shocking perceptions, how we, if we see some impediment, automatically throw the body back to prevent ourselves from hitting it. These sorts of sudden movements of the pilot holding the steering rod can easily lead to dangerous disturbances of the vestibular system. Similarly, the irritation of the eye by dust and small alien bodies can lead to considerable disturbances to the vestibular system; for this reason, the safe protection of the eye ... is of special significance.*
>
> *(Zuntz, 1912f, pp. 9–13)*

In his treatise, Zuntz ultimately came to realize the significance of the vestibular organ for the pilot, using comparative anatomical studies on the structure and function of the labyrinth for crustaceans, fish, amphibians, and birds:

> *The pilot in the air is by no means dependent to the same degree on the otolith apparatus; apart from a visual sense of orientation, significant assistance*

> *is provided by the sense of pressure on the skin for determining every fluctuation to the centroid in his physical bearing. Experienced pilots are not wrong when they say that they perceive with great accuracy every shift of their airplane with their buttocks. Since, as we saw earlier, our skin's sense of pressure reacts to small alterations of pressure with the greatest sensitivity, we thus become aware of small fluctuations in our physical bearing. But by no means does a so fine connection of the nervous centers innervating the muscles-to-skin sensitiveness exist in the derriere region as it does, for example, in the soles of the feet. They are constantly used, while walking and in a variety of positions of the body, as organs that assess the relationship of the human to the ground, the floor. This is why the regulation mechanisms, which react to any unevenness in the load of the soles of the feet by purposeful, compensating movements of the muscles, are much more fully formed than those which are in the skin of the buttocks. Since all these regulations, governed both by the skin as well as by the ear labyrinth, occur more or less unconsciously, the pilot cannot discern how many of his movements are triggered by the ear labyrinth and how many by the sensitivity of his skin, when he automatically executes appropriate regulatory movements to compensate for the movements of his seat. In any case, anyone whose otolith apparatus does not function normally is taking a great risk when he attempts to fly an airplane. He is most certainly missing a key means of assistance which conveys the rapid and half-automatic reaction to every change and movement of his seat. This warning does not only apply to people whose otolith apparatus is completely destroyed, instead it is particularly important for those with marginal disruptions of the apparatus, such as an inflammation of the eardrum. Sometimes this may arise even when the eustachian tube is blocked and will manifest itself in slight dizziness, which is particularly noticeable given fluctuations of pressure in the ear drum, that is, when sneezing, blowing the nose or in similar processes. The decreased ability to perform caused by such relatively insignificant interferences is noticeable in particular where the steering of the airplane is adjusted to the greatest extent possible to the normal compensative movements, which we execute instinctively when our body shifts positions and such an adjustment is being implemented – and rightly so – to all new apparatuses. You can still find some airplanes with steering equipment which is asymmetrical, arranged towards the right-hand side. I find this downright alarming since it can give rise to interferences with the pilot's state of equilibrium precisely in critical moments.*
>
> *(Zuntz, 1912f, pp. 27–28)*

As illustrated by the above statement, Zuntz not only recognized and described the significance of this organ for the pilot; he also worked out those requirements that, from the perspective of medical physiology, are particularly important for suitability tests for pilots – which did not exist at the time. The considerations expressed by Zuntz here were directly applied in the guidelines proposed by Koschel (1913) at the main assembly of the *Wissenschaftliche Gesellschaft für Flugtechnik* (Scientific Association for Flight Technology), regarding pilot qualifications, that had been drawn up by the subcommittee for medical physiological issues (Koschel, 1914, pp. 143–156).

In the subsequent chapter, Zuntz also addressed the aspects of occupational medicine in the field of aviation. He pursued the potential hazards of poisoning when filling and deflating balloons with gas. Using descriptions of accidents, Zuntz listed those events in the course of this work that had represented a hazard, providing suggestions for accident prevention and general rules for saving someone who was poisoned that are still applicable today (Dietl, 2003, p. 2838):

> *Much more acute than the danger posed by the exhaust of balloon gas during flight is another, which results when the people entrusted with filling the balloon sometimes breathe in almost pure or very highly concentrated balloon gas. If it is pure hydrogen, then this may only lead to temporary suffocation and unconsciousness, which would not result in dire consequences if the injured party is quickly carried out to fresh air. But one must keep in mind when filling the balloons that unconsciousness while breathing pure hydrogen or hydrogen mixed with small quantities of air can occur without the person doing the filling observing any shortness of breath or other subjective warnings. Because of this, no person should be filling the balloon without another person constantly supervising the work close by. When filling a balloon with coal gas, the danger is greater because carbon monoxide, which, depending on its origins, is contained in coal gas in quantities of 5–10% or even more (in so-called water gas up to almost 50%), will combine with red hemoglobin to form carboxyhemoglobin, making oxygen supply to the organs impossible for a longer period of time [even] after the injured party has been removed from the area where poisonous gas abounds.*
>
> *(...)*
>
> *The carbon monoxide bound to the hemoglobin disappears again when air is inhaled that is completely free of the aforementioned toxin. But this regeneration of the blood can occur only slowly because, after the injured party has been removed from the carbon monoxide-containing atmosphere, the air in the lungs still contains the poisonous gas for a long time and furthermore, new quantities of the poison continue to enter the lungs from the blood. A lack of oxygen for three is enough suffices to disable the nervous respiratory center. Thus, simply exposing the party poisoned with carbon monoxide to fresh air seldom results in revitalization. The disabled respiration system has to be substituted by pumping air into the respiratory tracts artificially, applying the methods of artificial respiration, primarily consisting of strongly compressing the chest cage in a regular rhythm and then decreasing the pressure. These methods should thus be practiced by all persons who are employed with filling balloons or the supervision thereof (Zuntz, 1912f, pp. 30–32).*
>
> *(...)*
>
> *However, if the blood has been completely saturated with carbon monoxide, it often proves necessary to introduce pure oxygen. The balloon filling areas should be equipped for this anyway since it is appropriate that tanks with compressed oxygen be kept available to be taken on airships for high altitude flights. When a tank of this sort is already equipped with an inhalation mask, this should be held in front of the mouth and nose and the chest cage should be caused to expand forcefully by means of the introduction of oxygen through the tightly closed mask. Then the oxygen flow should be shut*

> *off for a moment and the mask be removed, whereupon the elasticity of the chest cage should allow the air to flow out of the lungs.*
>
> *(Zuntz, 1912f, pp. 30* et seq*)*[20]

In Chapters 5–8 of his treatise, Zuntz addressed the high altitude physiological aspects of aeronautics in detail, essentially drawing on his own studies that he had pursued at the turn of the century. However, since his text was not solely intended for physicians and physiologists but also for flight technicians (Zuntz, 1912f, p. 67), Zuntz explained the influence of great altitudes on the circulatory and respiratory systems in terms that could be generally understood (Zuntz, 1912f, pp. 35–54).

Again, Zuntz succeeded in deriving a series of guidelines on conduct for safety and accident prevention while flying, based on his description of the pathophysiological circumstances. They included, to cite but a few examples, an examination of the Eustachian tube before commencing a flight, and a thought-out dietary plan that takes into account non-flatulent foods prior to and during the flight:

> *As already mentioned, the gases collected in the digestive apparatus will expand when the air becomes thinner as the balloon ascends. Here, external compensation for the pressure is not possible since the many bends and twists of the intestinal tract and the temporary blockage due to the walls pressing against one another will enable compensation by the natural exit passageways of the intestine only very gradually. However, during an ascent to 5,000 m altitude, that is, in an atmosphere with half the [normal] pressure, the gases filling the digestive apparatus will expand to twice their normal volume in accordance with Mariotte's Law. This can lead to an unpleasant pressure affecting the diaphragm because of the existing heavy flatulence of the intestinal tract, and can thus impede breathing.*
>
> *Any respiratory difficulty, even if extremely marginal, will notably increase the risk of oxygen deficiency, which is already present at this high altitude anyway.*
>
> *(...)*
>
> *But there is another factor, which may aggravate this interference even further. A large part of the gases forming in the intestinal tract, namely the carbonic acids which form the major bulk of the fermentation process, are absorbed into the blood circulating in the intestinal wall and excreted by the blood into the lungs. Now the absorption of gas into the blood depends on the partial pressure and, supposing that a given quantity of gas is present, this will be lower if the absolute air pressure is lower. Thus, when the air pressure sinks to half [of normal] at an altitude of 5,000 m, the absorption of gases from the intestines into the blood sinks by half, which of course will considerably promote the disruptive accumulation in the intestinal tract. Therefore, rational nourishment must be regarded as a factor which should by no means be neglected if a high altitude flight is to be a success. The absolute quantity of the food consumed is to be low, and flatulent vegetables, sweets, easily fermenting bread high in fiber like pumpernickel, or sparkling drinks and the like must be avoided.*
>
> *(Zuntz, 1912f, p. 39)*

The patho-physiological processes described here by Zuntz and the nutritional recommendations he derives therefrom can be recognized in the guidelines on special nourishment issued to airplane and airship pilots on the front in World War I (Figure 3.9).

A further requirement often expressed by Zuntz was that additional oxygen be taken along for extremely high ascents. He believed that artificial oxygen

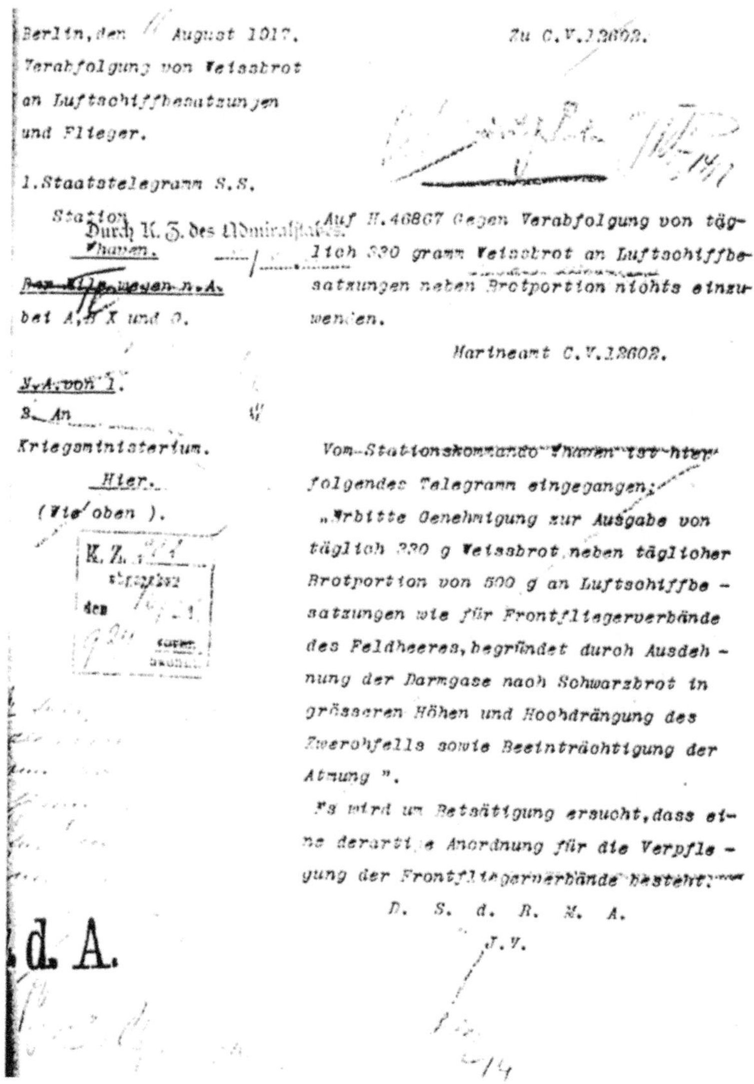

Berlin, den 16 August 1917.

Zu C.V.12602.

Verabfolgung von Weissbrot an Luftschiffbesatzungen und Flieger.

1. Staatstelegramm S.S.

Station ...haven.

Durch K. Z. des Admiralstabes.

bei A, B X und O.

Auf H.46867 Gegen Verabfolgung von täglich 330 gramm Weissbrot an Luftschiffbesatzungen neben Brotportion nichts einzuwenden.

Marineamt C.V.12602.

2. An Kriegsministerium.

Hier.

(Wie oben).

K. Z.

den

Vom Stationskommando ...haven ist hier folgendes Telegramm eingegangen:

„Erbitte Genehmigung zur Ausgabe von täglich 330 g Weissbrot neben täglicher Brotportion von 500 g an Luftschiffbesatzungen wie für Frontfliegerverbände des Feldheeres, begründet durch Ausdehnung der Darmgase nach Schwarzbrot in grösseren Höhen und Hochdrängung des Zwerchfells sowie Beeinträchtigung der Atmung".

Es wird um Betätigung ersucht, dass eine derartige Anordnung für die Verpflegung der Frontfliegerverbände besteht.

D. S. d. R. M. A.

J. V.

d. A.

Figure 3.9 Permission for the "Administration of white bread to airship personnel and pilots" dated August 16, 1917.
(German Federal Archives, Military Archive, Freiburg/i. Breisgau, RM 3/V 8970, p. 121)

inhalation was indispensable for people with normal physical constitutions at altitudes of over 4000 m above sea level. This is "the superior means for overcoming high altitude sickness" (Zuntz, 1912f, p. 53). If the subjects were older and had arteriosclerotic vessels and/or cardiovascular diseases, Zuntz did not rule out complications arising at even lower altitudes (Zuntz, 1912f, p. 52):

> *This appears to be the case especially in individual parts of the brain, so that in this organ even a small reduction of oxygen content in the arterial blood can cause damages. Probably, what has been easily proven for the muscles applies for the brain as well, namely that the active parts of the organ exploit the oxygen of the flowing blood to a much higher degree than the passive parts. In a recently published work, Tangl revealed that an intense excitement of the brain, which he achieved by means of intermittent violent lighting, significantly increases oxygen consumption in animals sedated by curare. He also discovered that oxygen content was much lower than in the blood flowing from the cerebral veins during the light stimulus than that flowing before or after such stimuli. Now, the normal functioning of the senses, normal control of our muscles and rapid consideration all entail an intensive activity in certain parts of the brain which specially perform the relevant functions. In this way, we are able to understand why, even at relatively moderate altitudes, different brain functions will suffer with different individuals. While one will have trouble maintaining physical balance, will feel dizzy and unsteady while walking, the other will be unable to process sensory impressions; to read the instruments precisely, to observe acoustic signals exactly and the like. For some observations in the balloon, it is of particular import that the sense of color is among those brain functions which are easily damaged by oxygen deficiency. Readings of the spectroscope or the interferometer and similar instruments appear to quickly become flawed at great altitudes.*
>
> *(Zuntz, 1912f, p. 47)*

In this context, Zuntz criticized the manner in which airplane designers restricted the free movement of the pilot's chest cage by the arrangement of the seat and steering equipment:

> *But here another aspect comes into consideration, resulting from the fact that we cannot breathe unrestrictedly given the permanent fixed tensing of the arm muscles, which to this day is forced on pilots by most airplane models. The muscles stretching from the chest cage to the upper arm are fixed, and as a result the chest cage has to remain relatively tensed, which naturally is a hindrance to inhalation and exhalation. It is possible to train and practice overcoming this involuntary breathing inhibition that can easily develop when the arms are tensed, and in particular so when attention is also intensely fixed in a certain direction. I believe it is not inconsequential that in training pilots for high altitude flights, particular attention be given to preventing these respiratory inhibitions.*
>
> *(Zuntz, 1912f, p. 54)*

Based on the characterization of these dangers caused by a relative or absolute oxygen deficiency, Zuntz determined the maximum necessary oxygen quantity for high altitude flights to be about 420 l per hour per person using his own calculation method (Zuntz, 1912f, p. 60). Zuntz strongly promoted his own, safer design of a mouthpiece for oxygen inhalation that was to replace the tube stuck loosely into the mouth that was usual practice at the time. Clearly, he recognized what dangers high altitude pilots were exposed to:

> *When unconsciousness or even half-sleep occurs, the mouthpiece can slip away from the lips completely, causing the certain death of the aeronaut. Therefore, it should be that the respiratory tracts are permanently connected to the oxygen source for high altitude flights going up higher than 8,000 m. The mouthpieces can be used in the manner in which I and my colleagues have deployed them for many years during respiration experiments. A soft rubber disk with a hole drilled in the middle is inserted between the lips and teeth, which has a tube attached connecting it to the oxygen source. The mouthpiece is kept airtight by means of the elasticity of the lips and the wetness of the mouth. But you can secure it even further by placing an elastic swathe with a slit in the front over the lips, similar to the moustache-trainer, and tying it behind the head. The nose is hermetically sealed using a clip. Since the nose is not used, respiration does not occur completely normally; nevertheless it does occur in a manner which can be tolerated by most people for hours without any difficulty.*
>
> *(Zuntz, 1912f, p. 58)*

Zuntz believed the future lay in the use of oxygen masks made of hard rubber or metal, such as had been developed by his student von Schroetter (Figures 2.16, 3.5). He stated that he was not yet able to evaluate the advantages of masks which were worn in connection with a cap, as designed by the Berlin physician and balloonist Flemming in 1911 (Figure 3.4), since at the time his publication went to the presses he still had no personal experience with this design. When aeronauts wanted to set new high altitude records during balloon ascents,[21] Zuntz recommended the use of mouthpieces because the damaging "dead space volumes," a term coined by Zuntz,[22] were significantly lower (approximately 150 ccm) than when using mask systems (*ca.* 290 ccm) (Zuntz, 1912f, p. 64).

After estimating and weighing the dangers from medical, physiological, and technical perspectives, Zuntz maintained the belief that ascents to extreme heights were hazardous despite all general safety measures taken. Because of this, Zuntz recommended requiring that, prior to commencing a height ascent, respiratory devices and oxygen supply should be tested in the pneumatic chamber to ensure they are intact (Zuntz, 1912f, p. 64). The German pilot Linnekogel tested the functionality of the oxygen device and his respiratory mask, of the type designed by von Schroetter, prior to his flight with which he set the high altitude world record (6300 m above sea level) on March 31, 1914 (Figure 3.10).

Zuntz calculated the maximum altitude attainable while breathing artificial oxygen to be 14,300 m above sea level. He stated that flights exceeding this altitude would have to be undertaken in an enclosed capsule, as proposed by

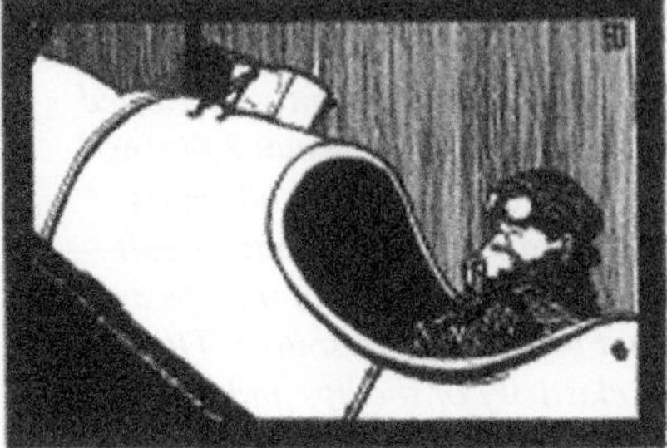

Figure 3.10 The German pilot Linnekogel and the Dräger Company advertising the oxygen device and a respiratory mask to obtain the world record of high-altitude flight on March 31, 1914 (6300 m above sea level).
(Kehrt, 2006, p. 139)

his student von Schroetter (Figure 3.7) (Zuntz, 1912f, p. 64). However, at the end of his treatise, he revealed his skepticism about a project such as the one proposed by von Schroetter, since he doubted the scientific value of such an undertaking (Zuntz, 1912f, pp. 66–67):

> *The realization of this project will depend on whether physical or meteorological problems of sufficient import exist that would justify the travel costs and the danger to the pilot's life, which should not be underestimated. In view of the successes achieved by unmanned recording balloons, it still appears doubtful whether this question can be answered in the affirmative.*
>
> *(Zuntz, 1912f, pp. 66–67)*

Using these numerous examples from his publication *Zur Physiologie und Hygiene der Luftfahrt*, it can be proven that, from today's perspective, Zuntz had recognized and described significant aeronautic-medical issues as early as in 1912. Thus, he made a decisive contribution to the development of this new field of medical research. In addition, Zuntz's dedicated intervention at this point in time regarding the laws regulating physicians' insurance presumably strengthened the self-confidence and self-conception of the "aeronautic physicians in Germany" just prior to the outbreak of World War I (Figure 3.11).[23]

OPEN LETTER

from the Flying Doctors to the *Versicherungskasse für die Aerzte Deutschlands* *(German Insurance Fund for Doctors)*

We protest the fact that the *Versicherungskasse für die Aerzte Deutschlands* (German Insurance Fund for Doctors) – whose regulations treat accidents as diseases, do not exclude any kind of accident, do not prohibit insured individuals from engaging in any athletic or other kind of activity or raise premiums for individuals who engage in such activities, and do not require any declaration from the insured regarding his lifestyle, athletic activities, or recreational activities – has refused to comply with the justified claims submitted by Dr. Halben, an associate professor and ophthalmologist from Berlin who was seriously injured in a balloon accident, such refusal being issued on the grounds that his illness was the result of an accident during a hot-air-balloon trip he led (organized by the Berlin Airborne Travel Association and fully authorized). We consider it an injustice if the fund bases its rejection of the claim solely on a stipulation of the Prussian *Allgemeines Landrecht* (General Code) according to which the insured individual loses his entitlement to compensation if he undertakes any activity or allows third parties to undertake any activity which changes the circumstances under which the insurance policy was originally taken out in a way that harms the insured individual or increases the risks that he faces.

By submitting this as its grounds, the fund admits that it has no hold for rejecting such a claim if the insured individual was already a hot-air-balloon rider or pilot at the time the insurance policy was taken out. As already stated, an insured individual is neither asked nor required to make any declaration in this regard, and the regulations do not require him to pay any additional premiums on account of such circumstances. Thus, the fund is also admitting that it wants to use this provision of the law to deny the claim, based on an allegation that the insured individual in the case at hand did not pilot hot air balloons when he joined the insurance fund. The resulting inequality in the treatment of individuals who engage in any kind of sport when they join the plan and those who begin a sport after joining the plan, as well as the transferability of this principle to almost any other change in lifestyle (motor-vehicles and other means of transportation, moving away from a healthy and safe location to a different one, assuming dangerous medical duties, working in the department of infectious disease, quarantine barracks, plague laboratory, etc.,) creates insecurity and inequity for the insured.

However, we must also protest emphatically against the notion that air travel is an exceptionally dangerous activity that doctors have no business engaging in. On the contrary, we believe that ballooning is well-suited for the relaxation and rejuvenation of doctors, who have more difficulty finding free time for daily sport activities (riding, tennis, etc.,) than people in other professions, and can seldom get away for a short vacation. Each and every balloon trip, which begins with an ascent on Sunday morning or Saturday evening and allows us to return to duty early on Monday, refreshes our spirits and toughens our constitutions with lasting effect.

Even if the numerous statistics published show a balloon accident rate of approximately 0.5 % of all riders, the following should also be taken into account: First, even the most enthusiastic balloon rider hardly undertakes more than six flights per year. Second, the beneficial effects on his health this has, which also benefit his insurance company, balance the scales considerably with respect to the risk of an accident. This risk is not considerably greater for a doctor who takes six balloon trips in a year than for one who goes mountain climbing for six weeks a year or drives his automobile for work and pleasure trips all year long.

Furthermore, the fund is overlooking the fact that doctors in particular have important tasks and obligations to perform in the area of airborne travel. For German doctors who are proud of their faithful fulfillment of duty on land and water (ship's doctors, marine doctors), it is also a matter of honor not to remain behind the curve in the area of aeronautics, but also to contribute enthusiastically and expertly to the promotion, regulation, and security of airborne travel. Only balloon-riding doctors can research and develop physiology, hygiene, prevention, and therapy, as well as insurance matters, with regard to aeronautics, and they are called upon first and foremost to contribute to the development of legal regulations with regard to medical requirements for airborne travelers. For example, an assembly of balloonist doctors has already met under Dr. Halben's chairmanship to consult on the principle that admission to pilot training and the use of a pilot's license should be dependent on the results of a doctor's examination and supervision, as well as to discuss minimum health requirements regarding various areas of physical health. The assembly submitted the results of its discussions to the *Berliner Verein für Luftfahrt* (Berlin Aeronautical Association), which presented the assembly's views at the *Deutscher Luftfahrertag* (German Aeronautical Congress).

Furthermore, those of us who are active or inactive military or marine doctors confirm that the military administration treats all injuries we suffer as a result of unofficial balloon trips undertaken with official permission (which is almost never denied for domestic trips) during our terms of service (trips for pleasure, sport, and competition) as injuries suffered in the line of duty. In addition, we all confirm that the balloons and pilots of the *Deutsche Luftfahrer-Verband* (German Aeronautical Association) will be placed at the disposal of the military leadership in case of war. Given this, every flight by a pilot who is not a military doctor should be viewed as a "military exercise," or, respectively, an exercise that should be treated as equally important. Thus, the insurance fund's actions run contrary to the intent of its regulations, according to which, "the conscription of a member into military exercises while on leave shall have no influence on his membership." It is also acting contrary to the aforementioned praiseworthy policy of the military authorities and the national sentiment of doctors at a time in which the entire nation is convinced of the importance of promoting and expanding aeronautics.

As long as the fund persists in its current application of the regulations to the detriment of balloonists, we cannot recommend our friends to join it, as much as we support a collegial union to protect against the risks of disease and invalidity in principle – especially in our high-risk profession. If the intended changes to the regulations are carried out, we must also protest emphatically against any disadvantage to flying doctors. An exclusion of aeronauts would be anti-national and out of step with the times; raising the premiums would only be permissible if the increases were graduated according to risk categories – that means for cyclists, hunters, marksmen, and equestrian sportsmen, coach-drivers, mountain-climbers, ice skaters, tobogganists, bobsledders, skiers, motorists, and so on. It would be difficult to implement such distinctions and maintain a just system if the different risk levels associated with various professional medical activities were not also taken into account by dividing them into risk classes.

Figure 3.11 "Open letter of aeronautic physicians to the insurance fund for Germany's physicians."
(*Deutsche Luftfahrer-Zeitschrift* (German Aeronauts Journal) 17 (1913), pp. 475 *et seq.*)

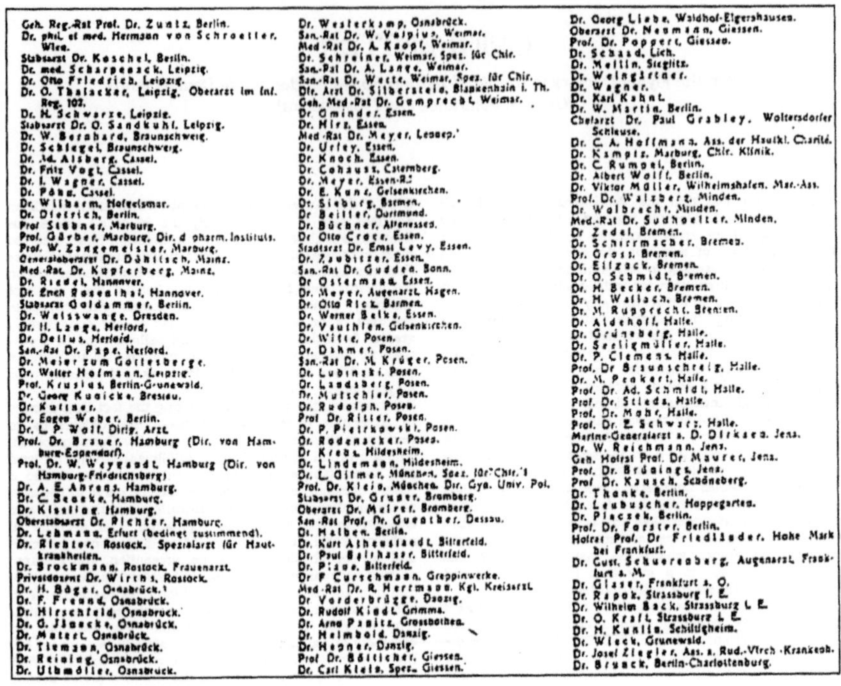

Geh. Reg.-Rat Prof. Dr. Zuntz, Berlin.
Dr. phil. et med. Hermann von Schroetter, Wien.
Stabsarzt Dr. Koschel, Berlin.
Dr. med. Scharpenack, Leipzig.
Dr. Otto Friedrich, Leipzig.
Dr. O. Thalacker, Leipzig. Oberarzt im Inf. Reg. 107.
Dr. H. Schwarze, Leipzig.
Stabsarzt Dr. O. Sandkuhl, Leipzig.
Dr. W. Bernhard, Braunschweig.
Dr. Schlegel, Braunschweig.
Dr. Ad. Alsberg, Cassel.
Dr. Fritz Vogt, Cassel.
Dr. I. Wagner, Cassel.
Dr. Pöhn, Cassel.
Dr. Wilbarm, Hofgeismar.
Dr. Dietrich, Berlin.
Prof. Stübner, Marburg.
Prof. Gürber, Marburg, Dir. d. pharm. Instituts.
Prof. W. Zangemeister, Marburg.
Generaloberarzt Dr. Dähitsch, Mainz.
Med.-Rat Dr. Kupferberg, Mainz.
Dr. Riedel, Hannover.
Dr. Erich Rosenthal, Hannover.
Stabsarzt Goldammer, Berlin.
Dr. Weisswange, Dresden.
Dr. H. Lange, Herford.
Dr. Dellus, Herford.
San.-Rat Dr. Pape, Herford.
Dr. Meier zum Gottesberge.
Dr. Walter Hofmann, Leipzig.
Prof. Krusius, Berlin-Grunewald.
Dr. Georg Kuoicke, Breslau.
Dr. Kuttner.
Dr. Eugen Weber, Berlin.
Dr. L. P. Wolf, Dirig. Arzt.
Prof. Dr. Brauer, Hamburg (Dir. von Hamburg-Eppendorf).
Prof. Dr. W. Weygandt, Hamburg (Dir. von Hamburg-Friedrichsberg)
Dr. A. E. Ahrens, Hamburg.
Dr. C. Seeke, Hamburg.
Dr. Kissling, Hamburg.
Oberstabsarzt Dr. Richter, Hamburg.
Dr. Lehmann, Erfurt (bedingt zustimmend).
Dr. Richter, Rostock, Spezialarzt für Hautkrankheiten.
Dr. Brockmann, Rostock, Frauenarzt.
Privatdozent Dr. Wirths, Rostock.
Dr. H. Böger, Osnabrück.
Dr. F. Freund, Osnabrück.
Dr. Hirschfeld, Osnabrück.
Dr. G. Jänecke, Osnabrück.
Dr. Materr, Osnabrück.
Dr. Tiemann, Osnabrück.
Dr. Reining, Osnabrück.
Dr. Uthmöller, Osnabrück.
Dr. Westerkamp, Osnabrück.
San.-Rat Dr. W. Vulpius, Weimar.
Med.-Rat Dr. A. Knopf, Weimar.
Dr. Schreiner, Weimar, Spez. für Chir.
San.-Rat Dr. A. Lange, Weimar.
San.-Rat Dr. Wette, Weimar, Spez. für Chir.
Dir. Arzt Dr. Silberstein, Blankenhain i. Th.
Geh. Med.-Rat Dr. Gumprecht, Weimar.
Dr. Gminder, Essen.
Dr. Hirz, Essen.
Med.-Rat Dr. Meyer, Lennep.
Dr. Urfey, Essen.
Dr. Knoch, Essen.
Dr. Cohausz, Osternberg.
Dr. Meyer, Essen-R.
Dr. E. Kunz, Gelsenkirchen.
Dr. Sieburg, Barmen.
Dr. Beitter, Dortmund.
Dr. Büchner, Altenessen.
Dr. Otto Croce, Essen.
Stadtarzt Dr. Ernst Levy, Essen.
Dr. Zaubitzer, Essen.
San.-Rat Dr. Gudden, Bonn.
Dr. Ostermann, Essen.
Dr. Meyer, Augenarzt, Hagen.
Dr. Otto Rick, Barmen.
Dr. Werner Belke, Essen.
Dr. Vauthlen, Gelsenkirchen.
Dr. Witte, Posen.
Dr. Dahmer, Posen.
San.-Rat Dr. M. Krüger, Posen.
Dr. Lubinski, Posen.
Dr. Landsberg, Posen.
Dr. Mutschler, Posen.
Dr. Rudolph, Posen.
Prof. Dr. Ritter, Posen.
Dr. P. Pietrkowski, Posen.
Dr. Rodenacker, Posen.
Dr. Krebs, Hildesheim.
Dr. Lindemann, Hildesheim.
Dr. L. Gilmer, München, Spez. für Chir.
Prof. Dr. Klein, München, Dir. Gyn. Univ. Pol.
Stabsarzt Dr. Gruner, Bromberg.
Oberarzt Dr. Meirer, Bromberg.
San.-Rat Prof. Dr. Guenther, Dessau.
Dr. Halben, Berlin.
Dr. Kurt Atheustaedt, Bitterfeld.
Dr. Paul Balthasar, Bitterfeld.
Dr. Pisoe, Bitterfeld.
Dr. F. Curschmann, Greppinwerke.
Med.-Rat Dr. R. Hertmann, Kgl. Kreisarzt.
Dr. Vorderbrügge, Danzig.
Dr. Rudolf Kiedl, Grimma.
Dr. Arno Paaitz, Grossbothen.
Dr. Helmbold, Danzig.
Dr. Hepner, Danzig.
Prof. Dr. Bötticher, Giessen.
Dr. Carl Klein, Spez. Giessen.
Dr. Georg Liebe, Waldhof-Elgershausen.
Oberarzt Dr. Neumann, Giessen.
Prof. Dr. Poppert, Giessen.
Dr. Schaad, Lich.
Dr. Mellin, Steglitz.
Dr. Weingärtner.
Dr. Wagner.
Dr. Karl Kahnt.
Dr. W. Martin, Berlin.
Chefarzt Dr. Paul Grabley, Woltersdorfer Schleuse.
Dr. C. A. Hoffmann, Ass. der Hautkl. Charité.
Dr. Kampis, Marburg, Chir. Klinik.
Dr. C. Rumpel, Berlin.
Dr. Albert Wolff, Berlin.
Dr. Viktor Müller, Wilhelmshafen, Mar.-Ass.
Prof. Dr. Walzberg, Minden.
Dr. Wolbrecht, Minden.
Med.-Rat Dr. Sudhoelter, Minden.
Dr. Zedel, Bremen.
Dr. Schirrmacher, Bremen.
Dr. Gross, Bremen.
Dr. Eilzack, Bremen.
Dr. O. Schmidt, Bremen.
Dr. H. Becker, Bremen.
Dr. H. Wallach, Bremen.
Dr. M. Rupprecht, Bremen.
Dr. Aldehoff, Halle.
Dr. Grüneberg, Halle.
Dr. Seeligmüller, Halle.
Dr. P. Clemens, Halle.
Prof. Dr. Braunschweig, Halle.
Dr. M. Penkert, Halle.
Prof. Dr. Ad. Schmidt, Halle.
Prof. Dr. Stieda, Halle.
Prof. Dr. Mohr, Halle.
Prof. Dr. E. Schwarz, Halle.
Marine-Generalarzt a. D. Dirksen, Jena.
Dr. W. Reichmann, Jena.
Geh. Hofrat Prof. Dr. Maurer, Jena.
Prof. Dr. Brünings, Jena.
Prof. Dr. Kausch, Schöneberg.
Dr. Thanke, Berlin.
Dr. Leubuscher, Hoppegarten.
Dr. Placzek, Berlin.
Prof. Dr. Forster, Berlin.
Hofrat Prof. Dr. Friedländer, Hohe Mark bei Frankfurt.
Dr. Gust. Schuerenberg, Augenarzt, Frankfurt a. M.
Dr. Glaser, Frankfurt a. O.
Dr. Rapok, Strassburg i. E.
Dr. Wilhelm Back, Strassburg i. E.
Dr. O. Kraft, Strassburg i. E.
Dr. H. Kunlin, Schiltigheim.
Dr. Wieck, Grunewald.
Dr. Josef Ziegler, Ass. a. Rud.-Virch.-Krankenh.
Dr. Brasck, Berlin-Charlottenburg.

Figure 3.11 (*Continued*)

Notes

1. A more detailed description is given in the following publications: *Berlinische Nachrichten* (Berlin News) (1788) No. 111 dated September 13, 1788.

   cf. Berlinische Nachrichten (1788) No. 117 dated September 27, 1788.

   cf. Berlinische Nachrichten (1788) No. 118 dated September 30, 1788.

   cf. Günther, P. E.: *Jean Pierre Blanchard's 33. Luftfahrt 1788 – Zeitgenössische Berichte der ersten bemannten Luftfahrt in Berlin, Beiträge zur Geschichte der Ballonfahrt 3* (Jean Pierre Blanchard's 33rd Flight in 1788 – Contemporary Reports of the First Manned Flight in Berlin, Contributions to the History of Talloon Flight) (1988).
2. The term "high altitude flights" (*Höhenfahrten*) is to be distinguished from "height ascents" (*Hochfahrten*). High altitude flights are flights which take place at a certain, specified altitude, in which context this, normally relatively low, altitude is to be maintained by the balloon pilot for as long a time as possible. In contrast, height ascents intend to set absolute altitude records.
3. Otto Lilienthal was born on May 23, 1848, in Anklam. He was an engineer, and had been interested in issues of flight technology since his childhood. In 1890 he constructed his first glider, and one year later he conducted his first gliding flight over a length of 15m. Over the following years, he was successful in extending the flights to 350m. He died on August 10, 1896, from injuries sustained in a serious crash during one of these

experimental flights. Today, Otto Lilienthal is viewed as the "Nestor" of aviation (*cf.* Brockhaus dictionary, volume 7, Brockhaus 1952–1958, pp. 246 *et seq.*; Hörnes, H.: *Buch des Fluges* (Book of Flight), volume 3, G. Szelinski 1911–1912, p. 292).

The history of the technical development of aviation in Berlin was readdressed in 1987 in a new work (*cf.* Koelle, H. H.: *Der Beitrag Berlins zur Entwicklung der Luft- und Raumfahrt* (Berlin's Contribution to the Development of Aeronautics and Space Flight), Yearbook 1987 I of the Deutsche Gesellschaft für Luft- und Raumfahrt e.V. (DGLR German Association for Aeronautics and Space Flight), pp. 431–434).

4. *cf.* On this item:

   Die Umschau 6 (Survey) (1902), pp. 176–177.

   Die Umschau 6 (Survey) (1902), pp. 862–864.

   Deutsche Zeitschrift für Luftschiffahrt (German Journal for Airship Flight) 15 (1911) No. 2, p. 9.

   Deutsche Zeitschrift für Luftschiffahrt (German Journal for Airship Flight) 15 (1911) No 13, p. 18.

5. *cf.* Charts "Opfer des Fluges" (Victims of Flight), *Deutsche Zeitschrift für Luftschiffahrt* (German Journal for Airship Flight) Berlin 15 (1911) No. 26, pp. 22–25; von Schroetter counted the number of victims since the start of motorized flying up until November 1911 at 115 (*cf.* von Schroetter, H.: *Hygiene der Aeronautik und Aviatik* (Hygiene of Aeronautics and Aviation), Braumüller 1912, p. 112).
6. On September 23, 1910, the American pilot Chavez crashed just before landing after flying over the Simplon Pass and died a few days later from serious injuries sustained in the accident (cf. Hörnes, H.: *Buch des Fluges* (Book of Flight), volume 3, G. Axelinski 1912, p. 368).
7. Weitlaner, F.: Die psychologischen, physiologischen und pathologischen Momente beim Simplonflug Chavez (The psychological, physiological and pathological factors of the Chavez flight over Simplon), *Flug- und Motortechnik* 4 (1910), p. 754. In this work, Weitlaner refers to the work by von Schroetter on the occasion of the First International Airship Flight Exhibition in Frankfurt a. M. 1909 (*cf.* von Schroetter, H.: Hygiene der Aeronautik und Aviatik (Hygiene of aeronautics and aviation), in *Denkschrift zur ersten internationalen Luftschiffahrtsausstellung* (Commemorative Work on the First International Airship Flight Exhibition), Frankfurt a. M. 1909, pp. 203–233.
8. *Berliner Lokal-Anzeiger* (Berlin daily paper) dated May 24, 1902, No. 237, column "Aus der Reichshauptstadt" (From the Capital of the Reich).

   cf. Similarly worded commentary in: *Berliner Lokal-Anzeiger* (Berlin daily paper) of May 22, 1902, No. 234, column "Aus der Reichshauptstadt" (From the Capital of the Reich).

   Vossische Zeitung (daily newspaper) dated May 23, 1902, No. 235, column "Luftschiffahrt-Kongress" (Airship Flight Congress).

9. *cf.* the following treatises by Flemming and Koschel: Flemming, –.: Unfälle und Rettungsmassnahmen auf dem Gebiet der Luftschiffahrt (Accidents and rescue measures in the field of airship flight), *Illustrierte Aeronautische Mitteilungen* (Illustrated Aeronautical News) 12 (1908), pp. 489–497.

   Flemming, –.: *Der Arzt im Ballon* (The Doctor in the Balloon), in Bröckelmann, K.: *Wir Luftschiffer* (We Aeronauts), Ullstein 1909b, pp. 172–187.

   Flemming, –.: Physiologische und physikalische Messungen und Beobachtungen bei einer Hochfahrt (Physiological and physical measurements and observations during a height ascent), *Illustrierte Aeronautische Mitteilungen* (Illustrated Aeronautical News) 12 (1909a), pp. 1019–1028.

Flemming, –.: Freiballonhygiene (Free balloon hygiene), In: Mehl, A.: Der *Freiballon in Theorie und Praxis* (The Free Balloon in Theory and Practice), Franckh'sche Verlagshandlungt 1911, pp. 219–236 *et seq.*

Flemming, –.: *Bewusstlosigkeit im Luftschiff* (Unconsciousness in an airship), Umschau (Survey) 16 (1912), pp. 960 *et seq.*

Koschel, E.: Eine wissenschaftliche Hochfahrt mit dem Ballon "Berlin" (A scientific height ascent with the balloon "Berlin"), *Jahrbuch des Berliner Vereins für Luftschiffahrt* (Yearbook of the Berlin Society for Airship Flight) (1912), pp. 116–132.

Koschel, E.: Welche Anforderungen müssen an die Gesundheit der Führer von Luftfahrzeugen gestellt werden? (What health requirements should aircraft pilots meet?), *Jahrbuch der Wissenschaftlichen Gesellschaft für Flugtechnik* (Yearbook of the Scientific Association for Flight Technology) 2 (1914), pp. 143–156.

10. For further information, see: Koschel, E.: Eine wissenschaftliche Hochfahrt mit dem Ballon "Berlin" (A scientific height ascent with the balloon "Berlin"), *Jahrbuch des Berliner Vereins für Luftschiffahrt* (Yearbook of the Berlin Society for Airship Flight) (1912), p. 116; Harsch, V: *Die Deutsche Gesellschaft für Luft- und Raumfahrtmedizin e.v. (DGLRM)*, Books on Demand GmbH 2001, p. 1).
11. Krusius later became one of the physicians to sign the open letter [from] "the aeronaut physicians to the insurance fund for physicians" in Germany dated October 1, 1913.
12. The Privatdozent (junior professor) and ophthalmologist Halben studied the ophthalmologic problems facing aeronautics and aviation. He effectively promoted the use of a monocle in place of fixated glasses while in a balloon or flying a plane, and provided the following interesting justification for this:

    "Perhaps the monocle ... will also prove to be advantageous for pilots who are able to steer with their eyes open at least temporarily; its manageability has truly proven itself in almost every other sport. It guarantees almost the same degree of corrected vision as spectacles, it can be serviced with one hand, can be easily replaced and cleaned if dirty, and combines the advantages of not wearing glasses with those of wearing glasses. The "unarmed" eye remains free from all disadvantages posed by glasses, the "armed" eye offers all advantages of the lens. The free eye [can be] fairly protected against cold, wind and foreign objects by completely and moderately squeezing it shut, [it] only needs to be opened every few moments to fully exploit the visual performance unclouded by any lens. The monocle may prove useful especially when landing. It can be spontaneously removed or dropped if there is danger of crashing and offers protection against dust-raising wind and foreign objects when landing ... My specially constructed glasses broke every time I landed the 3 times I used them, whereas my monocle never broke in 20 different landings. During my most recent landing, my monocle and that of my companion remained intact while we all three suffered serious bone breakages and concussions." (Halben, –.: Die Augen der Luftfahrer (The eyes of the aeronaut), *Jahrbuch der Wissenschaftlichen Gesellschaft für Flugtechnik* (Yearbook of the Scientific Association for Flight Technology) 2 (1914), p. 163).
13. *cf. Jahrbuch der Wissenschaftlichen Gesellschaft für Flugtechnik* (Yearbook of the Scientific Association for Flight Technology) 2 (1914), p. 218.
14. Arthur Berson was born on August 6, 1859, in Neusandez in Galicia. He studied physiogeography and meteorology. In 1890 he arrived at the Prussian Meteorological Institute in Berlin, and later followed Assmann to the aeronautic observatory in

Berlin-Reinickendorf and Lindenberg. Berson conducted about 100 balloon ascents for scientific purposes between 1891 and 1910. On July 31, 1901, he set a world record together with his colleague Süring during a height ascent in Berlin; they reached 10,800 m above sea level. This record was to remain unbroken for 15 years. In 1902, Berson accompanied von Schroetter and Zuntz on one of their balloon ascents for physiological purposes. He died on January 10, 1943, in Berlin. (*cf.* Brockhaus dictionary, volume 2, Brockhaus 1952–58, p. 40; Hörnes, H.: *Buch des Fluges* (Book of Flight), volume 3, G. Szelinski 1912, pp. 286 *et seq*).

15. Reinhard Süring was born on May 15, 1866, in Hamburg. He studied mathematics and natural sciences in Göttingen, Marburg, and Berlin. In 1889, he completed his doctorate in Berlin with the dissertation entitled *Die vertikale Temperaturabnahme in Gebirgsgegenden in ihrer Abhängigkeit von der Bewölkung* (The Vertical Decrease of Temperatures in Mountain Regions in Relation to Cloud Formation). In 1890, Süring became head of the climate department of the Prussian Meteorological Institute. From 1892 to 1901 he was employed at the meteorological observatory. During this time, the scientific ascents *Wissenschaftliche Luftfahrten* with Berson took place. In 1901 Süring returned to the Prussian Meteorological Institute, and remained there until 1909, when he became the director of the meteorological observatory in Potsdam. Except for an interruption from 1932–1945, he held this position almost until the end of his life. He authored well over 100 publications, and in 1915 edited the third edition of the standard work *Lehrbuch der Meteorologie* (Textbook of Meteorology) together with Hann. He was also the editor of the *Meteorologische Zeitschrift* (Meteorological Journal) for many years. He died on December 29, 1950, in Potsdam.

    (*cf.* Brockhaus dictionary, volume 11, Brockhaus 1952–58, p. 351; König, W.: Obituary of Professor Dr. Reinhard Süring, *Zeitschrift für Meteorologie* (Journal for Meteorology) 5 (1951), pp. 33 *et seq.*)

16. This source can be found in: *Civil-Luftschiffahrt, Wissenschaftliche Fahrten 1895–99* (Civil Airship Flight, Scientific Flights), files of the Bundesarchiv (German Federal Archive) in Freiburg, PH 18/17, unnumbered pages.

17. *cf.* de Schroetter, E.[H.]: Communication d'éxperiences physiologiques faites pendant un voyage en ballon [à] 7500 m et rapport sur différents essais concernant l'altitude de l'influence de l'air raréfi sur l'organisme humain (Presentation of physiological experiments made during a balloon voyage at [a height of] 7500 m and report on various altitude-related examinations of the impact of thin air on the human body), *Archivio Italiano de Biologia* (Italian Archive of Biology) (Pisa) 36 (1901), pp. 66–88.

    However, in a summary of the *Ballonfahrten des Deutschen Vereins für Luftschiffahrt im Jahr 1901* (Balloon Flights of the German Society for Airship Flight in 1901), Süring is mentioned neither as a passenger nor as the pilot of the balloon (*cf. Illustrierte Aeronautische Mitteilungen* (Illustrated Aeronautical News) 6 (1902), p. 28).

18. For further details compare: Hallion, –. and Tissot, –.: Recherches expérimentales sur l'influence des variations rapides d'altitude sur les phénomènes chimiques et physiques de la respiration a l'état de repos (recherches faites au cours d'une ascension en ballon) (Experimental research on the impact of rapid changes in altitude on the chemical and physical phenomena of respiration in a state of rest (research performed during a balloon flight)), *Comptes rendus hebdomadaires des séances et mémoires de la Société de Biologie* (Weekly reports of the meetings and records of the [French] Biological Society) 53 (1901a), pp. 1030–1032.

cf.: Hallion, –. and Tissot, –.: Recherches expérimentales sur l'influence des variations rapides d'altitude sur les gaz du sang et sur la pression artérielle (Experimental research on the impact of rapid changes in altitude on blood gas and arterial pressure), *Comptes rendus hebdomadaires des séances et mémoires de la Société de Biologie* (Weekly reports of the meetings and records of the [French] Biological Society) volume 53 (1901b), pp. 1032–1034.

19. *cf. Illustrierte Aeronautische Mitteilungen* (Illustrated Aeronautical News) 8 (1904), p. 18.

    In Assmann's *Wissenschaftliche Luftfahrten* (Scientific Flights) (1899), Gross also referred to the necessity of a closed basket for flights to extreme altitudes without providing more detailed construction plans (*cf.* Assmann, R., Berson, A. and Gross, H.: *Wissenschaftliche Luftfahrten*, 1 (1899), p. 141).

20. In 1918, Zuntz once more delved more deeply into this problem (*cf.* Zuntz, N.: Über künstliche Atmung mit und ohne Zufuhr hochprozentigen Sauerstoffes (On artificial respiration with and without high-percentage oxygen supply), *Deutsche militärärztliche Zeitschrift* (German Military Journal) 47 (1918), pp. 311–317).

21. The elevation record set in Berlin, 1901, by Berson and Süring, of 10,800 m above sea level, was not broken until 1926. Loewy reports this was done by Vaccice (*cf.* Loewy, A: *Physiologie des Höhenklimas* (Physiology of High Altitude Climate), Springer 1932, p. 360).

22. *cf.* the following sources: Zuntz, N.: Physiologie der Blutgase und des respiratorischen Gaswechsels (Physiology of Blood Gas and the Respiratory Gas Exchange), p. 100. In: Herrmann, L.: *Handbuch der Physiologie des Kreislaufs, der Athmung und der thierischen Wärme* (Handbook of Physiology of the Circulatory System, Respiration and Animal Heat), volume 4, F. C. W. Vogel 1882.

    cf. Edholtz, O. G. and Weiner, J. S.: *Principles and Practice of Human Physiology*, London 1981, p. 4.

23. In the archive of the Berliner Ärzteversorgung (Berlin Physicians' Pension Provision [offices]) and the Kassenärztliche Vereinigung (Association of Panel Physicians), there are no documents on the Versicherungskasse für die Ärzte Deutschlands (Insurance Fund for Germany's Physicians). Thus, the question as to the reaction of the insurance fund to this letter has remained unresolved to this day.

4 Conclusion

Based on the reconstruction of Zuntz's personal and scientific development of the investigation of the historical construction and expansion of the fields of study and institutions connected with aviation, it is possible to gain an understanding of the evolution of scientific aviation in Berlin, as shown in Figure 4.1. This diagram illustrates how closely the various organizations of the aviation association, the military and the scientific institutes in Berlin were both interrelated and mutually dependent. If the *Deutscher Verein zur Förderung der Luftschiffahrt* (German Association for the Encouragement of Aviation) was the starting point for the development of aviation in Berlin, then the decisive force behind the logistical expansion of this association lay in Assmann's plans for meteorologically equipped scientific air expeditions. It became possible to realize these plans thanks to both a donation to the association from the German Kaiser in 1892, and the support of personnel from the Berlin Aviation Department, later the Aviation Battalion (1899).

As the technical and logistical possibilities grew, the medical-physiological problems for manned high altitude flights in aeronautics increased as well. The studies in high altitude physiology started by Zuntz at the beginning of the 1890s in Lazarus's pneumatic cabinet at the *Jüdisches Krankenhaus* (Jewish Hospital), which were complemented in the middle of the decade by the laboratory and field physiological studies at the *Tierphysiologisches Institut* (Veterinary Physiological Institute) at the *Landwirtschaftliche Hochschule* (Agricultural College) and on Monte Rosa, contributed decisively to the understanding and solution of the questions that aeronautics and aviation were facing. Thus, at the turn of the twentieth century Berlin had access to resources in the scientific and technological fields of aviation that were unique in the world. Examples include the altitude record of 10,800 meters above sea level, set on the Berlin balloon trip by Berson and Süring in 1901. Such achievements, which also met with high international acclaim, were made possible only by cooperation among the abovementioned disciplines and institutions. No less important were the considerable contributions made by physicians and physiologists such as Zuntz, Loewy, and von Schroetter. It was Zuntz and his students who, by their high altitude physiology studies in high mountains, in the pneumatic chamber and in balloons, established that high altitude travel required the inhalation of artificial oxygen, and who developed the appropriate equipment together with the meteorologists Assmann, Berson, and Süring, and the *Deutsche Verein zur Förderung der Luftschiffahrt*. Shortly after Zuntz and von Schroetter had undertaken their first physiological experiments with balloons, the military physicians Koschel and Flemming were occupied with related problems in military aeronautics. From 1902 onwards, Zuntz and von

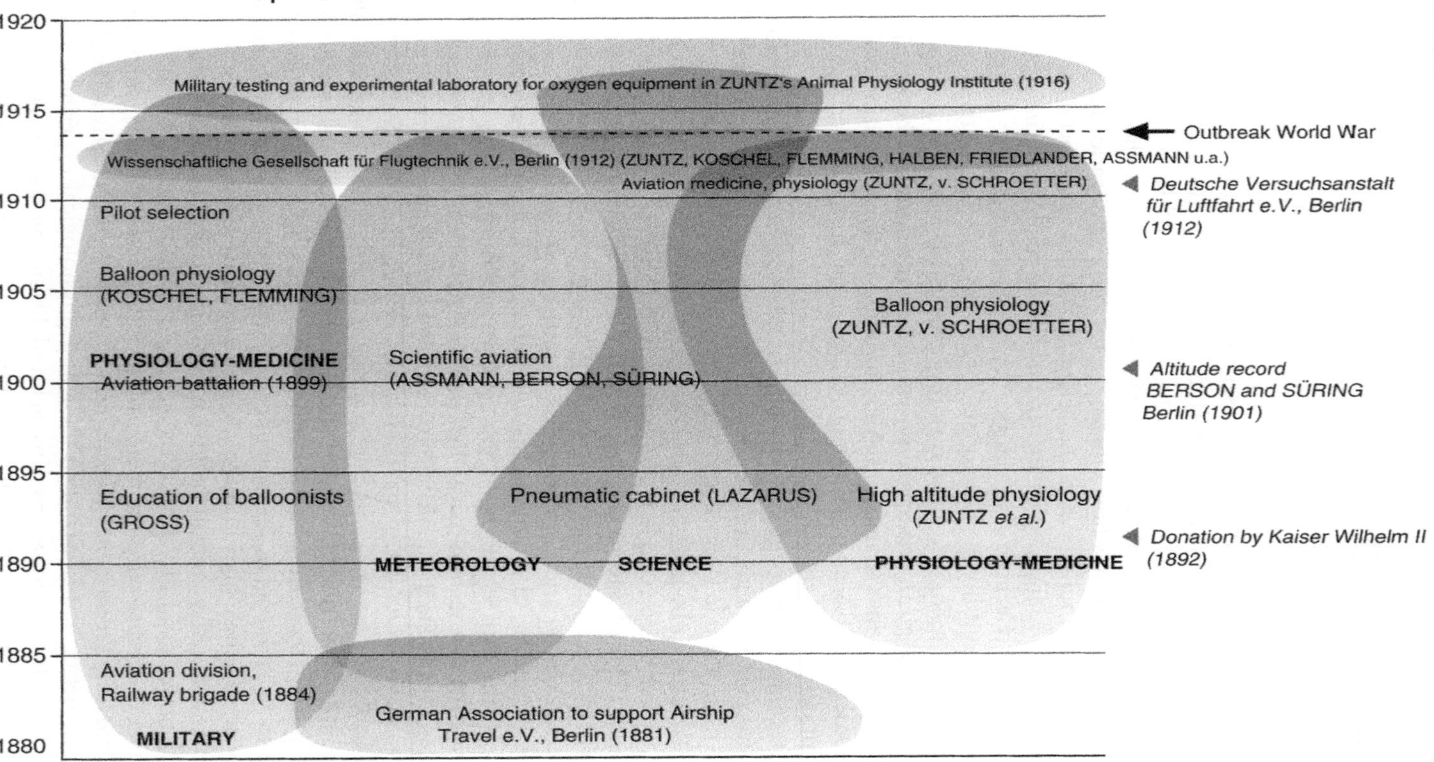

Figure 4.1 Development of scientific aviation in Berlin (1880–1918).

Schroetter were concerned primarily with questions of pilots' suitability, and determining criteria for selecting aviation personnel. Until 1912, it was primarily von Schroetter, Zuntz's student, who worked on subjects relating to the physiology of balloon travel. The strong influence of terrestrial high altitude research is clearly evident in these works as well. With the foundation of the *Deutsche Versuchsanstalt für Luftfahrt* (German Experimental Institute for Aviation) in Berlin-Adlershof and the *Wissenschaftliche Gesellschaft für Flugtechnik* (Scientific Society for Aviation Technology; known from 1914 as the *Wissenschaftliche Gesellschaft für Luftfahrt* (Scientific Society for Aviation) in Berlin in 1912, the groundwork was laid for the further development of aviation technology in Germany. The studies in the physiology of high altitudes and balloon travel that Zuntz and von Schroetter pursued between 1891 and 1912 made it possible for Germany to make an almost entirely smooth transition to similar kinds of research in aviation, since there was already an awareness of the problems in terms of medical, physiological, and psychological issues. The significance and necessity of research in aviation medicine was reflected by the establishment of a subcommittee for questions of medicine and psychology in the Scientific Society for Aviation Technology. The leading civilian and military scientists in this field in Germany were all members of this subcommittee.

In the brief period before the outbreak of World War I, during which the Scientific Society for Aviation had to discontinue its work, the subcommittee was where all research into aviation medicine in Germany was concentrated. Once again, Zuntz played a decisive role in this development. His treatise *Zur Physiologie und Hygiene der Luftfahrt*, published in 1912, marks, together with the work of von Schroetter (1912) and the French scientists Cruchet and Mouliner (1911, pp. 589–592), the final establishment of aviation medicine as an autonomous branch of research within medicine. These authors recognized that aviation posed a series of new demands in medicine, physiology, and psychology for the human organism that went beyond the high altitude phenomena that had dominated previous research in aeronautics. "The Physician in the Balloon" (Figure 4.2) evolved into the aviation physician.

With the outset of World War I, all civil research into aviation medicine was de-prioritized in favor of military research. When, in 1916, the German military leadership transferred the *Militärische Forschungs- und Versuchstelle für Sauerstoffgeräte* (Military Research and Experimental Laboratory for Oxygen Equipment) to the ground floor of Zuntz's *Tierphysiologisches Institut*, and Koschel was assigned to do research on aviation medicine there, it demonstrated that both scientists and the German state recognized the work that Zuntz had achieved in the field.

If it is the task of modern aviation medicine to "explore the mechanisms of the body's capability to adapt to the requirements of flight, to recognize the risk limits of human capability, and to find means and ways to stretch these limits as far as possible using appropriate measures" (Müller, 1973, p. 46), then Zuntz was pursuing precisely the answers to these questions as early as 1912.

Figure 4.2 "The Physician in the Balloon" in a contemporary photograph from 1909. (Bröckelmann, 1909, p. 179)

The early history of aviation medicine has been closely associated, both in Germany and abroad, with the life and influence of the Berlin physiologist Zuntz and his school. It is thus all the more difficult to understand why the leading specialists in aviation medicine in the 1930s, 1940s and later did not study Zuntz's work in adequate detail. His works regarding high altitude physiology and his publication *Zur Physiologie und Hygiene der Luftfahrt* are not cited at all in a comprehensive work on aviation medicine dating from 1950, *German Aviation Medicine: World War II* (Department of the Air Force, 2nd volume, 1950). This treatise was compiled in Heidelberg over a period of three years, at the request of the Americans, following the defeat of the German Reich in 1945, by the leading specialists in German aviation medicine.[1]

In particular, Strughold's article[2] on the historical development of aviation medicine in Germany gives the impression that any systematic research of aviation medicine had begun only as late as in 1930 – and, as we have seen, this view is not tenable. The situation is similar regarding Schubert's monograph from the mid-1930s, *Physiologie des Menschen im Flugzeug* (Schubert, 1935), in which not a single citation indicates Zuntz's early work in this field. One can only speculate about the background for such behavior. The obvious explanation of anti-Semitism, based on Zuntz's Jewish origins, is refuted by the fact that both works repeatedly refer to and examine in detail the publications of his Jewish colleague and friend Loewy (*German Aviation Medicine: World War II*, volume 1, pp. 195 *et seq.*, p. 348).

The structure of thought, the methodological approach, and the comprehensive interpretation of research results that formed the basis of Zuntz's research into high altitude physiology and aviation medicine can still be considered exemplary today as an approach to applied and integrated physiology. Zuntz practiced a physiology that was not, as Adolph aptly described it in his monograph *Physiology of Man in the Desert,* based on "armchair postulations" (Adolph, 1947, p. 5) but rather achieved by the "sweat of the brow" (Adolph, 1947, p. 5). Indeed, his emphasis on practically oriented research did not permit Zuntz to conceive, pursue, and describe his veterinary and human physiology solely from his desk. By this empirical approach to physiological research, Zuntz's scientific work represented a strong counterpoint to the well-established formal reductionism of mathematics and physics of a scientist like Frank.[3] Since the turn of the century, physiological research in Germany was influenced (not entirely positively) by this movement. While the reduction to essentials was indeed able to contribute to developing new methods, its influence on the conceptual level remained minimal. The turn to reductionist-oriented research at German universities was quite in keeping with the larger trend of European science around the turn of the century, since the *Zeitgeist* was considerably influenced by the epochal discoveries in the fields of physics and chemistry. French- and English-speaking physiologists followed this trend only in part (Rothschuh, 1969, pp. 160–162). In France and England, from the beginning, applied physiology was seen not merely as a means of applying discoveries from basic physiological research; instead, scientists were aware that applied physiology could make its own contribution to basic research. Ultimately, this enabled the concept and subject matter of applied physiology to gain a footing in the Anglo-Saxon realm starting in the 1930s (McArdle *et al.*, 1996, pp. xxvii–xl; Tipton, 2003, pp. 106, 153, 204). Hence, Zuntz's research and teaching in America in 1908 was of great importance for the evolutionary path as applied physiology emerged as an autonomous specialized discipline within physiology. Zuntz had achieved an international reputation with the methods and approaches he had developed in Berlin. His method was not yet known as applied physiology, but should be understood as precisely that, and it was appreciated as such by physiologists in the United States and England. In this way, then, Zuntz doubtlessly contributed to shaping the early developmental phase of applied physiology in the United States, especially in the field of research into the physiology of nutrition.[4] This can be seen, among other places, in the fact that the first biography and assessment of Zuntz's influence after his death – apart from the published obituary – was not published in Germany but in the United States (Forbes, 1955, pp. 3–15), and that he is still well recognized in current English textbooks (McArdle *et al.*, 1996, p. xl; Tipton, 2003, pp. 106–331).

Zuntz's physiological oeuvre was one of the most comprehensive ever published in the German-speaking world. He was one of the last scientists to have a complete and full understanding of his field. For Plesch, Zuntz was consequently "the last great physiologist of the world" (Plesch, 1949, p. 62); for Barnard and Holloszy, "Zuntz was a remarkably insightful scientist who was far ahead of his time" (Barnard and Holloszy, 2003, p. 293). Zuntz's universality went hand in hand with an intelligence that was both highly analytical and practically inclined, and was characterized by his ability to elaborate skillfully as a teacher on what he had discovered. In his publications he used an esthetically advanced German language in describing experiments and observations in the laboratory and field. His book *Höhenklima und Bergwanderungen* exemplifies this, because it is readable by the layman as well as by experts, and links the German language of medical research with literature and the true spirit of discoverers such as Alexander von Humboldt at the frontier of sciences. This connection was a special achievement of Zuntz's work that merits further study in a broader discussion of his life and work.[5] As a researcher, he pursued new themes tirelessly without abandoning earlier ones. Instead, he strove to connect everything new to his own work in ways that made sense – always in search of a "well-rounded view of nature" (Armsby, 1917, pp. 211–343). High altitude physiology and aviation medicine owe their basic tenets to this researcher. The methodological approaches and rapid practical applications mean that Zuntz is not only a model for the working philosophy of basic applied environmental and occupational physiology at the Center of Space Medicine Berlin; he is also a modern mastermind of translational or integrative-organismic physiology at the turn of the twentieth century.

Notes

1. *cf.* Kirsch, K. and Winau, R.: The early days of space medicine in Germany, *Aviation Space Environ. Med.* 57 (1986) p. 634.
2. *cf.* Strughold, H.: Development of aviation medicine in Germany. In: *German Aviation Medicine: World War II*, volume 1, Dept of the Air Force 1950, pp. 3–11.

   Years before, Strughold had already concerned himself with the history of this specialty in another article: see Strughold, H.: Geschichtliches zur Luftfahrtmedizin (Historical notes on aviation medicine), *Luftfahrtmedizinische Abhandlungen* 1 (1936–37), pp. 16–22. Although in the latter treatise Strughold emphasizes the importance of Zuntz's research for research in high altitude physiology, there is no reference to Zuntz's *Zur Physiologie und Hygiene der Luftfahrt.*
3. Otto Frank was born on June 21, 1865, in Grossumstadt, Odenwald, Germany. He completed his medical studies in 1889, and in 1891 went to work for Ludwig in Leipzig. In 1894, he was Voit's assistant in Munich. He remained there until 1905, when he accepted an appointment to the chair of physiology in Giessen. In 1908, he returned to Munich to take over the chair held by Voit (1831–1908). Frank's scientific work is marked above all by a physical and mathematical analysis of circulation. His research in that field had already received great recognition and respect during his lifetime. Frank died in 1944 in Munich (*cf.* Rothschuh, K. E.: *Geschichte der Physiologie* (A History of Physiology), Springer 1953, pp. 184–186).

   By contrast, however, the Berlin physiologist Gauer (1909–1979), remarked that Frank "tyrannized the Deutsche Physiologische Gesellschaft (German Physiological Society) for decades with his mathematical chatter" (oral statement by K. Kirsch to the author in December 1987).
4. *cf.* Armsby, H. P.: *The Nutrition of Farm Animals*, Macmillan 1917, pp. 90, 211–343.

   cf. Benedict, F.: An apparatus for studying the respiratory exchange. *Am. J. Physiol.* 24 (1909) pp. 345–374.
5. The role of the German language as part of the "glory of the German university" at the turn of the twentieth century was recently discussed by W. Frühwald; *cf.* in this context his lecture *Die Sprachen der anderen* (Learning languages to understand others) given at the Annual Meeting of the Alexander von Humboldt Foundation on June 14, 2007.

List of works by Nathan Zuntz

1867

Zuntz, Nathan. Über den Einfluss des Partiardrucks der Kohlensäure auf die Vertheilung dieses Gases im Blute. *Zentralblatt für die medizinischen Wissenschaften* 5 (1867a): 529–533.

Zuntz, Nathan. Zur Kenntniss des Stoffwechsels im Blute. *Zentralblatt für die medizinischen Wissenschaften* 5 (1867b): 801–804.

1868

Zuntz, Nathan. *Beiträge zur Physiologie des Blutes*. Dissertation, Bonn, 1868. Reviewed in *Zentralblatt für die medizinischen Wissenschaften* 6 (1868): 630–631.

Pflüger, E., and Nathan Zuntz. Über den Einfluss der Säuren auf die Gase des Blutes. *Pflügers Archiv* 1 (1868): 361–374.

1870

Zuntz, Nathan. Über die Bindung der Kohlensäure im Blute. *Berliner klinische Wochenschrift* 7 (1870): 185–186.

1871

Zuntz, Nathan. Über eine Untersuchung der Ursachen der constanten Eigenwärme bei den warmblütigen Thieren. *Berliner klinische Wochenschrift* 8 (1871): 321.

Roehrig, A., and Nathan Zuntz. Zur Theorie der Wärmeregulation und der Balneotherapie. *Pflügers Archiv* 4 (1871): 57–90.

1872

Zuntz, Nathan. Ist Kohlenoxydhämoglobin eine feste Verbindung? *Pflügers Archiv* 5 (1872): 584–588.

1873

Zuntz, Nathan. Vergleichende Untersuchung der Wirksamkeit verschiedener im Handel vorkommender Pepsinsorten. *Berliner klinische Wochenschrift* 10 (1873): 403–404.

1875

Zuntz, Nathan. Über eine Untersuchung der Gase der Lippspringer Arminius-Quelle. *Berliner klinische Wochenschrift* 12 (1875a): 225, 241.

Zuntz, Nathan. Über im hiesigen Physiologischen Institute angestellte Versuche, welche den Einfluss des veränderten Athemdrucks auf den Kreislauf des Blutes betrafen. *Berliner klinische Wochenschrift* 12 (1875b): 510.

Zuntz, Nathan. Über einen Fall von Arsenvergiftung durch einen grünen Lampenschirm. *Berliner klinische Wochenschrift* 12 (1875c): 587–588.

1876

Zuntz, Nathan. Über den Einfluss der Curarevergiftung auf den thierischen Stoffwechsel. *Pflügers Archiv* 12 (1876): 522–528.

1877

Zuntz, Nathan. Über die Respiration des Säugethier-Fötus. *Pflügers Archiv* 14 (1877): 605–627.

von Mering, J., and Nathan Zuntz. In wiefern beeinflusst Nahrungszufuhr die thierischen Oxydationsprocesse? *Pflügers Archiv* 15 (1877): 634–636.

1878

Zuntz, Nathan. Über den Einfluss der Innervation auf den Stoffwechsel ruhender Muskeln. *Berliner klinische Wochenschrift* 15 (1878a): 141.

Zuntz, Nathan. Über die Quelle und Bedeutung des Fruchtwassers. *Pflügers Archiv* 16 (1878b): 548–550.

Zuntz, Nathan. Über die Wirkungen des Stickoxydulgases. *Pflügers Archiv* 17 (1878c): 135–136.

Zuntz, Nathan. Beiträge zur Kenntniss der Einwirkungen der Athmung auf den Kreislauf. *Pflügers Archiv* 17 (1878d): 374–412.

1879

Zuntz, Nathan. Gesichtspunkte zum kritischen Studium der neueren Arbeiten auf dem Gebiete der Ernährung. *Landwirthschaftliches Jahrbuch* 8 (1879): 67–117.

1882

Zuntz, Nathan. Physiologie der Blutgase und des respiratorischen Gaswechsels. In *Handbuch der Physiologie des Kreislaufs, der Athmung und der thierischen Wärme*, edited by L. Herrmann. Vogel: Leipzig, 1882a.

Zuntz, Nathan. Über den Stoffwechsel fiebernder Tiere. *Archiv für Anatomie und Physiologie/Physiologische Abteilung* (1882b): 115–118.

Zuntz, Nathan. Über die Bedeutung der Amidsubstanzen für die thierische Ernährung. *Archiv für Anatomie und Physiologie/Physiologische Abteilung* (1882c): 424–425.

Zuntz, Nathan. Kritik an der Arbeit von Winternitz: Über die Bedeutung der Hautfunction für die Körpertemperatur und Wärmeregulation. *Archiv für Anatomie und Physiologie/Physiologische Abteilung* (1882d): 122–123.

Zuntz, Nathan. Bemerkung zu der "Entgegnung" des Hrn. Professor W. Winternitz. *Archiv für Anatomie und Physiologie/Physiologische Abteilung* (1882e): 568–570.

1883

Zuntz, Nathan, and J. von Mering. Inwiefern beeinflusst Nahrungszufuhr die thierischen Oxydationsprocesse? *Pflügers Archiv* 32 (1883): 173–221.

1884

Cohnstein, J., and Nathan Zuntz. Untersuchungen über das Blut, den Kreislauf und die Athmung beim Säugethier-Fötus. *Pflügers Archiv* 34 (1884): 173–233.

1885

Zuntz, Nathan. Über den Nährwerth der sogenannten Fleischpeptone. *Pflügers Archiv* 37 (1885): 313–324.

1886

Zuntz, Nathan. Über die Ursache der Apnoe des Fötus. *Berliner klinische Wochenschrift* 45 (1886a): 785.

Zuntz, Nathan. Über den wechselnden Gehalt des strömenden Blutes an geformten Bestandtheilen und seine Ursachen. *Berliner klinische Wochenschrift* 45 (1886b): 785.

Zuntz, Nathan. Zur Richtigstellung gegen Herrn Professor Hitzig. *Pflügers Archiv* 39 (1886c): 473–475.

Zuntz, Nathan, and J. Geppert. Über die Natur der normalen Athemreize und den Ort ihrer Wirkung. *Pflügers Archiv* 38 (1886): 337–338.

1887

Zuntz, Nathan. Über die Einwirkung des Alkohols auf den Stoffwechsel des Menschen. *Archiv für Anatomie und Physiologie/Physiologische Abteilung* (1887): 178.

Zuntz, Nathan, and C. Lehmann. Über Respiration und den Gaswechsel. *Berliner klinische Wochenschrift* 24 (1887): 428–430.

1888

Zuntz, Nathan. Über die Kräfte, welche den respiratorischen Gasaustausch in den Lungen und in den Geweben des Körpers vermitteln. *Pflügers Archiv* 42 (1888): 408–418.

Cohnstein, J., and Nathan Zuntz. Untersuchungen über den Flüssigkeits-Austausch zwischen Blut und Geweben unter verschiedenen physiologischen Bedingungen. *Pflügers Archiv* 42 (1888a): 303–341.

Cohnstein, J., and Nathan Zuntz. Weitere Untersuchungen zur Physiologie des Säugethier-Fötus. *Pflügers Archiv* 42 (1888b): 342–392.

Geppert, J., and Nathan Zuntz. Über die Regulation der Athmung. *Pflügers Archiv* 42 (1888): 189–245.

1889

Zuntz, Nathan. Über die Wärmeregulation beim Menschen, nach Versuchen des Herrn A. Loewy. *Archiv für Anatomie und Physiologie/Physiologische Abteilung* (1889): 558–561.

Zuntz, Nathan, and C. Lehmann. Untersuchungen über den Stoffwechsel des Pferdes bei Ruhe und Arbeit. *Landwirthschaftliches Jahrbuch* 18 (1889): 1–156.

1890

Zuntz, Nathan. Über die Leistungen der menschlichen Muskulatur als Arbeitsmaschine. *Naturwissenschaftliche Rundschau* 5 (1890a): 337–341.

Zuntz, Nathan. Einige Versuche zur diätetischen Verwendung des Fettes. *Therapeutische Monatshefte* 4 (1890b): 471–474.

Zuntz, Nathan. Über die Einwirkung der Muskelthätigkeit auf den Stoffverbrauch des Menschen. Nach Versuchen des Cand. med. G. Katzenstein. *Archiv für Anatomie und Physiologie/Physiologische Abteilung* (1890c): 367–376.

Zuntz, Nathan, and C. Lehmann. Remarks on the chemistry of respiration in the horse during rest and exercise. *Journal of Physiology* 11 (1890): 396–398.

1891

Zuntz, Nathan. Über die Unwirksamkeit des Curare vom Magen her. *Pflügers Archiv* 49 (1891a): 437.

Zuntz, Nathan. Bemerkungen über die Verdauung und den Nährwerth der Cellulose. *Pflügers Archiv* 49 (1891b): 477–483.

Zuntz, Nathan. Stoffwechselversuche am Pferde. *Landwirthschaftliches Jahrbuch* 20 (1891c): 267.

Zuntz, Nathan, and J. Geppert. Nochmals über den Einfluss der Muskelthätigkeit auf die Athmung. *Deutsches Archiv für klinische Medizin* 43 (1891/92): 444.

Zuntz, Nathan, and A. Magnus-Levy. Beiträge zur Kenntniss der Verdaulichkeit und des Nährwerthes des Brodes. *Pflügers Archiv* 49 (1891): 438–460.

1892

Zuntz, Nathan. Die Ernährung des Herzens und ihre Beziehung zu seiner Arbeitsleistung. *Deutsche medizinische Wochenschrift* 18 (1892a).

Zuntz, Nathan. Nachruf an Ernst von Brücke. *Berliner klinische Wochenschrift* 29 (1892b): 59–60.

Zuntz, Nathan. Die Ergebnisse der jüngsten Arbeiten über Herzthätigkeit und Kreislauf. *Deutsche Zeitschrift für Thiermedicin und vergleichende Pathologie* 18 (1892c): 261–277.

Zuntz, Nathan. Beitrag zur Physiologie des Geschmacks. *Archiv für Anatomie und Physiologie/Physiologische Abteilung* (1892d): 556.

Zuntz, Nathan. Bemerkungen zu der Abhandlung von B. Werigo: Zur Frage über die Wirkung des Sauerstoffs auf die Kohlensäureausscheidung in den Lungen. *Pflügers Archiv* 52 (1892e): 191–193.

Zuntz, Nathan. Zusatz zu meinen Bemerkungen über die Wirkung des Sauerstoffs auf die Kohlensäureausscheidung in den Lungen. *Pflügers Archiv* 52 (1892f): 198–200.

Zuntz, Nathan, and J. von Mering. Über die Stellung des Stimmbandes bei Lähmung des Nervus recurrens. *Archiv für Anatomie und Physiologie/Physiologische Abteilung* (1892): 163–164.

1893

Zuntz, Nathan. Welche Mittel stehen uns zur Hebung der Ernährung zu Gebote? *Deutsche medizinische Wochenschrift* 19 (1893a): 466–468.

Zuntz, Nathan. Über die Natur und die Bindung der Basen und Säuren im Blute. *Archiv für Anatomie und Physiologie/Physiologische Abteilung* (1893b): 556–559.

Zuntz, Nathan, and –. Vogelius. Über die Neubildung von Kohlehydraten im hungernden Organismus. *Archiv für Anatomie und Physiologie/Physiologische Abteilung* (1893): 378–380.

Lehmann, C., F. Müller, I. Munk, H. Senator, and Nathan Zuntz. Untersuchungen an zwei hungernden Menschen. *Archiv für pathologische Anatomie und Physiologie und für klinische Medicin* 131 Suppl (1893): 1–229.

1894

Zuntz, Nathan. Über eine neue Methode zur Messung der circulirenden Blutmenge und der Arbeit des Herzens. *Pflügers Archiv* 55 (1894): 521–524.

Zuntz, Nathan, J. Frenzel, and W. Loeb. Über die Bedeutung der verschiedenen Nährstoffe als Erzeuger der Muskelkraft. *Archiv für Anatomie und Physiologie/Physiologische Abteilung* (1894): 541–543.

Zuntz, Nathan, F. Lehmann, and O. Hagemann. Über Haut- und Darmathmung. *Archiv für Anatomie und Physiologie/Physiologische Abteilung* (1894): 351–355.

Lehmann, F., O. Hagemann, and Nathan Zuntz. Zur Kenntniss des Stoffwechsels beim Pferde. *Landwirthschaftliches Jahrbuch* 23 (1894): 125–165.

Loewy, A., and Nathan Zuntz. Einige Beobachtungen über die Alkalescenzveränderungen des frisch entleerten Blutes. *Pflügers Archiv* 58 (1894a): 507–510.

Loewy, A., and Nathan Zuntz. Über die Bindung der Alkalien in Serum und Blutkörperchen. *Pflügers Archiv* 58 (1894b): 511–522.

von Noorden, H., and Nathan Zuntz. Über die Einwirkung des Chinins auf den Stoffwechsel des Menschen. *Archiv für Anatomie und Physiologie/Physiologische Abteilung* (1894): 203–209.

1895

Zuntz, Nathan. Zur Kenntniss des Phlorhizindiabetes. *Archiv für Anatomie und Physiologie/Physiologische Abteilung* (1895a): 570–574.

Zuntz, Nathan. Vom Internationalen Physiologen-Congress in Bern. *Berliner klinische Wochenschrift* 32 (1895b): 863.

Zuntz, Nathan, and E. Frank. Studien über Wundheilung mit besonderer Berücksichtigung der Jodpräparate. *Dermatologische Zeitschrift* 13 (1895): 305–311.

Zuntz, Nathan, and –. Schumburg. Vorläufiger Bericht über die zur Gewinnung physiologischer Merkmale für die zulässige Belastung der Soldaten auf Märschen im thierphysiologischen Laboratorium der landwirthschaftlichen Hochschule angestellten wissenschaftlichen Versuche. *Deutsche militärärztliche Zeitschrift* 24 (1895a): 49–81.

Zuntz, Nathan, and W. Schumburg. Einwirkung der Belastung auf Stoffwechsel und Körperfunctionen des marschirenden Soldaten. *Archiv für Anatomie und Physiologie/Physiologische Abteilung* (1895b): 378–382.

Zuntz, Nathan, and P. Strassmann. Über das Zustandekommen der Athmung beim Neugeborenen und die Mittel zur Wiederbelebung Asphyktischer. *Berliner klinische Wochenschrift* 32 (1895): 264–265.

Bintz, N., and Nathan Zuntz. Über Wirkungen und Verhalten der Nosophens im Thierkörper. *Fortschritte der Medizin* 13 (1895): 517.

1896

Zuntz, Nathan. Über Prüfung des Gesetzes von der Erhaltung der Energie im Thierkörper. Bemerkungen zu den bezüglichen Versuchen des Herrn Chauveau (Paris). *Archiv für Anatomie und Physiologie/Physiologische Abteilung* (1896a): 358–361.

Zuntz, Nathan. Über die Wärmeregulirung bei Muskelarbeit. *Berliner klinische Wochenschrift* 33 (1896b): 709–710.

Zuntz, Nathan. Über die Bedeutung der Galle und des Pankreassecretes für die Resorption der Fette. *Archiv für Anatomie und Physiologie/Physiologische Abteilung* (1896c): 344–345.

Zuntz, Nathan. Zur Frage über den Umfang der zuckerbildenden Function der Leber. *Zentralblatt für Physiologie* 10 (1896d): 561–564.

Zuntz, Nathan. Das Ergebniss der Fütterungsversuche mit Rübenblättern. *Zeitschrift für die Rübenzuckerindustrie des Deutschen Reiches* (1896e): 559.

Zuntz, Nathan. Über die Rolle des Zuckers im thierischen Stoffwechsel. *Archiv für Anatomie und Physiologie/Physiologische Abteilung* (1896f): 538–542.

Zuntz, Nathan, and J. Geppert. Zur Frage von der Athemregulation bei Muskelthätigkeit. *Pflügers Archiv* 62 (1896): 295–303.

Zuntz, Nathan, and W. Schumburg. Physiologische Versuche mit Hilfe der Röntgen-Strahlen. *Archiv für Anatomie und Physiologie/Physiologische Abteilung* (1896): 550.

Schumburg, W., and Nathan Zuntz. Zur Kenntniss der Einwirkungen des Hochgebirges auf den menschlichen Organismus. *Pflügers Archiv* 63 (1896): 461–494.

1897

Zuntz, Nathan. Über den Stoffverbrauch des Hundes bei Muskelarbeit. *Pflügers Archiv* 68 (1897a): 191–211.

Zuntz, Nathan. Über den Werth der wichtigsten Nährstoffe für die Muskelarbeit nach Versuchen am Menschen, ausgeführt von Prof. Newton Heynemann aus New York. *Archiv für Anatomie und Physiologie/Physiologische Abteilung* (1897b): 535–544.

Zuntz, Nathan. Zur Pathogenese und Therapie der durch rasche Luftdruckänderungen erzeugten Krankheiten. *Fortschritte der Medizin* (1897c): 632–639.

Zuntz, Nathan. Ausstellung von Apparaten zur Messung der Athmung. *Verhandlungen des 15. Congresses für innere Medicin in Berlin vom 9.–12. Juni 1897* (1897d): 561–562.

Zuntz, Nathan, and L. Zuntz. Über die Wirkungen des Hochgebirges auf den menschlichen Organismus. *Himmel und Erde* 9 (1897): 289–316.

Zuntz, Nathan, and E. Bogdanow. Über die Fette des Fleisches. *Archiv für Anatomie und Physiologie/Physiologische Abteilung* (1897): 149–150.

Zuntz, Nathan, and J. Frentzel. Die Elementaranalyse nach gasanalytischen Methoden mit Hilfe der Berthelot'schen Bombe. *Berichte der deutschen Chemischen Gesellschaft* 30 (1897): 380–382.

Zuntz, Nathan, and A. Loewy. Über die Bedeutung des Sauerstoffmangels und der Kohlensäure für die Innervation der Athmung. *Archiv für Anatomie und Physiologie/Physiologische Abteilung* (1897): 379–390.

1898

Zuntz, Nathan. Die Aufgaben der Wissenschaft für die Förderung der Teichwirtschaft. *Fischerei-Zeitung* 1 (1898a): 624–627.

Zuntz, Nathan. Die Bedeutung der Fische in Natur und Menschenleben. *Fischerei-Zeitung* 1 (1898b): 153–156.

Zuntz, Nathan. Über die Beziehung zwischen Wärmewerth und Nährwerth der Kohlenhydrate und Fette. *Archiv für Anatomie und Physiologie/Physiologische Abteilung* (1898c): 267–270.

Zuntz, Nathan, and E. Cavazzani. Über die Zuckerbildung in der Leber. *Archiv für Anatomie und Physiologie/Physiologische Abteilung* (1898): 539–542.

Zuntz, Nathan, and O. Hagemann. Untersuchungen über den Stoffwechsel des Pferdes bei Ruhe und Arbeit. *Landwirtschaftliche Jahrbücher* 27 Ergänzungsband 3 (1898): 1–338 mit Anhang.

Zuntz, Nathan, and K. Knauthe. Gesichtspunkte zur Beurteilung praktischer Fütterungsversuche an Fischen. *Fischerei-Zeitung* 1 (1898a): 480–483.

Zuntz, Nathan, and K. Knauthe. Über die Verdauung und den Stoffwechsel der Fische. *Archiv für Anatomie und Physiologie/Physiologische Abteilung* (1898b): 149–154.

Zuntz, Nathan, and O. Polimanti. Über die Fettbildung aus Eiweiss. *Archiv für Anatomie und Physiologie/Physiologische Abteilung* (1898): 261–262.

Tangl, F., and Nathan Zuntz. Über die Einwirkung der Muskelarbeit auf den Blutdruck. *Pflügers Archiv* 70 (1898): 544–558.

1899

Zuntz, Nathan. Leistungen und Aufgaben der Tierphysiologie im Dienste der Landwirtschaft. Festrede zur Feier des Geburtstages Sr. Majestät des Kaisers am 26.1.1899 zu Berlin. Berlin: Parey, 1899a.

Zuntz, Nathan. Zwei Apparate zur Dosirung und Messung menschlicher Arbeit. *Archiv für Anatomie und Physiologie/Physiologische Abteilung* (1899b): 372.

Zuntz, Nathan. Über eine Methode zur Aufsammlung und Analyse von Darm- und Gährungsgasen. *Archiv für Anatomie und Physiologie/Physiologische Abteilung* (1899c): 579.

1900

Zuntz, Nathan. Nach von Herrn Ossow angestellten Versuchen: Über die Einwirkung der Galle auf die Verdauungsvorgänge. *Archiv für Anatomie und Physiologie/ Physiologische Abteilung* (1900a): 380–382.

Zuntz, Nathan. Über die Herkunft der flüchtigen Fettsäuren in der Butter. *Archiv für Anatomie und Physiologie/Physiologische Abteilung* (1900b): 382–384.

Zuntz, Nathan. Über den Kreislauf der Gase im Wasser. *Archiv für Anatomie und Physiologie/Physiologische Abteilung* Suppl. (1900c): 311–315.

Zuntz, Nathan, and K. Knauthe. Bemerkungen zu den Fütterungsversuchen des Herrn von Schrader auf Sunder und den anschließenden Karpfenanalysen von Prof. Dr. Franz Lehmann. *Fischerei-Zeitung* 3 (1900): 194–197.

Zuntz, Nathan, and S. Kostin. Versuche, die Methode des Nachweises von Kohlenoxyd in der Luft zu verfeinern. *Archiv für Anatomie und Physiologie/Physiologische Abteilung* Suppl. (1900): 315–316.

Zuntz, Nathan, and L. Sternberg. Über den Einfluss des Labfermentes auf die Verdauung des Milcheiweisses. *Archiv für Anatomie und Physiologie/Physiologische Abteilung* (1900): 362–363.

1901

Zuntz, Nathan. Über die Bedeutung der verschiedenen Nährstoffe als Erzeuger der Muskelkraft. *Pflügers Archiv* 83 (1901a): 557–571.

Zuntz, Nathan. Bemerkungen zur therapeutischen Verwertung der Muskelthätigkeit. *Zeitschrift für diätetische und physikalische Therapie* (1901b): 99–103.

Zuntz, Nathan. Sind kalorisch äquivalente Mengen von Kohlehydraten und Fetten für Mast und Entfettung gleichwerthig?. *Berliner klinische Wochenschrift* 39 (1901c): 26–27.

Zuntz, Nathan. Gymnastik. In *Handbuch der Physikalischen Therapie,* edited by A. Goldschneider, and P. Jacob. Leipzig: Thieme, 1901d.

Zuntz, Nathan. Über den Stoffwechsel des Pferdes. *Die Landwirtschaftlichen Versuchs-Stationen* 55 (1901e): 117–128.

Zuntz, Nathan, and W. Schumburg. Studien zu einer Physiologie des Marsches. In *Bibliothek v. Coler. Sammlung von Werken aus dem Bereiche der medicinischen Wissenschaften*, 6. Band, edited by O. Schjerning. Berlin: Hirschwald, 1901.

1902

Zuntz, Nathan. Über neuere Nährpräparate in physiologischer Hinsicht. *Deutsche Pharmazeutische Gesellschaft* 12 (1902a): 363.

Zuntz, Nathan. Abwehr gegen Herrn Prausitz. *Zeitschrift für Biologie* 43 (1902b): 113–116.

Zuntz, Nathan. Der Mensch als calorische Maschine und der zweite Hauptsatz. *Physikalische Zeitschrift* (1902c): 184–185.

Zuntz, Nathan, and O. Hagemann. Bemerkungen zu vorstehender Kritik Pfeiffer's. *Die Landwirtschaftlichen Versuchs-Stationen* 56 (1902): 289–292.

Zuntz, Nathan, and H. von Schroetter. Über zwei Ballonfahrten, bei welchen die Hauptaufmerksamkeit dem Studium der Athmung gewidmet war. *Archiv für Anatomie und Physiologie/Physiologische Abteilung* Suppl. (1902): 436.

Zuntz, Nathan. Zur therapeutischen Verwerthung der Muskelthätigkeit. *Zeitschrift für diätetische und physikalische Therapie* 5 (1902): 99.

von Schroetter, H., and Nathan Zuntz. Ergebnisse zweier Ballonfahrten zu physiologischen Zwecken. *Pflügers Archiv* 92 (1902): 479–520.

1903

Zuntz, Nathan. Über die Wärmeregulation bei Muskelarbeit. *Deutsche Medicinal Zeitung* 25 (1903a): 265–267.

Zuntz, Nathan. Eine Methode zur Schätzung des Eiweiss- und Fettgehaltes im lebenden Thierkörper. *Archiv für Anatomie und Physiologie/Physiologische Abteilung* (1903b): 205–208.

Zuntz, Nathan. Über Beziehungen zwischen Körpergrösse und Stoffverbrauch beim Gehen. *Archiv für Anatomie und Physiologie/Physiologische Abteilung* (1903c): 380.

Zuntz, Nathan. Einfluss der Geschwindigkeit, der Körpertemperatur und der Übung auf den Stoffverbrauch bei Ruhe und bei Muskelarbeit. *Pflügers Archiv* 95 (1903d): 192–208.

Zuntz, Nathan. Über die Frage der Sauerstoffaufspeicherung in den thierischen Geweben. *Archiv für Anatomie und Physiologie/Physiologische Abteilung* Suppl. (1903e): 492–498.

Zuntz, Nathan. Die Merkmale der Ermüdung. *Die Umschau* 7 (1903f): 741–744.

Loewy, A., and Nathan Zuntz. Über den Mechanismus der Sauerstoffversorgung des Thierkörpers. *Deutsche medizinische Wochenschrift* 29 (1903): 365–366.

1904

Zuntz, Nathan. Über den Mechanismus der Zuckerbildung des hepatischen Glykogens. *Archiv für Anatomie und Physiologie/Physiologische Abteilung* (1904a): 220–225.

Zuntz, Nathan. Zur Technik der künstlichen Geflügelzucht. *Deutsche Landwirtschaftliche Presse* 32 (1904b): 281–282.

Zuntz, Nathan. Zum Ausbau der fischereilichen Wissenschaft. *Fischerei-Zeitung* 7 (1904c): 581–583.

Durig, A., and Nathan Zuntz. Beiträge zur Physiologie des Menschen im Hochgebirge. *Archiv für Anatomie und Physiologie/Physiologische Abteilung* Suppl. (1904a): 417–456.

Durig, A., and Nathan Zuntz. Bericht über einige Untersuchungen zur Physiologie des Menschen im Hochgebirge. In *Sitzungsbericht der Königlichen Preussischen Akademie der Wissenschaften*, Berlin: Königliche Akademie der Wissenschaft (1904b) pp. 1041–1042.

Loewy, A., and Nathan Zuntz. Über den Mechanismus der Sauerstoffversorgung des Körpers. *Archiv für Anatomie und Physiologie/Physiologische Abteilung* (1904): 166–216.

1905

Zuntz, Nathan. Über den Winterschlaf der Tiere. *Naturwissenschaftliche Wochenschrift* 4 (1905a): 145–148.

Zuntz, Nathan. Zur Kritik der Blutkörperchenzählung. *Archiv für Anatomie und Physiologie/Physiologische Abteilung* (1905b): 441–444.

Zuntz, Nathan. Über die Wirkungen des Sauerstoffmangels im Hochgebirge. *Archiv für Anatomie und Physiologie/Physiologische Abteilung* Suppl. (1905c): 416–430.

Zuntz, Nathan. Zur Bedeutung des Blinddarmes für die Verdauung beim Kaninchen. *Archiv für Anatomie und Physiologie/Physiologische Abteilung* (1905d): 403–412.

Zuntz, Nathan. Besonderheiten eines von ihm nach dem Princip von Regnault und Reiset gebauten Respirationsapparates. *Archiv für Anatomie und Physiologie/ Physiologische Abteilung* Suppl. (1905e): 431–434.

1906

Zuntz, Nathan. Gesichtspunkte für die Mästung unserer Haustiere. *Illustrierte Landwirtschaftliche Zeitung* 26 (1906a): 52–55.

Zuntz, Nathan. Wissenschaftliche und praktische Studien zur Teichwirtschaft. *Fischerei-Zeitung* 9 (1906b): 401–406.

Zuntz, Nathan. Wissenschaftliche und praktische Studien zur Teichwirtschaft. *Mitteilungen des Fischerei-Vereins für die Provinz Brandenburg* (1906c): 48–57.

Zuntz, Nathan. Die Bedeutung der "Verdauungsarbeit" im Gesamtstoffwechsel des Menschen und der Tiere. *Naturwissenschaftliche Rundschau* 21 (1906d): 501–503.

Zuntz, Nathan. Das Tierphysiologische Institut. In *Die Königliche Landwirtschaftliche Hochschule in Berlin. Festschrift zur Feier des 25 jährigen Bestehens*, edited by L. Wittmack. Berlin: Parey, 1906e.

Zuntz, Nathan. Teichmeliorationen und Fütterung in Teichen, erläutert an Ergebnissen eigener Untersuchungen. *Jahrbuch der Deutschen Landwirtschafts-Gesellschaft* 21 (1906f): 352–358.

Zuntz, Nathan. Die Tiere im Dienste der Wissenschaft und der Heilkunde. In *Mensch und Erde*, 2 Vol., edited by H. Krämer. Berlin-Leipzig-Wien-Stuttgart: Deutsches Verlagshaus Bong, 1906g.

Zuntz, Nathan, A. Loewy, F. Müller, and W. Caspari. *Höhenklima und Bergwanderungen in ihrer Wirkung auf den Menschen.* Berlin: Deutsches Verlagshaus Bong, 1906.

Loewy, A., and Nathan Zuntz. Die physikalischen Grundlagen der Sauerstofftherapie. In *Handbuch der Sauerstofftherapie*, edited by M. Michaelis. Berlin: Hirschwald, 1906.

1907

Zuntz, Nathan. Das neue internationale Institut für Hochgebirgsforschungen, Laboratorio scientifico Angelo Mosso. *Internationale Wochenschrift* 1 (1907): 1180–1186.

Zuntz, Nathan, R. Ostertag, –. Strigel, and –. Hempel. Zur Milchsekretion des Schweines. *Zentralblatt für Physiologie* 21 (1907): 609–610.

Ostertag, R., and Nathan Zuntz. Studien über die Lecksucht der Rinder. *Zeitschrift für Infektionskrankheiten, parasitäre Krankheiten und Hygiene der Haustiere* 2 (1907): 409–424.

1908

Zuntz, Nathan. Über künstliche Parthenogenese nach eigenen Beobachtungen im Laboratorium von Jacques Loeb, Berkeley. *Zentralblatt für Physiologie* 22 (1908a): 708–709.

Zuntz, Nathan. Zur Erklärung der spezifisch dynamischen Wirkung der Eiweißstoffe. *Zentralblatt für Physiologie* 22 (1908b): 67–69.

Zuntz, Nathan. Die Kraftleistungen des Tierkörpers. Festrede zur Feier des Geburtstages Sr. Majestät des Kaisers am 26. Januar 1908 in Berlin. Berlin: Parey, 1908c.

Zuntz, Nathan, and C. Oppenheimer. Über verbesserte Modelle eines Respirationsapparates nach dem Prinzip von Regnault und Reiset. *Biochemische Zeitschrift* 14 (1908): 361–406.

Ostertag, R., and Nathan Zuntz. Untersuchungen über die Milchsekretion und die Ernährung der Ferkel. *Landwirthschaftliches Jahrbuch* 37 (1908a): 201–260.

Ostertag, R., and Nathan Zuntz. Nachtrag zu den Untersuchungen über die Milchsekretion des Schweines. *Landwirthschaftliches Jahrbuch* 37 (1908b): 1051–1052.

1909

Zuntz, Nathan, and J. Plesch. Methode zur Bestimmung der zirkulierenden Blutmenge beim lebenden Tiere. *Biochemische Zeitschrift* 11 (1909a): 47–60.

Zuntz, Nathan. Das neue Tierphysiologische Institut der Königl. Landwirtschaftlichen Hochschule. *Landwirthschafliches Jahrbuch* 38 Ergänzungsband 5 (1909b): 473–491.

Zuntz, Nathan. Die Verhütung der Erkrankungen nach Aufenthalt in komprimierter Luft nach J. U. Haldane und –.Boycott. *Fortschritte der Medizin* 27 (1909c): 561–563.

Zuntz, Nathan. Zu Darwins 100. Geburtstag. *Zentralblatt für Physiologie* 23 (1909d): 199–200.
Zuntz, Nathan. Charles Darwin. *Medizinische Klinik* 5 (1909e): 298.
Zuntz, Nathan. Beobachtungen der Wirkung des Höhenklimas. *Medizinische Klinik* 5 (1909f): 396–399.
Zuntz, Nathan. Über künstliche Parthenogenese nach eigenen Beobachtungen im Laboratorium von Jacques Loeb. *Zentralblatt für Physiologie* 22 (1909g): 710.
Zuntz, Nathan, and A. Loewy. *Lehrbuch der Physiologie des Menschen.* Leipzig: Vogel, 1909.
Frentzel, J. and Nathan Zuntz. Ernährung und Volksnahrungsmittel. Berlin: Teubner, 1909.

1910

Zuntz, Nathan. Klima und Mensch. In *Bericht des Vorstandes der Gesellschaft Urania* 1910a: 10.
Zuntz, Nathan. Ein Universalrespirationsapparat für grosse Tiere. *Zentralblatt für Physiologie* 24 (1910b): 809–810.
Zuntz, Nathan. Ergebnisse von Gaswechseluntersuchungen an Wiederkäuern. *Zentralblatt für Physiologie* 24 (1910c): 810.
Zuntz, Nathan. Verdauungsarbeit und spezifisch-dynamische Wirkung der Nahrungsmittel. *Medizinische Klinik* 6 (1910d): 351–354.
Zuntz, Nathan. Über die chemischen Sinne. *Chemisch-technisches Repetitorium* 34 (1910e): 146.
Zuntz, Nathan. Respiratorischer Stoffwechsel und Atmung während der Gravidität. *Archiv für Gynäkologie* (1910f): 452.
Zuntz, Nathan. Beruht die Stoffwechselsteigerung nach Kohlehydratzufuhr auf Verdauungsarbeit? *Zentralblatt für Physiologie* 24 (1910g): 714–715.
Zuntz, Nathan. Der Respirationsapparat des Tierphysiologischen Instituts. *Jahrbuch der Deutschen Landwirtschafts-Gesellschaft* 27 (1910h): 180.
Zuntz, Nathan. Sportliche und hygienische Eindrücke einer Amerikareise. *Körperkultur. Künstlerische Monatsschrift für Hygiene und Sport* V (1910i): 9.

1911

Zuntz, Nathan. Einleitung zum Kapitel Gaswechsel and Betrachtungen über die Beziehungen zwischen Nährstoffen und Leistungen des Körpers. In *Handbuch der Biochemie des Menschen und der Tiere*, 4. Band edited by C. Oppenheimer. Jena: Fischer, 1911a.
Zuntz, Nathan. Gärungsprozesse bei der Verdauung der Wiederkäuer. *Medizinische Klinik* 26 (1911b): 1028.
Zuntz, Nathan. Einige Ergebnisse von Gaswechselversuchen an Wiederkäuern. *Archivo di Fisiologia* 9 (1911c): 236.
Zuntz, Nathan. Physiologische und hygienische Wirkungen der Seereisen. *Zeitschrift für Balneologie, Klimatologie und Kurort-Hygiene* 4 (1911d): 165–168.
Zuntz, Nathan. Der neue Respirationsapparat in der Berliner Landwirtschaftlichen Hochschule. *Umschau* 15 (1911e): 92–97.

Zuntz, Nathan. Nachruf für Christian Bohr. *Zentralblatt für Physiologie* 25 (1911f): 755f.

Zuntz, Nathan. Über die Wechselwirkung der Organe im menschlichen Körper. *Naturwissenschaftliche Wochenschrift* 10 (1911g): 356–361.

Zuntz, Nathan. Zur Methodik der Klimaforschung. *Medizinische Klinik* 22 (1911h): 854–855.

Zuntz, Nathan. Beiträge zur Physiologie der Klimawirkungen. *Zentralstelle für Balneologie* 1 (1911i): 1–9.

Zuntz, Nathan. Künstliches Klima für Versuche am Menschen. *Zeitschrift für Balneologie, Klimatologie und Kurort-Hygiene* 3 (1911j): 643–644.

Zuntz, Nathan. Leistungsfähigkeit und Sauerstoffbedarf bei maximaler Arbeit. *Medizinische Klinik* 1 (1911k): 21–23.

Zuntz, Nathan. Zur Physiologie der Spiele und Leibesübungen. *Blätter für Volksgesundheitspflege* 11 (1911l): 241–246.

Zuntz, Nathan. Angelo Mosso einige Worte des Gedenkens. *Zentralblatt für Physiologie* 25 (1911m): 93–94.

Zuntz, Nathan. Angelo Mosso. *Deutsche medizinische Wochenschrift* 27 (1911n): 31.

Zuntz, Nathan. Bibliographie des gesamten Sports. Internationale Hygiene-Exhibition Dresden 1911. *Verlag der Internationalen Hygiene-Ausstellung*. Leipzig: Veit & Comp, 1911o.

Zuntz, Nathan, C. Brahm, and A. Mallwitz. *Sonderkatalog der Abteilung Sportausstellung der Internationalen Hygieneausstellung*. Internationale Hygiene Exhibition Dresden 1911. *Verlag der Internationalen Hygiene-Ausstellung*, Dresden: 1911.

Zuntz, Nathan, and W. Crohnheim. Die Bedeutung der Naturnahrung für die Ernährung der Teichfische. In: *Aus deutscher Fischerei*. Neudamm: Fischerei-Verein für die Provinz Brandenburg, 1911.

Zuntz, Nathan, and A. Loewy. Remarques sur les derniers travaux de M. Tissot relatifs à la genèse du mal de montagne. *Journal de Physiologie* 13 (1911): 1.

Caspari, W., and Nathan Zuntz. Stoffwechsel. In *Handbuch der physiologischen Methodik*. 1. Band, 3. Abteilung, edited by R. Tigerstedt. Berlin: Hirzl, 1911.

Markoff, I., F. Müller, and Nathan Zuntz. Neue Methode zur Bestimmung der im menschlichen Körper umlaufenden Blutmenge. *Zeitschrift für Balneologie* 4 (1911): 373, 409, 441.

1912

Zuntz, Nathan. Zur Physiologie der Spiele und Leibesübungen. *Körper und Geist* 20 (1912a): 145–155.

Zuntz, Nathan. Physiologische Bedeutung des nach Witterung und Boden wechselnden Mineralgehaltes der Futtermittel. *Jahrbuch der Deutschen Landwirtschafts-Gesellschaft* 27 (1912b): 570–593.

Zuntz, Nathan. Gas- und Stoffwechsel bei eiweißarmer Ernährung. *Zentralblatt für Physiologie* 26 (1912c): 725–730.

Zuntz, Nathan. Neue Methode zur Bestimmung der umlaufenden Blutmenge im lebenden Körper. *Zentralblatt für Physiologie* 26 (1912d): 87–89.

Zuntz, Nathan. Körperkultur und Sport. *Salonblatt* (1912e): 1716–1721.

Zuntz, Nathan. Zur Physiologie und Hygiene der Luftfahrt. *Luftfahrt und Wissenschaft* 3 (1912f): 1–67.

Zuntz, Nathan. Beiträge zur Physiologie der Klimawirkungen. *Zeitschrift für Balneologie, Klimatologie und Kurort-Hygiene* 4 (1912g): 523–525.

Zuntz, Nathan. Zur Klärung der Versuchsergebnisse von Chauveau über die Minderwertigkeit der Fette Kohlenhydraten gegenüber als Energiespender bei Muskelarbeit. *Biochemische Zeitschrift* 44 (1912h): 290–291.

Zuntz, Nathan. Vorführung des Respirationsapparates des Tierphysiologischen Instituts der Kgl. Landwirtschaftlichen Hochschule Berlin. *Jahrbuch der Deutschen Landwirtschaftlichen Gesellschaft* 27 (1912i): 180–188.

Zuntz, Nathan. Gibt es einen nennenswerten intrapulmonalen Sauerstoffverbrauch? *Zeitschrift für klinische Medizin* 74 (1912j): 347–351.

Zuntz, Nathan. Über die Einwirkungen des Lichts auf den menschlichen Organismus. *Medizinische Klinik* 8 (1912k): 587.

Zuntz, Nathan. Verdauungsarbeit und spezifisch-dynamische Wirkung der Nährstoffe. *15. Internationaler Hygiene-Kongress zu Washington* 2 (1912): 390–394.

Zuntz, Nathan, R. von der Heide, and W. Klein. Zum Studium der Respiration und des Stoffwechsels der Wiederkäuer. *Landwirtschaftliche Versuchsstationen* 79/80 (1912): 781.

Durig, A., H. von Schroetter, and Nathan Zuntz. Über die Wirkung intensiver Belichtung auf den Gaswechsel und die Atemmechanik. *Biochemische Zeitschrift* 39 (1912): 469–495.

Durig, A., and Nathan Zuntz. Zur physiologischen Wirkung des Seeklimas. *Biochemische Zeitschrift* 39 (1912a): 428–434.

Durig, A., and Nathan Zuntz. Beobachtungen über die Wirkung des Höhenklimas auf Teneriffa. *Biochemische Zeitschrift* 39 (1912b): 435–468.

1913

Zuntz, Nathan. Über einige Arbeiten zur Physiologie der Verdauung und des Stoffwechsels. *Berliner klinische Wochenschrift* 50 (1913a): 2132–2135.

Zuntz, Nathan. Vergleich der Verdauungs- und Ernährungsverhältnisse des Pferdes und des Rindes. *Arbeiten der Landwirtschafts-Kammer für die Provinz Hannover* 34 (1913b): 154–156.

Zuntz, Nathan. Einfluss chronischer Unterernährung auf den Stoffwechsel. *Biochemische Zeitschrift* 55 (1913c): 341–354.

Zuntz, Nathan. Zur Kenntnis der Einwirkung des winterlichen Höhenklimas auf den Menschen. *Zeitschrift für Balneologie, Klimatologie und Kurort-Hygiene* 6 (1913d): 509–511.

Zuntz, Nathan. Die Einwirkung der Salze und ihrer Ionen auf die Oxydationsprozesse in unserem Körper. *Zentralstelle für Balneologie* 2 (1913e): 333–334.

Zuntz, Nathan. Einiges über die Teichdüngungsstation Sachsenhausen-Oranienburg. *Fischerei-Zeitung* 16 (1913f): 347–354.

Zuntz, Nathan. Die Beziehungen der Mikroorganismen zur Verdauung. *Die Naturwissenschaften* 1 (1913g): 7–11.

Zuntz, Nathan. Antwort auf Anfrage über Teichdüngung. *Fischerei-Zeitung* 16 (1913h): 597.

Zuntz, Nathan, and A. Loewy. *Lehrbuch der Physiologie des Menschen*. 2. Auflage. Leipzig: Vogel, 1913.

Durig, A., and Nathan Zuntz. Die Nachwirkung der Arbeit auf die Respiration in grösseren Höhen. *Skandinavisches Archiv für Physiologie* 29 (1913): 133–148.

Voeltz, W., and Nathan Zuntz. Untersuchungen über den Nährwert der Kartoffelschlempe und ihres Ausgangsmaterials. Leitende Gesichtspunkte. *Landwirthschaftliches Jahrbuch* 44 (1913): 681–684.

von der Heide, R., W. Klein, and Nathan Zuntz. Untersuchungen über den Nährwert der Kartoffelschlempe und ihres Ausgangsmaterials. Respirations- und Stoffwechselversuche am Rinde. Über den Nährwert der Kartoffelschlempe und ihrer Ausgangsmaterialien. *Landwirthschaftliches Jahrbuch* 44 (1913): 765–832.

1914

Zuntz, Nathan. Meinungsaustausch. Wie begegnen wir dem Ausfall der ausländischen Futtermittel? *Deutsche landwirtschaftliche Tierzucht* 37 (1914a): 428.

Zuntz, Nathan. Physiologie des Sportes und der Leibesübungen. *Himmel und Erde* 26 (1914b): 439–453.

Zuntz, Nathan. Zur Kenntnis des Stoffwechsels und der Atmung von Wassertieren. *Berliner klinische Wochenschrift* 51 (1914c): 621–622.

Zuntz, Nathan. Erfahrungen und Gesichtspunkte für das Studium des tierischen Stoffwechsels mit Hilfe von Respirationsapparaten. *Internationale Agrartechnische Rundschau* 5 (1914d): 465–476.

Zuntz, Nathan. Neuere Forschungen, betreffend die Verfütterung zuckerhaltiger Nährmittel. *Zeitschrift des Vereins der Deutschen Zuckerindustrie* 64 (1914e): 485–498, 643–656, 658–659, 667–668.

Nicolai, G. F., and Nathan Zuntz. Füllung und Entleerung des Herzens bei Ruhe und Arbeit. *Berliner klinische Wochenschrift* 51 (1914): 821–824.

1915

Zuntz, Nathan. Tier-Ernährung und Fütterung. In *Volksernährung im Kriege*. Berlin: Hobbing, 1915a.

Zuntz, Nathan. Einwirkung der Kriegslage auf die Teichwirtschaft. *Fischerei-Zeitung* 18 (1915b): 217–218.

Zuntz, Nathan. Ausnutzbarkeit eines neuartigen Vollbrotes. *Berliner klinische Wochenschrift* 52 (1915c): 91–92.

Zuntz, Nathan. Über Ernährungsfragen. *Berliner klinische Wochenschrift* 52 (1915d): 507–508.

Zuntz, Nathan. Über die Ausnutzung des Strohmehls. *Berliner klinische Wochenschrift* 52 (1915e): 586–587.

Zuntz, Nathan. Unsere Ernährung im Kriege. *Neue Rundschau* 26 (1915f): 405–411.

Zuntz, Nathan. Die Einwirkung der Kriegslage auf die Teichwirtschaft. *Fischerei-Zeitung* 18 (1915g): 217–218.

Zuntz, Nathan. Einfluss des Krieges auf Ernährung und Gesundheit des deutschen Volkes. *Medizinische Klinik* 43/44 (1915h): 1–24.

Zuntz, Nathan. Krieg und die deutsche Landwirtschaft. *Arbeiten der Landwirtschafts-Kammer für die Provinz Brandenburg* 38 (1915i): 22–35.

Zuntz, Nathan. Nachrufe. *Berliner klinische Wochenschrift* 52 (1915j): 1291–1292.
Zuntz, Nathan. Zur Physiologie der Schweisssekretion. *Berliner klinische Wochenschrift* 52 (1915k): 1292.
Zuntz, Nathan. Gesichtspunkte zur Bemessung des Umfanges des Zuckerrübenanbaues und zur Hebung des Zuckerverbrauches. *Correspondenz der Vereinigung zur Hebung des Zuckerverbrauches e.V.* 18 (1915l): 429–438.
Zuntz, Nathan. Ersatzfuttermittel. *Mitteilungen der Deutschen Landwirtschafts-Gesellschaft* 30 (1915m): 226–228.
Zuntz, Nathan, and R. von der Heide. Untersuchungen am Schaf über die Verdaulichkeit und Verwertung des Birkenholzschliffes. *Sitzungsberichte der Königlichen Preussischen Akademie der Wissenschaften* 41 (1915): 695.
Haberlandt, G., and Nathan Zuntz. Über die Verdaulichkeit der Zellwände des Holzes. *Sitzungsberichte der Kgl. Preussischen Akademie der Wissenschaften* 41 (1915): 686–708.
Hehl, H., and Nathan Zuntz. Die fettarme Küche. *Flugschriften zur Volksernährung* (1915): 1–16.
Kuczynski, R., and Nathan Zuntz. *Unsere bisherige und künftige Ernährung im Kriege.* Braunschweig: Vieweg, 1915.

1916

Zuntz, Nathan. Über Knochenweiche. *Deutsche Landwirtschaftliche Presse* 43 (1916a): 471.
Zuntz, Nathan. Zur Einwirkung der Kriegslage auf die Teichwirtschaft. *Fischerei-Zeitung* 19 (1916b): 127–129.
Zuntz, Nathan. Bedeutung des Sportes in der zukünftigen Jugenderziehung. *Blätter für Volksgesundheitspflege* 16 (1916c): 60–64.
Zuntz, Nathan. Ergänzung des Futtereiweißes durch Ammonsalze. *Illustrierte Landwirtschaftliche Zeitung* 36 (1916d): 13.
Durig, A., C. Neuberg, and Nathan Zuntz. Ergebnisse der unter Führung von Prof. Pannwitz ausgeführten Teneriffaexpedition 1910. *Biochemische Zeitschrift* 72 (1916): 253–284.
Loewy, A., and Nathan Zuntz. Einfluss der Kriegskost auf den Stoffwechsel. *Berliner klinische Wochenschrift* 53 (1916): 825–837.
von der Heide, R., M. Steuber, and Nathan Zuntz. Untersuchungen über den Nährwert des Strohstoffs. *Biochemische Zeitschrift* 73 (1916): 161–192.

1917

Zuntz, Nathan. Sportlaboratorien. *Die Umschau* 21 (1917a): 206.
Zuntz, Nathan. Die Aufgaben des Arztes beim gegenwärtigen Stande der Ernährungsfragen. *Deutsche medizinische Wochenschrift* 43 (1917b): 1409–1412.
Zuntz, Nathan. Zur Vervollkommnung des Leims als Eiweissersatz. *Berliner klinische Wochenschrift* 54 (1917c): 540.
Zuntz, Nathan. Einfluß des Krieges auf Ernährung und Gesundheit des Deutschen Volkes. *Medizinische Klinik* 11 (1917d): 1176, 1204.

Zuntz, Nathan. Bemerkungen zu der von Gad-Andresen beschriebenen "neuen" Methode zur Bestimmung von Kohlenoxyd im Blute. *Biochemische Zeitschrift* 78 (1917e): 231–232.

Zuntz, Nathan. *Gesichtspunkte zur Anpassung des Landwirtes an die Kriegslage.* Berlin: Parey, 1917e.

Brahm, C., and Nathan Zuntz. Wert der Abbauprodukte des Horns als Nähr- und Genussmittel. *Deutsche medizinische Wochenschrift* 43 (1917): 1061–1062.

Brahm, C., R. von der Heide, M. Steuber, and Nathan Zuntz. Untersuchungen über den Einfluß mechanischer und chemischer Einwirkungen auf den Nährwert von Futterstoffen. *Biochemische Zeitschrift* 79 (1917): 389–441.

von der Heide, R., M. Steuber, and Nathan Zuntz. Untersuchungen über den Nährwert des Strohstoffs. *Biochemische Zeitschrift* 73 (1917): 161–192.

1918

Zuntz, Nathan. *Ernährung und Nahrungsmittel.* Leipzig-Berlin: Teubner, 1918a.

Zuntz, Nathan. Über künstliche Atmung mit und ohne Zufuhr hochprozentigen Sauerstoffes. *Deutsche militärärztliche Zeitschrift* 47 (1918b): 311–317.

Zuntz, Nathan. Ernährung und Nahrungsmittel. *Aus Natur und Geisteswelt* 19 (1918c): 1–136.

Zuntz, Nathan. Hugo Thiel. *Deutsche Landwirtschaftliche Presse* 45 (1918d): 161–163.

Zuntz, Nathan. Bilanzbestimmung des tierischen Stoffwechsels mit Hilfe der kalometrischen Bombe. *Berliner klinische Wochenschrift* 55 (1918e): 393–395.

Zuntz, Nathan. Bedeutung der Lupinen für die Fütterung der landwirtschaftlichen Nutztiere nach dem Kriege. *Mecklenburgische Landwirtschaftliche Wochenschrift* (1918f): 403–410.

Zuntz, Nathan. Der hungrige Riese. *Berliner Tageblatt* Nr. 321 (1918g).

Zuntz, Nathan. Die Ernährungsverhältnisse Deutschlands nach dem Kriege. *Zeitschrift für ärztliche Fortbildung* 15 (1918h): 535–540.

Zuntz, Nathan. Ernährungsfragen. *Land und Frau* 2 (1918i): 338–339.

Zuntz, Nathan, and A. Loewy. Weitere Untersuchungen über den Einfluß der Kriegskost auf den Stoffwechsel. *Biochemische Zeitschrift* 90 (1918): 244–264.

von der Heide, R., M. Steuber, and Nathan Zuntz. Versuche über den Nährwert von aufgeschlossenem Holz. *Deutsche Landwirtschaftliche Presse* 45 (1918): 67–68.

1920

Zuntz, Nathan. Beeinflussung des Wachstums der Horngebilde (Haare, Nägel, Epidermis) durch spezifische Ernährung. *Deutsche medizinische Wochenschrift* 46 (1920): 145–146.

Zuntz, Nathan, and A. Loewy. *Lehrbuch der Physiologie des Menschen*, 3. Auflage. Leipzig: Vogel, 1920.

List of sources

Archival sources

Archiv der Jerusalems- und Neuen Kirchengemeinde Berlin, *Berlin (Archives of the Jerusalem and New Church Parish Berlin)*

Serial No. 181 of baptisms, October 23, 1889, *Archiv der Jerusalems- u. Neuen Kirchengemeinde Berlin* (Archives of the Jerusalem and New Church Parish Berlin), Parish Register of the New Church, p. 173.

Archiv der Humboldt Universität zu Berlin *(Archives of the Humboldt University Berlin)*

Amtliches Verzeichnis des Personals und der Studierenden der Kgl. Landwirtschaftlichen Hochschule Berlin (Official Register of the Staff and Students of the Royal Agricultural University Berlin in the years 1896, 1915, and 1920).

Letter of Zuntz to Flügge dated February 3, 1915, NL Flügge, No. 172.

Archiv der Max-Planck-Gesellschaft *(Archives of the Max Planck Society)*

Letter from Harnack to the Prussian Minister of Education dated February 25, 1911, A 1 1218, fol. 35.

The list of scientists printed in the annex of the meeting's record , A 1 1218, 1911, fol. 76/77.

Letter of the Minister of Education dated April 18, 1911. U I K., A 1 1219, fol. 69.

This document was marked as Confidential, 1911, fol. 71–75

Expert report by Zuntz dated May 15, 1911. A 1 1220, September 8, 1911–October 6, 1911, fol. 144.

Report by Cohnheim dated May 14, 1911. Max Planck Society Archive: A 1 1220, fol. 143.

Archiv der Rheinischen Friedrich-Wilhelms-Universität Bonn, *Bonn (Archives of the Rheinische Friedrich-Wilhelm-University, Bonn)*

Zuntz, Certificate of having been awarded the degree of Doctor of Medicine (*Promotionsurkunde*), July 31, 1868; records kept by the office of the

Medizinisches Dekanat (Dean of the Medical Department) at the University of Bonn on the Zuntz *Habilitation*, non-paginated.

Handwritten *curriculum vitae* by N. Zuntz, July 6, 1870; records kept by the office of the *Medizinisches Dekanat* (Dean of the Medical Department) at the University of Bonn on the Zuntz *Habilitation*, non-paginated.

Pflüger's opinion on Zuntz (probably written in July of 1870); records kept by the office of the *Medizinisches Dekanat* (Dean of the Medical Department) at the University of Bonn on the Zuntz *Habilitation*, non-paginated.

Letter from the Curator of the University of Bonn to Zuntz, July 7, 1870; records kept by the office of the *Medizinisches Dekanat* (Dean of the Medical Department) at the University of Bonn on the Zuntz *Habilitation*, non-paginated.

Letter from Zuntz to Dean Rindfleisch of the Medical Department, July 9, 1870, non-paginated.

Letter from Dean Rindfleisch to the Curator of the University Beseler, October 1st, 1870; records kept by the office of the *Medizinisches Dekanat* (Dean of the Medical Department) at the University of Bonn on the Zuntz *Habilitation*, non-paginated.

Letter from Pflüger to the Medical Department in Bonn, February 7, 1874; records kept by the office of the *Medizinisches Dekanat* (Dean of the Medical Department) at the University of Bonn on the Zuntz *Habilitation*, non-paginated.

Letter from the Medical Department to the Minister for Intellectual, Instructional and Medical Matters, February 12, 1874; records kept by the office of the *Medizinisches Dekanat* (Dean of the Medical Department) at the University of Bonn on the Zuntz *Habilitation*, non-paginated.

Letter from the Minister for Intellectual, Instructional, and Medical Matters addressed to the Medical Department in Bonn, July 29, 1874; records kept by the office of the *Medizinisches Dekanat* (Dean of the Medical Department) at the University of Bonn on the Zuntz *Habilitation*, non-paginated.

Letter from Zuntz to the Dean of the Philosophy Department in Bonn, August 4, 1919; Archives of the Rheinische Friedrich-Wilhelm-University, Bonn; Ehrenpromotionsakte Zuntz (file on the honorary doctorate awarded to Zuntz).

Archiv der Trinitatis Gemeinde zu Berlin, *Berlin (Archives of the Trinitatis Parish of Berlin)*

Death Register of the Trinitatis Parish of the Protestant Church in Berlin, February 1919 until January 1923, No. 223, fol. 90.

Archiv der Universität Wien, *Wien (Archives of the University of Vienna)*

Personalblatt (staff record sheet) of Arnold Durig; Archives of the University of Vienna, non-paginated.

Personalblatt (staff record sheet) of Hermann von Schrötter; Archives of the University of Vienna, non-paginated.

Bancroft Library, University of California, Berkeley

Letter of Zuntz to the *Präsident des Kriegsernährungsamts* (President of the War Nurition Department) March 9, 1917. Emil Fischer papers, BANC MSS 71/95 z, unpaginated.

*Bundesarchiv (*Militärarchiv*) Freiburg, Freiburg/i.Breisgau (German Federal Archives, Military Archive, Freiburg)*

Response of the War Department to the Royal Inspectorate of the Transportation Troops No. 550/10.02.A6, November 13, 1902; records of the German Federal Archive (Military Archive) Freiburg, PH 9 V/25, p. 137.

Letter from the War Department to the Inspectorate of the Transportation Troops; records of the German Federal Archive (Military Archive) Freiburg, PH 9 V/25, p. 166.

Civil-Luftschiffahrt, Wissenschaftliche Fahrten 1895–99 (Civil Airship Flight, Scientific Flights 1895–99); records of the German Federal Archive (Military Archive) Freiburg, PH 18/17, non-paginated.

Letter from the Aviator Battalion to the Inspectorate of the Transportation Troops, December 13, 1903; records of the German Federal Archive (Military Archive) Freiburg, PH 9 V/25, p. 161.

Letter from the War Department to the Royal Inspectorate of the Transportation Troops, January 9, 1904; records of the German Federal Archive (Military Archive) Freiburg, PH 9 V/25, p. 166.

Staatstelegramm (State telegram) on the provision of white bread to aeronauts and aviators, August 16, 1917; records of the German Federal Archive (Military Archive) Freiburg, RM 3 V/8970, p. 121.

Geheimes Staatsarchiv (GStA) Preussischer Kulturbesitz, *Berlin (Secret Central Archives of the Berlin State Library Prussian Cultural Heritage)*

Letter from Thiel to Lucius, September 22, 1880; GStA PK, I. HA Rep. 87 B, records of the *Preussisches Ministerium für Landwirtschaft, Domänen und Forsten*, B *Landwirtschaftsabteilung* (Prussian Ministry for Agriculture, Crown Domains and Forests, B Agriculture Department), No. 20075, fol. 45.

Letter from Zuntz to Thiel, December 20, 1880; GStA PK, I. HA Rep. 87 B, records of the *Preussisches Ministerium für Landwirtschaft, Domänen und Forsten*, B *Landwirtschaftsabteilung* (Prussian Ministry for Agriculture, Crown Domains and Forests, B Agriculture Department), No. 20075, fol. 65 *et seq.*

Letter from Lucius to Zuntz, December 27, 1880; GStA PK, I. HA Rep. 87 B, records of the *Preussisches Ministerium für Landwirtschaft, Domänen und Forsten*, B *Landwirtschaftsabteilung* (Prussian Ministry for Agriculture, Crown Domains and Forests, B Agriculture Department), No. 20075, fol. 67.

Letter from Zuntz to Lucius, December 31, 1880; GStA PK, I. HA Rep. 87 B, records of the *Preussisches Ministerium für Landwirtschaft, Domänen und Forsten* (Prussian Ministry for Agriculture, Crown Domains and Forests, B Agriculture Department), B *Landwirtschaftsabteilung* (Agriculture Department) No. 20075, fol. 68.

Letter from Zuntz to Lucius, February 5, 1881; GStA PK, I. HA Rep. 87 B, records of the *Preussisches Ministerium für Landwirtschaft, Domänen und Forsten* (Prussian Ministry for Agriculture, Crown Domains and Forests, B Agriculture Department), B *Landwirtschaftsabteilung* (Agriculture Department), No. 20075, fol. 82.

Letter from Zuntz to Lucius, April 27, 1888; GStA PK, I. HA Rep. 87 B, records of the *Preussisches Ministerium für Landwirtschaft, Domänen und Forsten* (Prussian Ministry for Agriculture, Crown Domains and Forests, B Agriculture Department), B *Landwirtschaftsabteilung* (Agriculture Department), No. 20077, fol. 23 *et seq.*

Letter from Lucius to Zuntz, April 30, 1888; GStA PK, I. HA Rep. 87 B, records of the *Preussisches Ministerium für Landwirtschaft, Domänen und Forsten* (Prussian Ministry for Agriculture, Crown Domains and Forests, B Agriculture Department), B *Landwirtschaftsabteilung* (Agriculture Department), No. 20077, fol. 25.

Letter from Zuntz to the Ministry for Agriculture, Crown Domains and Forests, April 24, 1893; GStA PK, I. HA Rep. 87 B, records of the *Preussisches Ministerium für Landwirtschaft, Domänen und Forsten* (Prussian Ministry for Agriculture, Crown Domains and Forests, B Agriculture Department), B *Landwirtschaftsabteilung* (Agriculture Department), No. 20079, fol. 119.

Letter from Podbielski to Kaiser Wilhelm II, October 8, 1904; GStA PK, I. HA Rep. 89, *Geheimes Zivilkabinett, jüngere Periode* (Secret Civil Cabinet, more recent period) No. 31929, fol. 130 R.

Letter from Podbielski to Kaiser Wilhelm II, January 29, 1906; GStA PK, I. HA Rep. 89, *Geheimes Zivilkabinett, jüngere Periode* (Secret Civil Cabinet, more recent period), No. 31929, fol. 141.

Letter from Kaiser Wilhelm II to Podbielski, January 31, 1906; GStA PK, I. HA Rep. 89, *Geheimes Zivilkabinett, jüngere Periode* (Secret Civil Cabinet, more recent period), No. 31929, fol. 142.

Bericht über die zum Besuche des internationalen Kongresses in Wien ausgeführte Reise (Report on the Trip Made for the Purpose of Visiting the International Congress in Vienna) by Zuntz to the Minster for Agriculture, Crown Domains and Forests, June 3, 1907; GStA PK, I. HA Rep. 87 B, records of the *Preussisches Ministerium für Landwirtschaft, Domänen und Forsten* (Prussian Ministry for Agriculture, Crown Domains and Forests, B Agriculture

Department), B *Landwirtschaftsabteilung* (Agriculture Department), No. 200139, fol. 176 R.

Letter from Fuchs to Zuntz, January 18, 1919; GStA PK, I. HA Rep. 87 B, records of the *Preussisches Ministerium für Landwirtschaft, Domänen und Forsten* (Prussian Ministry for Agriculture, Crown Domains and Forests, B Agriculture Department), B *Landwirtschaftsabteilung* (Agriculture Department), No. 20090, fol. 205–205 R.

Letter from Zuntz to the Minister for Agriculture, Crown Domains and Forests, January 8, 1908; GStA PK, I. HA Rep. 87 B, records of the *Preussisches Ministerium für Landwirtschaft, Domänen und Forsten,* B *Landwirtschaftsabteilung* (Agriculture Department), No. 200139, fol. 192–193 R.

Letter from the Minister for Agriculture, Crown Domains and Forests to Kaiser Wilhelm II, January 19, 1908; GStA PK, I. HA Rep. 89, *Geheimes Zivilkabinett jüngere Periode* (Secret Civil Cabinet, more recent period), No. 31929, fol. 152.

Letter from the President of the Agricultural College of Berlin to the Minister for Agriculture, Crown Domains and Forests, March 6, 1918; GStA PK, I. HA Rep. 87 B, records of the *Preussisches Ministerium für Landwirtschaft, Domänen und Forsten* (Prussian Ministry for Agriculture, Crown Domains and Forests, B Agriculture Department), B *Landwirtschaftsabteilung* (Agriculture Department), No. 20090, fol. 237.

Petition of the Minster for Agriculture, Crown Domains and Forests to Kaiser Wilhelm II, July 21, 1918; GStA PK, I. HA Rep. 89, *Geheimes Zivilkabinett jüngere Periode* (Secret Civil Cabinet, more recent period), No. 31930, fol. 24.

Letter from Zuntz to the Minister for Agriculture, Crown Domains and Forests, November 3, 1918; GStA PK, I. HA Rep. 87 B, records of the *Preussisches Ministerium für Landwirtschaft, Domänen und Forsten* (Prussian Ministry for Agriculture, Crown Domains and Forests, B Agriculture Department), B *Landwirtschaftsabteilung* (Agriculture Department), No. 20090, fol. 204.

Pension certificate for Nathan Zuntz; GStA PK, I. HA Rep. 87 B, records of the *Preussisches Ministerium für Landwirtschaft, Domänen und Forsten* (Prussian Ministry for Agriculture, Crown Domains and Forests, B Agriculture Department), B *Landwirtschaftsabteilung* (Agriculture Department), No. 20090, fol. 215 *et seq.*

Museum des Heimatschutzvereins, Montafon *(Museum of the Association for the Preservation of the Homeland, Montafon)*

Estate of Arnold Durig, non-paginated.

Nordrhein-Westfälisches Hauptstaatsarchiv *(North-Rhine-Westphalian State Archives, Düsseldorf)*

Serial No. 23 of the roll of members of the synagogue in Bonn paying their dues, 1880/1881; records of the government of Cologne, North-Rhine-Westphalian State Archives, Düsseldorf.

Österreichisches Staatsarchiv (Kriegsarchiv), Wien *(Austrian State Archives, war archives, Vienna)*

Haupt-Grundbuchblatt (Main register folio): Hermann Ritter Schrötter von Kristelli; records of the Austrian State Archives, war archives, Vienna, non-paginated.

Österreichisches Staatsarchiv (Verwaltungsarchiv), Wien *(Austrian State Archives, Administration Archive, Vienna)*

Habilitationsverfahren (Habilitation procedure) of Hermann von Schrötter; Austrian State Archives, Administration Archive, Vienna, documents regarding *Unterricht* (instruction), non-paginated.

Standesausweis (official professional identification) of Hermann von Schrötter; Austrian State Archives, Administration Archive, Vienna), documents kept by the *Soziale Verwaltung* (welfare administration), non-paginated.

Staatsbibliothek Preussischer Kulturbesitz, *Berlin (Berlin State Library Prussian Cultural Heritage)*

Letter of Zuntz to the medical counselor Dr. Loebker (Bochum), February 20, 1908; Staatsbibliothek Preussischer Kulturbesitz; Darmstaedter collection on Nathan Zuntz, 3d. 1885, fol. 88–96.

Fragment of a letter from Zuntz to an unknown person, February 2, 1920; Staatsbibliothek Preussischer Kulturbesitz, Darmstaedter collection on Nathan Zuntz, 3d. 1885, fol. 3.

Letter from Leo Zuntz to Darmstaedter; Staatsbibliothek Preussischer Kulturbesitz, Darmstaedter collection on Nathan Zuntz, 3d. 1885, fol. 48–49.

Stadtarchiv Bonn (City Archive Bonn)

Birth Certificate No. 397, October 7, 1847; *Geburtenbuch* (Register of Births) of the *Standesamt* (Registry of Vital Statistics) of Bonn 1 1847, p. 199.

Birth Certificate No. 52, January 28, 1849; *Geburtenbuch* (Register of Births) of the *Standesamt* (Registry of Vital Statistics) of Bonn 1 1849, p. 26.

Birth Certificate No. 295, June 26, 1850; *Geburtenbuch* (Register of Births) of the *Standesamt* (Registry of Vital Statistics) of Bonn 1 1850, p. 149.

Birth Certificate No. 86, February 19, 1852; *Geburtenbuch* (Register of Births) of the *Standesamt* (Registry of Vital Statistics) of Bonn 1 1852, p. 45.

Birth Certificate No. 508, November 2, 1854; *Geburtenbuch* (Register of Births) of the *Standesamt* (Registry of Vital Statistics) of Bonn 1 1854, p. 254.

Birth Certificate No. 477, November 8, 1855; *Geburtenbuch* (Register of Births) of the *Standesamt* (Registry of Vital Statistics) of Bonn 1 1855, p. 240.

Birth Certificate No. 111, March 7, 1858; *Geburtenbuch* (Register of Births) of the *Standesamt* (Registry of Vital Statistics) of Bonn 1 1858, p. 56.

Birth Certificate No. 194, April 13, 1861; *Geburtenbuch* (Register of Births) of the *Standesamt* (Registry of Vital Statistics) of Bonn 1 1861, p. 97.

Birth Certificate No. 553, October 23, 1863; *Geburtenbuch* (Register of Births) of the *Standesamt* (Registry of Vital Statistics) of Bonn 1 1863, p. 278.

Birth Certificate No. 554, October 23 1863; *Geburtenbuch* (Register of Births) of the *Standesamt* (Registry of Vital Statistics) of Bonn 1 1863, p. 279.

Certificate of Death No. 374, June 13, 1874; *Sterbebuch* (Death Register) of the *Standesamt* (Registry of Vital Statistics) of Bonn 1 1874.

Birth Certificate No. 340, April 14, 1875; *Geburtenbuch* (Register of Births) of the *Standesamt* (Registry of Vital Statistics) of Bonn 1 1875.

Birth Certificate No. 449, May 14, 1877; *Geburtenbuch* (Register of Births) of the *Standesamt* (Registry of Vital Statistics) of Bonn 2 1877.

Birth Certificate No. 44, January 14, 1880; *Geburtenbuch* (Register of Births) of the *Standesamt* (Registry of Vital Statistics) of Bonn 1 1880.

Miscellaneous

Diary of Emma Zuntz.
Report from Heidelberg, unpaginated.

Bibliography

Adler, H. G. 1974. *Der verwaltete Mensch*. Tübingen: Mohr Siebeck.

Adolph, E. F. *et al.* 1947. (Reprint 1969). *Physiology of Man in the Desert*. New York, NY: Interscience Publishers.

Anonymous. 1901. Hochfahrten im Luftballon. *Die Umschau* 5: 689–694.

Anonymous. 1902a. Dritte Tagung der Internationalen Kommission für wissenschaftliche Luftfahrt. *Illustrierte Aeronautische Mitteilungen* 6: 138–149.

Anonymous. 1902b. Unfälle bei Luftschiffahrten. *Die Umschau* 6: 861–864.

Anonymous. 1913. Wissenschaftliche Ballonfahrten. *Deutsche Luftfahrer-Zeitschrift* 16: 391–392.

Anonymous. 1917. Geh. Rat Prof. Dr. N. Zuntz. *Die Umschau* 21: 760.

Anonymous. April 2, 1927. Zum Jubiläum der Firma Zuntz sel. Wwe. *Bonner Zeitung*.

Anonymous. 1963. Nathan Zuntz – October 7, 1847–March 23, 1920. *Journal of the American Dietetic Association* 43: 364.

Armsby, H. P. 1917. *The nutrition of farm animals*. New York, NY: Macmillan.

Artelt, W., E. Heischkel, and C. Wehmer 1952. *Periodica medica*. Stuttgart: Thieme.

Asen, J. 1955. *Gesamtverzeichnis des Lehrkörpers der Universität Berlin*, Vol. 1. Leipzig: Harrassowitz.

Assmann, R. 1915. *Das Königlich Preußische Aeronautische Observatorium Lindenberg*. Braunschweig: Friedrich Vieweg & Sohn.

Assmann, R. 1918. Nachruf von Berson. *Zeitschrift für Flugtechnik und Motor-Luftschiffahrt* 9: 62–64.

Assmann, R., A. Berson, and H. Gross. 1899. *Wissenschaftliche Luftfahrten*, Vol. 1. Braunschweig: Friedrich Vieweg & Sohn.

Auerbach, F. 1883. *Hundert Jahre Luftschiffahrt*. Breslau.

Auerbach, P. S. 2007. *Wilderness Medicine*. Oxford: Elsevier Books.

Barnard, R. J., and J. O. Holloszy. 2003. The metabolic systems: aerobic metabolism and substrate utilization in exercising skeletal muscle. In *Exercise Physiology*, edited by C. M. Tipton, pp. 292–321. Oxford: Oxford University Press.

Bast, T. H. 1931. Max Johann Sigismund Schultze. *Annals of Medical History* 3: 166–178.

Benedict, F. 1909. An Apparatus for Studying the Respiratory Exchange. *American Journal of Physiology* 24: 345–374.

Benedict, F. G. 1938. Biographical Memoir of Henry Prentiss Armsby 1853–1921. *National Academy of Sciences of the United States of America, Biographical Memoirs* 19: 271–284.

Berdrow, W. 1897. Die Ausrüstung wissenschaftlicher Ballon-Expeditionen. *Die Umschau* 1: 64–67.

Berg, A. 1909. Das Schrifttum der Luftschiffahrt im 20. Jahrhundert. *Illustrierte Aeronautische Mitteilungen* 13: 769–771.

Berghold, F., and W. Schaffert 2001. *Handbuch der Trekking- und Expeditionsmedizin*. Munich: DAV Summit Club.

Bergner, H. 1968. Zum Gedenken des 120. Geburtstages von Nathan Zuntz. *Archiv für Tierernährung* 18: 1–3.

Bergner, H. 1986. Nathan Zuntz and 100 years non-protein-nitrogen research in Berlin. *Archives of Animal Nutrition* 36: 127–130.

Berliner, B., and F. Müller 1911/12. Vergleichende meteorologische Beobachtungen am Strande und an der Binnenseite des Dünenwaldes in einem Ostseebade. *Veröffentlichungen der Zentralstelle für Balneologie* 1: 3–9.

Berson, A. 1895. Eine Reise in das Reich der Cirren. *Das Wetter* 22: 1–10.

Berson, A. 1902. Wissenschaftliche Hochfahrt vom 31.7.1901. *Illustrierte Aeronautische Mitteilungen* 6: 51–52.

Berson, A., and R. Süring 1901. Ein Ballonaufstieg bei 10 500 m. *Illustrierte Aeronautische Mitteilungen* 5: 117–119.

Bert, P. 1878. *La pression barométrique, Recherches de physiologie expérimentale.* Paris: Masson.

Blanchard, J. P. 1788. Nachricht über den Ballonaufstieg vom 27. September 1788. *Berlinische Nachrichten von Staats- und gelehrten Sachen*: 897.

Blasius, W. 1959. *Zur Geschichte der Deutschen Physiologischen Gesellschaft.* Giessen.

Bleker, J. 1972. *Die Geschichte der Nierenerkrankung.* Mannheim: Boehringer.

Blüm-Spieker, H. 1984. 600 Jahre Stadt Zons. *Katalog zum Anlass des Stadtjubiläums 1973.* Dormagen.

Bonnier, P. 1901. Recherches sur la compensation labyrinthique en ballon. *Comptes rendus hebdomadaires des séances et mémoires de la Société de Biologie* 53: 1034–1037.

Bower, T. 1987. *The paperclip conspiracy.* London: Michael Joseph.

Bretschneider, H. 1962. *Der Streit um die Vivisektion im 19. Jahrhundert.* Stuttgart: Gustav Fischer.

Brockhaus. 1952–1958. *Der grosse Brockhaus*, 12 volumes. Wiesbaden: Brockhaus.

Bröckelmann, K. 1909. *Wir Luftschiffer. Die Entwicklung der modernen Luftschifftechnik in Einzeldarstellungen.* Berlin: Ullstein.

Bröer, R. 2002. Legende oder Realität? – Werner Forssmann und die Herzkatheterisierung. *Deutsche Medizinische Wochenschrift* 127: 2151–2154.

Brooks, G. A., and L. B. Gladden 2003. The Metabolic Systems: Anaerobic Metabolism (Glycolytic and Phosphagen). In *Exercise Physiology*, edited by C. M. Tipton, pp. 322–360. Oxford: Oxford University Press.

Brooks, G. A., T. D. Fahey, and T. P. White 1996. *Exercise Physiology*, second edition. Human Bioenergetics and Its Applications. Mountain View: Mayfield Publishing Company.

Bruce, E. 1902. Die Entwickelung der Luftschiffahrt in Deutschland. *Illustrierte Aeronautische Mitteilungen* 6: 107–108.

Buskirk, E. R. 2003. The Temperature Regulatory System. In *Exercise Physiology*, edited by C. M. Tipton, pp. 423–451. Oxford: Oxford University Press.

Calugareanu, –., and V. Henri. 1901. Résultats des expériences faites pendant une ascension en ballon. *Comptes rendus hebdomadaires des séances et mémoires de la Société de Biologie* 53: 1037–1039.

Caspari, W. 1917. Nathan Zuntz zu seinem 70. Geburtstage. *Naturwissenschaften* 40: 87–89.

Clark, B. R. 1995. *Place of Inquiry. Berkeley.* Los Angeles, CA: University of California Press.

Cournand, A., and H. A. Ranges 1941. Cathetherization of the Right Auricle in Man. *Proceedings of the Society for Experimental Biology and Medicine* 46: 462–466.

Cournand, A. F. 1986. From roots to late budding. *The Intellectual Adventures of a Medical Scientist.* New York, London: Gardner Press.

Cruchet, R. 1911. Le vol en hauteur et le mal des aviateurs. *La Revue Scientifique*: 740–744.

Cruchet, R. 1911. L'air raréfié – ses méfaits – son emploi thérapeutique. *Le Journal Médical Français*: 507–515.

Cruchet, R., and R. Moulinier 1920a. *Air Sickness, Its Nature and Treatment*. London: J. Bale, Sons & Danielsson.

Cruchet, R., and R. Moulinier 1920b. *Le Mal des Aviateurs ses Causes et ses Remèdes*. Paris: J.B. Balliére et Fils.

Cüppers, S. 1994. *Die geschichtliche Entwicklung der Höhenphysiologie und ihre Bedeutung für die Luftfahrtmedizin bis 1961*. Dissertation. RWTH Aachen: Verlag Shaker.

Dempsey, J. A., and B. J. Whipp 2003. The respiratory system. In *Exercise Physiology*, edited by C. M. Tipton, pp. 138–187. Oxford: Oxford University Press.

de Schroetter, E. 1901/1902. Communication d'expériences physiologiques faites pendant un voyage en ballon à 7500m et rapport sur différents essais concernant l'étude de l'influence de l'air raréfié sur l'organisme humain. *Archives italiennes de biologie*: 86–88.

Deutsche Forschungsgemeinschaft. 1981. Reprint 1984. *Tierexperimentelle Forschung und Tierschutz*. Weinheim: Verlag Chemie.

Dietl, M., J. Dudenhausen, and N. Suttorp (eds) 2003. *Harrisons Innere Medizin 2*. Deutsche Ausgabe. Berlin: ABW Wissenschaftsverlag.

Dietrich, J. 1884. Die Wirkung comprimirter und verdünnter Luft auf den Blutdruck. *Archiv für experimentelle Pathologie und Pharmakologie* 18: 242–259.

Dietz, A. 1907. *Stammbuch der Frankfurter Juden, Geschichtliche Mitteilungen über die Frankfurter jüdischen Familien von 1349–1849*. Frankfurt am Main: J. St Goar.

Diezemann, N. 1989. *Die Kunst des Hungerns. Anorexie in literarischen und medizinischen Texten um 1900*. Dissertation, Universität Hamburg 2005.

Dräger, B. 1904. Über Sauerstoffinhalation bei Hochfahrten. *Illustrierte Aeronautische Mitteilungen* 8: 249–252.

Drosdoff, –., and –. Botschetschkaroff 1875. Die physiologische Wirkung der im Waldenburg'schen Apparate comprimirten Luft auf den arteriellen Blutdruck der Thiere. *Zentralblatt für die medicinischen Wissenschaften* 13: 65–67.

Du Bois-Reymond, E. 1912a. Über die Übung. In *Reden von Emil du Bois-Reymond*, edited by Estelle du Bois-Reymond, Vol. 2, pp. 99–140. Leipzig: Veit & Comp.

Du Bois-Reymond, E. 1912b. Über die wissenschaftlichen Zustände der Gegenwart. In *Reden von Emil du Bois-Reymond*, edited by Estelle du Bois-Reymond, Vol. 2, pp. 141–156. Leipzig: Veit & Comp.

Durig, A. 1906. Beiträge zur Physiologie des Menschen im Hochgebirge. *Pflügers Archiv* 113: 213–316.

Durig, A. 1911. Physiologische Ergebnisse der im Jahre 1906 durchgeführten Monte Rosa-Expedition: I.: Einleitung u. II.: Teilnehmer und Verlauf der Expedition. *Denkschriften der Kaiserlichen Akademie der Wissenschaften in Wien, Mathematisch-Naturwissenschaftliche Klasse* 86: 1–36.

Durig, A. 1920. N. Zuntz. *Wiener klinische Wochenschrift* 33: 344–345.

Ebel, G., and O. Wührs 1988. *Urania – eine Idee, eine Bewegung, eine Institution wird 100 Jahre alt–Festschrift*. Berlin: Urania.

Eckart, W. U. 2006. *Man, Medicine and the State. The Human Body as an Object of Government Sponsored Medical Research in the 20th Century*, Vol. 2. Stuttgart: Franz Steiner Verlag.

Eckart, W. U. 2008. *Medizingeschichte*. In *Forschergruppe zur Geschichte der Deutschen Forschungsgemeinschaft 1920–1970*. Bericht zur Abschlusskonferenz am 30. und 31. Januar 2008 in Berlin, pp. 62–108.

Edholtz, O. G., and J. S. Weiner. 1981. *Principles and Practice of Human Physiology*. London.

Eltzbacher, P. 1914. *Denkschrift – Die deutsche Volksernährung und der englische Aushungerungsplan*. Braunschweig: Vieweg.

Fangerau, H., and I. Müller 2005. National styles? Jacques Loeb's analysis of German and American science around 1900 in his correspondence with Ernst Mach. *Centaurus* 47: 207–225.

Feldkirch, B. 1909. 200 Jahre Luftschiffahrt. *Illustrierte Aeronautische Mitteilungen* 13: 735–753.

Felsch, P. 2007. *Laborlandschaften. Physiologische Alpenreisen im 19. Jahrhundert*. Göttingen: Wallstein.

Fischer, I. 1962a. *Biographisches Lexikon der hervorragenden Ärzte vor 1880*, 5 volumes. Munich: Urban & Schwarzenberg.

Fischer, I. 1962b. *Biographisches Lexikon der hervorragenden Ärzte der letzten fünfzig Jahre (1880–1930)*, 2 volumes. Munich: Urban & Schwarzenberg.

Fishman, A., and D. W. Richards (eds) 1964. *Circulation of the Blood – Men and Ideas*. New York: Oxford University Press.

Flemming, –. 1908. Unfälle und Rettungsmassnahmen auf dem Gebiete der Luftschiffahrt. *Illustrierte Aeronautische Mitteilungen* 12: 489–497.

Flemming, –.1909a. Physiologische und physikalische Messungen und Beobachtungen bei einer Hochfahrt. *Illustrierte Aeronautische Mitteilungen* 13: 1019–1028.

Flemming, –. 1909b. Der Arzt im Ballon. In *Wir Luftschiffer*, edited by K. Bröckelmann, pp. 172–187. Berlin–Vienna: Ullstein & Co.

Flemming, –. Freiballonhygiene. 1911. In *Der Freiballon in Theorie und Praxis*, edited by A. Mehl, pp. 219–236. Stuttgart: Franckh'sche Verlagshandlung.

Flemming, –. 1912. Bewusslosigkeit im Luftschiff. *Die Umschau* 14: 960–961.

Flemming, F. 2003. *Nach oben*. Munich: Piper-Verlag.

Forbes, R. M. 1955. Nathan Zuntz. *Journal of Nutrition* 57: 3–15.

Forssmann, W. 1929. Die Sondierung des rechten Herzens. *Klinische Wochenschrift* 8: 2085–2087, 2287.

Foucault, M. 1977. *Überwachen und Strafen*. Frankfurt/M.: Suhrkamp.

Fraenkel, A., and J. Geppert 1883. *Über die Wirkung der verdünnten Luft auf den Organismus*. Berlin: Springer-Verlag.

Franklin, K. J. 1938. A Short History of the International Congresses of Physiologists 1889–1938, *Annals of Sciences* 3: 254–265. Attached article in *History of the International Congresses of Physiological Sciences 1889–1968*, edited by Wallace O. Fenn. Baltimore, MD: Waverly Press, Inc.

Friedlaender, –. 1912. Zur Physiologie und Pathologie der Luftfahrt. *Jahrbuch der Wissenschaftlichen Gesellschaft für Flugtechnik* 1: 70–83.

Friedlaender, C., and E. Herter 1879. Über die Wirkung des Sauerstoffmangels auf den thierischen Organismus. *Zeitschrift für physiologische Chemie* 3: 19–51.

Gaule, J. 1902. Die Blutbildung im Luftballon. *Pflügers Archiv* 89: 119–153.

Geison, G. L. (ed.) 1987. *Physiology in the American Context. 1850–1940*. Bethesda, MD: American Physiological Society.

George, A., T. D. Fahey, and T. P. White 1996. *Exercise Physiology. Second Edition. Human Bioenergetics and Its Applications*. Mountain View: Mayfield Publishing Company.

German Aviation Medicine World War II 1950. Department of the Air Force. Vols 1 and 2. Washington, DC.

Germe, L. 1895. *Recherches sur les lois de la circulation pulmonaire sur la fonction hémodynamique de la respiration et l'asphyxie suivies d'une étude sur le mal de montagne et de ballon*. Paris: G. Masson.

Gimbel, J. 1986. US policy and German scientists. The early Cold War. *Political Science Quarterly* 101: 433–451.

Glatzer, N. 1964. *Leopold Zuntz*. Tübingen: Mohr (Siebeck).

Gorbatschew, P. 1891. Zur Frage über den Einfluss der Bergbesteigung auf Blutdruck, Körpertemperatur, Puls, Athem, Verluste durch Haut- und Lungenrespiration und Nahrungsmenge. *Österr.-ungar. Zentralblatt für die medicinischen Wissenschaften* 2: 1–2.

Granger, H. J. 1998. Cardiovascular Physiology in the Twentieth Century: Great Strides and Missed Opportunities. *American Journal of Physiology (Heart Circ. Physiol.)* 275: H1925–H1936.

Grawitz, E. 1895. Über die Einwirkung des Höhenklimas auf die Zusammensetzung des Blutes. *Berliner klinische Wochenschrift*: 713–740.

Greiner, E., and K.-H. Arndt 2004. Der erste deutsche Sportärztekongress 1912 – Programm für ein Jahrhundert. *Deutsche Zeitschrift für Sportmedizin* 55: 310–314.

Guenter, M. 1973. *Die Juden in Lippe von 1648 bis zur Emanzipation 1858*. Naturwissenschaftlicher und Historischer Verein für das Land Lippe: Detmold.

Günther, P. E. 1988. *Jean Pierre Blanchard's 33. Luftfahrt 1788 – Zeitgenössische Berichte der ersten bemannten Luftfahrt in Berlin*. Beiträge zur Geschichte der Ballonfahrt, Vol. 3. Berlin: PPE Günther.

Gunga, H.-C. 1989. *Leben und Werk des Berliner Physiologen Nathan Zuntz (1847–1920) unter besonderer Berücksichtigung seiner Bedeutung für die Frühgeschichte der Höhenphysiologie und Luftfahrtmedizin*. Dissertation, Freie Universität Berlin.

Gunga, H.-C., and K. Kirsch 1995a. Nathan Zuntz (1847–1920). A German pioneer in high altitude physiology and aviation medicine, Part I: Biography. *Aviation Space and Environmental Medicine* 66: 168–171.

Gunga, H.-C., and K. Kirsch 1995b. Nathan Zuntz (1847–1920). A German pioneer in high altitude physiology and aviation medicine, Part II: Scientific work. *Aviation Space and Environmental Medicine* 66: 172–176.

Gurtner, V. 1971. *Jungfrauexpress*. Zürich: Orell Fuessli.

Guy-Lussac, L., and J. Biot 1804. Extrait de la relation d'un voyage aérostatique. *Journal de physique, de chimie et d'histoire naturelle* 59: 317.

Hackenberger, W. 1915. *Deutschlands Eroberung der Luft*, Vol. 1. Siegen: Hermann Montanus.

Hagemann, O. 1914. *Anatomie des Pferdes, der Wiederkäuer, Schweine, Fleischfresser und des Hausgeflügels, mit besonderer Berücksichtigung des Pferdes*. Stuttgart: Eugen Ulmer.

Halben, –. 1914. Die Augen der Luftfahrer. *Jahrbuch der Wissenschaftlichen Gesellschaft für Flugtechnik* 2: 158–168.

Hallion, –., and –. Tissot 1901. Recherches expérimentales sur l'influence des variations rapides d'altitude sur les phénomènes chimiques et physiques de la respiration à l'état de repos (recherches faites au cours d'une ascension en ballon). *Comptes rendus hebdomadaires des séances et mémoires de la Société de Biologie* 53: 1030–1032.

Hallion, –., and –. Tissot 1901. Recherches expérimentales sur l'influence des variations rapides d'altitude sur les gaz du sang et sur la pression artérielle. *Comptes rendus hebdomadaires des séances et mémoires de la Société Biologie* 53: 1032–1034.

Hamilton, W. F. 1953. The physiology of the cardiac output. *Circulation* 8: 527–543.

Harsch, V. 2000. Aerospace medicine in Germany: from the very beginnings. *Aviation Space and Environmental Physiology* 71: 447–450.

Harsch, V. 2001. *Die Deutsche Gesellschaft für Luft- und Raumfahrtmedizin e.V. (DGLRM) 1961–2001*. Norderstedt: Books on Demand GmbH.

Harsch, V. 2004. *Leben, Werk und Zeit des Physiologen Hubertus Strughold (1898–1986)*. Flugmedizinische Reihe: Vol. 4. Neubrandenburg: Rethra.

Haushofer, H. 1972. *Die deutsche Landwirtschaft im technischen Zeitalter*. Stuttgart: Eugen Ulmer.

Hecker, C. 1894. Entwicklung der Körperoberfläche in ihrer Beziehung zur Wärmeproduktion. *Zeitschrift f. Veterinärkunde*: 97.

Heindel, W. 1971. *Personalbibliographien von Professoren und Dozenten des Histologisch-Embryologischen Institutes der Universität Wien im ungefähren Zeitraum von 1848–1968*. Dissertation, Universität Erlangen-Nürnberg.

Heller, R., W. Mager, and H. von Schroetter 1900. *Luftdruckerkrankungen*. Vienna: Springer.

Henderson, Y. 1920. *The library of the late professor Zuntz*. Science 51: 589.

Helwig, O., and F. Müller 1912. Beiträge zur Physiologie der Klimawirkungen. *Zentralstelle für Balneologie* I: 1–4.

Henneberg, W., and Th. Pfeifer 1890. Ueber den Einfluss eines einseitig gesteigerten Zusatzes von Eiweißstoffen zum Beharrungsfutter auf den Gesammtstoffwechsel des ausgewachsenen Thieres. *Journal für Landwirtschaft* 38: 258–259.

Henocque, –. 1901. Étude de l'activité de la réduction de l'oxyhémoglobine, dans les ascensions en ballon. *Comptes rendus hebdomadaires des séances et mémoires de la Société de Biologie* 53: 1003–1006.

Herken, H. 1980. Tierexperimentelle Prüfung von Arzneimitteln. *Deutsches Ärzteblatt* 77: 2617–2628.

Hermann, L. 1882. *Handbuch der Physiologie des Kreislaufs, der Athmung und der thierischen Wärme*, Vol. 4, Part 2. Leipzig: F. C. W. Vogel.

Hinterstoisser, F. 1908. Über Luftschiffahrt-Journalismus. *Illustrierte Aeronautische Mitteilungen* 12: 627–629.

Hinterstoisser, F. 1915. *Fünfundzwanzig Jahre Luftfahrt*. Vienna: Verlag von Streffleurs Militärisches Zeitschrift.

Hirsch, A. 1886. *Biographisches Lexikon der hervorragenden Ärzte aller Zeiten und Völker*, Vol. 4. Vienna-Leipzig: Urban & Schwarzenberg.

Hirsch, A. 1887. *Biographisches Lexikon der hervorragenden Ärzte aller Zeiten und Völker*, Vol. 5. Vienna-Leipzig: Urban & Schwarzenberg.

Hochachka, P. W., H.-C. Gunga, and K. Kirsch 1998. Our Ancestral Physiological Phenotype: An Adaptation for Hypoxia Tolerance and for Endurance Performance. *Proceedings of the National Academy of Sciences* 95: 1915–1920.

Hölder, H. 1960. *Geologie und Paläontologie in Texten und ihrer Geschichte*. Freiburg: Karl Alber.

Hörnes, H. 1912. *Buch des Fluges*, 3 volumes. Vienna: G. Szelinski.

Hultgren, H. 1997. *High Altitude Medicine*. Stanford, CA: Hultgren.

Hunt, Linda 1991. *Secret Agenda: The United States Government, Nazi Scientists, and Project Paperclip, 1945 to 1990*. New York, NY: St. Martin's Press.

Internationale Kommission für wissenschaftliche Luftschiffahrt 1903. Protokoll über die vom 20. bis 25. Mai 1902 zu Berlin abgehaltene Versammlung der Internationalen Kommission für wissenschaftliche Luftschiffahrt. Strassburg i.E.

Jackson, D. 1985. *Die Ballonfahrer*. Amsterdam: Time-Life.

Jacobsen, H., and –. Lazarus 1877. Über den Einfluss des Aufenthaltes in comprimirter Luft auf den Blutdruck. Zentralblatt für die medicinischen Wissenschaften: 929–930.

Jahrbuch der wissenschaftlichen Gesellschaft für Flugtechnik, 1912–1920. Bände 1–5. Berlin: Springer.

Jaquet, A., and F. Suter 1898. Über die Veränderungen des Blutes im Hochgebirge. *Correspondenzblatt für Schweizer Ärzte* 28: 104–116.

Jokl, E. 1967. Zur Geschichte der Höhenphysiologie. Forschung und Fortschritte 41: 321–328. Berlin: Akademie Verlag.

Jokl, E. 1975. Unveröffentlichte, eigenhändig zusammengestellte Bibliographie und selbstverfasster curriculum vitae.

Jokl, E. 1984. Aus der Frühzeit der Deutschen Sportmedizin, Deutscher Sportärztekongress vom 27.-29.September 1984 in Berlin. Berlin.

Jolly, J. 1901. Examens histologiques du sang, au cours d'une ascension en ballon. *Comptes rendus hebdomadaires des séances et mémoires de la Société de Biologie* 53: 1039–1040.

Katzenstein, G. 1891. Über die Einwirkung der Muskelthätigkeit auf den Stoffverbrauch des Menschen. *Pflügers Archiv* 49: 330–404.

Kaulen, –. 1917. Über den Einfluß des Fliegens auf den Blutdruck bei Menschen, Kaninchen und Mäusen. *Deutsche medizinische Wochenschrift*: 1562–1563.

Kiell, P. 1987. How Sports Medicine Began. *American Medical Athletic Association Newsletter* 2: 10.

Kirchner, M. 1896. *Grundriss der Militär-Gesundheitspflege*. Braunschweig.

Kirchhoff, A. 1910. *Die Erschließung des Luftmeeres. Luftschiffahrt und Flugtechnik in ihrer Entwicklung und ihrem heutigen Stande*. Leipzig: Otto Spamer.

Kirsch, K. and H.-C. Gunga 1988. Der Mensch in extremen Umwelten. In *100 Jahre Urania Berlin-Festschrift-Wissenschaft heute für morgen*, edited by Urania, pp. 155–161. Berlin-Bonn.

Kirsch, K., and R. Winau 1986. The Early Days of Space Medicine in Germany. *Aviation, Space and Environmental Medicine* 57: 633–635.

Kitano, H. 2002. Systems Biology: A Brief Overview. *Science* 295: 1662–1664.

Klingeberg-Kraus, S. 2001. *Beitrag zur Ernährungsforschung bei Pferden bis 1950 (Verdauungsphysiologie, Energie- und Eiweißstoffwechsel)*. Inaugural Dissertation, Hannover.

Koelle, H. H.: *Der Beitrag Berlins zur Entwicklung der Luft- und Raumfahrt* (Berlin's Contribution to the Development of Aeronautics and Space Flight), Yearbook 1987 I of the Deutsche Gesellschaft für Luft- und Raumfahrt e.V. (DGLR German Association for Aeronautics and Space Flight), pp. 431–434.

Kohler, R. E. 2002. *Landscapes and Labscapes*. Chicago, IL: University of Chicago Press.

Kolb, E. 1974. *Lehrbuch der Physiologie der Haustiere*. Stuttgart: Fischer.

König, W. 1951. Nachruf auf Professor Dr. Reinhard Süring. *Zeitschrift für Meteorologie* 5: 33–34.

Koschel, E. 1912. Eine wissenschaftliche Hochfahrt mit dem Ballon Berlin am 6. Oktober 1912. *Jahrbuch des Berliner Vereins für Luftschiffahrt*: 116–132.

Koschel, E. 1914. Welche Anforderungen müssen an die Gesundheit der Führer von Luftfahrzeugen gestellt werden? *Jahrbuch der Wissenschaftlichen Gesellschaft für Flugtechnik* 2: 143–156.

Koschel, E. 1922. Hygiene des Ersatzes bei den Luftstreitkräften. In *Handbuch der ärztlichen Erfahrungen im Weltkriege 1914/1918*, edited by O. von Schjerning, 7, 10–33. Leipzig: Barth.

Kraemer, H. 1906. *Der Mensch und die Erde*, Vol. 2. Berlin: Bong.

Krogh, A., and M. Krogh 1995. *Lives in Science*. New York, NY: Oxford University Press.

Kronecker, H. 1903. *Die Bergkrankheit*. Munich: Urban und Schwarzenberg.

Kronfeld, A. 1919. Eine experimentell-psychologische Tauglichkeitsprüfung zum Flugdienst. *Zeitschrift für angewandte Psychologie* 15: 193–235.

Langendorff, O. 1899. Zur Physiologie der Luftschiffahrt und des Alpensports. *Deutsche Revue* 24: 304–316.

Lazarus, J. 1895. Bergfahrten und Luftfahrten in ihrem Einfluß auf den menschlichen Organismus. *Berliner klinische Wochenschrift*: 672–673, 702–705.

Lebegott, W. 1882. *Die Ausathmung in verdünnter Luft*. Dissertation, Freie Universität Berlin.

Lehmann-Brune, M. 1997. *Der Koffer des Karl Zuntz. Fünf Jahrhunderte einer jüdischen Familie*. Düsseldorf: Droste.

Loewy, A. 1891. Die Wirkung ermüdender Muskelarbeit auf den respiratorischen Stoffwechsel. *Pflügers Archiv* 49: 405–422.

Loewy, A. 1895. *Untersuchungen über die Respiration und Circulation bei Änderung des Druckes und des Sauerstoffgehaltes der Luft*. Berlin: Springer.

Loewy, A. 1916a. Über den Stoffwechsel im Wüstenklima. *Zeitschrift für Balneologie, Klimatologie und Kurort-Hygiene* 7/8: 43–48.

Loewy, A. 1916b. Über den Stoffwechsel im Wüstenklima. *Veröffentlichungen der Zentralstelle für Balneologie* 3: 1–6.

Loewy, A. 1917. N. Zuntz zum 70. Geburtstage. *Deutsche medizinische Wochenschrift* 43: 1270–1272.

Loewy, A. 1920. Dem Gedächtnis an N. Zuntz. *Berliner Klinische Wochenschrift* 57: 433–435.

Loewy, A. 1922. Dem Andenken an Nathan Zuntz. *Pflügers Archiv* 194: 1–19.

Loewy, A. 1932. *Physiologie des Höhenklimas*. Berlin: Springer.

Loewy, A., and F. Müller 1904. Über den Einfluss des Seeklimas und der Seebäder auf den Stoffwechsel des Menschen. *Pflügers Archiv* 103: 450–475.

Loewy, A., and S. Placzek 1914. Die Wirkung der Höhe auf das Seelenleben des Luftfahrers. *Berliner klinische Wochenschrift*: 1020–1023.

Loewy, A., and H. von Schroetter 1905. Untersuchungen über Blutcirculation beim Menschen. *Zeitschrift für experimentelle Pathologie und Therapie* 1: 197–311.

Loewy, A., and H. von Schroetter 1926. Über den Energieverbrauch bei musikalischer Betätigung. *Pflügers Archiv* 211: 1–63.

Loewy, A., J. Loewy, and L. Zuntz 1897. Über den Einfluss der verdünnten Luft und des Höhenklimas auf den Menschen. *Pflügers Archiv* 66: 477–538.

Mamlock, –. 1920. Professor N. Zuntz. *Berliner Tageblatt und Handelszeitung* vom 25. März 1920.

Marquis, R. 1912. *Hygiène pratique de l'aviateur et de l'aéronaute*. Paris: A. Malione.

Martin, R. 1912. Millionäre in der Provinz Brandenburg. *Jahrbuch des Vermögens und Einkommens der Millionäre in der Provinz Brandenburg einschließlich Charlottenburg, Wilmersdorf und alle anderen Vororte*. Berlin.

McArdle, William D., Frank I. Katch, and Victor L. Katch 1996. *Exercise Physiology*. Baltimore, MD: Williams & Wilkins.

Mehl, A. 1911. *Der Freiballon in Theorie und Praxis*, 2 volumes. Stuttgart: Franck'hsche Verlagshandl.

Michaelis, M. 1906. *Sauerstofftherapie*. Berlin: Hirschfeld.

Miescher, F. 1893. Über die Beziehungen zwischen Meereshöhe und Beschaffenheit des Blutes. *Correspondenzblatt für Schweizer Ärzte* 23: 809–830.

Mitchell, J. H., and B. Saltin 2003. The Oxygen Transport System and Maximal Oxygen Uptake. In *Exercise Physiology*, edited by C. M. Tipton, pp. 255–291. Oxford: Oxford University Press.

Moedebeck, H. 1898a. Luftschiffahrt und Meteorologie, ihre Entwicklung und ihre Zukunft. *Die Umschau* 2: 819–822, 845–847.

Moedebeck, H. 1898b. Die Entwicklung der Luftschiffahrt in Deutschland. *Illustrierte Aeronautische Mittheilungen* 6: 107–109.

Moedebeck, H. W. L. 1886. Wie können die Fahrten der Berufsluftschiffer wissenschaftlich genutzt werden? *Zeitschrift des Deutschen Vereins zur Förderung der Luftschiffahrt* 5: 6–11.

Moedebeck, H. W. L. 1897. Der Sport in der Luftschiffahrt. *Illustrierte Mittheilungen des Oberrheinischen Vereins für Luftschiff-Fahrt* 1: 56, 58.

Moedebeck, H. W. L. 1906. 25 Jahre Geschichte des Berliner Vereins für Luftschifffahrt. *Illustrierte Aeronautische Mitteilungen* 10: 329.

Mosso, A. 1899. *Der Mensch auf den Hochalpen*. Leipzig: Veit.

Müllenhoff, K. 1891. Über die Wirkung der Luftverdünnung auf den menschlichen Körper. *Archiv für Anatomie und Physiologie, Physiologische Abteilung*: 344–350.

Müller, B. 1973. *Flugmedizin für die ärztliche Praxis*. Bonn: Kirschbaum.

Müller, E. A., and H. Franz 1952. Energieverbrauchsmessungen bei beruflicher Arbeit mit einer verbesserten Respirations-Gasuhr. *Arbeitsphysiologie* 14: 499–504.

Munk, F. 1956. *Das Medizinische Berlin um die Jahrhundertwende*. Munich: Urban & Schwarzenberg.

Nachmansohn, D., and R. Schmid 1988. *Die grosse Ära der Wissenschaft in Deutschland 1900 bis 1933*. Stuttgart: Wissenschaftliche Verlagsgesellschaft.

Neuberg, C. 1920. Nekrolog Nathan Zuntz. *Biochemische Zeitschrift* 105: Anhang.

Nöding, R., U. Gabler, L. Kipke, and R. Jankowski 1975. Die Anwendung einer modifizierten Anordnung der Atmung und des Gasstoffwechsels im Schwimmen. *Medizin und Sport* 15: 238–241.

Oppenheimer, C. 1911. *Handbuch der Biochemie des Menschen und der Tiere*. Jena: Gustav Fischer.

Pace, N. 1986. Tides of Physiology at Berkeley. *The Physiologist* 29 (Suppl.): 7–20.

Pagel, J. 1901. *Biographisches Lexikon hervorragender Ärzte des neunzehnten Jahrhunderts*. Berlin: Urban & Schwarzenberg.

Pauly, M. 1917. Zum 70. Geburtstage von Geheimrat Professor Dr. Zuntz. *Fischerei-Zeitung* 20: 349–351.

Petersen, K. 2002. *Die Entwicklungsgeschichte der Überdruckkammer und Indikationen für Hyperbare Sauerstoff-Therapie*. Inaugural-Dissertation, Heinrich-Heine Universität Düsseldorf.

Pflüger, E. 1877. Die Physiologie und ihre Zukunft. *Pflügers Archiv* 15: 361–365.

Pflüger, E. 1878. Wesen und Aufgaben der Physiologie. *Pflügers Archiv* 18: 427–442.

Plesch, J. 1949. *Ein Arzt erzählt sein Leben*. Munich: List.

Polis, P. 1896. *Über Wissenschaftliche Ballonfahrten und deren Bedeutung für die Physik der Atmosphäre*. Aachen.

Rasch, –. 1913. Die Entwicklung der Luftfahrt in Deutschland. *Deutsche Luftfahrer-Zeitschrift* 16: 241–242.

Raymond, P. 1986. *Literarische Entdeckung einer Landschaft: Die Romantisierung der Alpen*. Münster.

Reichel, H. 1911. Physiologische Ergebnisse der im Jahre 1906 durchgeführten Monte Rosa-Expedition. *Denkschriften der Kaiserlichen Akademie der Wissenschaften in Wien, Mathematisch-Naturwissenschaftliche Klasse* 86: 79–114.

Reymond, –. 1913. The hygiene and the physiology of the airman. *Journal of State Medicine*: 500–503.

Rheinberger, H.-J. 2006. *Experimentalsysteme und epistemische Dinge*. Frankfurt a. M.: Suhrkamp.

Rosenberg, G. 1901. Die zivil- und strafrechtliche Haftung des Luftschiffers. *Illustrierte Aeronautische Mittheilungen* 5: 89–93, 123–135.

Rothschuh, K. E. 1953. *Geschichte der Physiologie*. Berlin: Springer.

Rothschuh, K. E. 1954. 50 Jahre Deutsche Physiologische Gesellschaft (1904–1954). *Klinische Wochenschrift* 32: 271–273.

Rothschuh, K. E. 1968. *Physiologie – Der Wandel ihrer Konzepte, Probleme und Methoden vom 16. bis 19. Jahrhundert*. Freiburg im Breisgau: K. Alber.

Rothschuh, K. E. 1969. *Physiologie im Werden*. Stuttgart: Medizin Gesch. Kultur.

Rowell, L. B. 2003. The Cardiovascular System. In *Exercise Physiology*, edited by C. M. Tipton, pp. 98–137. Oxford: Oxford University Press.

Ruff, S., and H. Strughold. 1939, 1944, 1957 (1., 2. und 3. Aufl.). *Grundriss der Luftfahrtmedizin*. Munich: Barth.

Scheunert, A., and A. Trautmann. 1965. *Lehrbuch der Veterinär-Physiologie*. Berlin: Parey.

Schmidt, I. 1938. *Bibliographie der Luftfahrtmedizin*. Berlin: Springer.

Schmidt, R. F., F. Lang, and G. Thews 2005. *Physiologie des Menschen*. Berlin: Springer.

Schmidt-Nielsen, Bodil 1995. *August & Maria Krogh. Lives in Science*. New York: Oxford University Press.

Schmuck, H., and W. Gorzny 1982. *Gesamtverzeichnis des deutschsprachigen Schrifttums (GV) 1700–1910* 45: 227.

Schmuck, H., and W. Gorzny 1983. *Gesamtverzeichnis des deutschsprachigen Schrifttums (G.V.) 1700–1910* 90: 385.

Schubert, F. 1911. Wissenschaftliche Fahrt des Ballons Harburg III. *Deutsche Zeitschrift für Luftschiffahrt* 15: 16–17.

Schubert, G. 1935. *Physiologie des Menschen im Flugzeug*. Berlin: Springer.

Schulte, K. H. S. 1976. Bonner Juden und ihre Nachkommen bis um 1930. *Veröffentlichungen des Stadtarchivs Bonn* 16: 549.

Schulthess, H. 1915. *Schulthess' europäischer Geschichtskalender 56:* 56, 113.

Schwalbe, J. 1908. Reichs-Medizinal-Kalender für Deutschland auf das Jahr 1908, Part II. Leipzig.

Schwalbe, J. 1926. *Reichs-Medizinal-Kalender 1926–27*, Part 2. Leipzig.

Schwartz, E., and R. Chambers 1947. Wilhelm Caspari 1872–1944. *Science* 105: 613.

Seehaus, –. 1915. Jahrestafeln zur Geschichte der Akademie. *Die Königlich Preussische Landwirthschaftliche Akademie*, Bonn-Poppelsdorf: 123.

Simons, E., and O. Oelz 2001. Kopfwehberge. *Eine Geschichte der Höhenmedizin*. Zürich: AS Verlag & Buchkonzept.

Sittmann, G. 1900. Hochfahrten im Dienste der medizinischen Wissenschaft. *Jahresbericht des Münchener Vereins für Luftschiffahrt:* 36–43.

Skalweit, A. 1927. Die deutsche Kriegsernährungswirtschaft. 96. Stuttgart: Deutsche Verlagsanstalt.

Soubies, M. J. 1907. *Physiologie de l'aéronaute*. Dissertation, Paris.

Steffen, R. 1984. *Reisemedizin*. Berlin: Springer.

Stegemann, J. 1984. *Leistungsphysiologie*. Stuttgart: Thieme.

Strasser, P., and R. Andreas (eds) 2006. *Montafon 1906–2006*. Schruns: Montafoner Schriftenreihe.

Strughold, H. 1936/37. Geschichtliches zur Luftfahrtmedizin. *Luftfahrtmedizinische Abhandlungen* 1: 16–22.

Strughold, H. 1950. Development of aviation medicine in Germany. In *German Aviation Medicine World War II*, German Aviation Medicine, Vol. 1, pp. 3–11. Washington, DC: Dept. of the Air Force.

Süring, R. 1909. Wissenschaftliche Ballonfahrten. In *Wir Luftschiffer*, edited by K. Bröckelmann, pp. 48–65. Berlin: Ullstein & Co.

The New York Times 1920. Offers Hope to the Bald. German scientist says certain food elements make hair grow. February 11.

Tigerstedt, R. 1911. *Handbuch der physiologischen Methodik*. Leipzig: S. Hirzel.

Tipton, C. 2003a. *Exercise Physiology: People and Ideas*. New York, NY: Oxford University Press.

Tipton, C. 2003b. The autonomic nervous system. In *Exercise Physiology*, edited by C. M. Tipton, pp. 188–254. Oxford: Oxford University Press.

Tissandier, G. 1887–90 (Reprint 1980). *Histoire des ballons et des Aéronautes célèbres*. Paris. (Reprint Molsheim).

Tittel, K. 2004. Leistungen Deutschlands für die internationale Sportmedizin – Historische Reminiszenzen. *Deutsche Zeitschrift für Sportmedizin* 55: 315–321.

Tögel, C. 2006. *Freud und Berlin*. Berlin: Aufbau Verlag.

von Basch, S. 1877. Über den Einfluss der Athmung von comprimirter und verdünnter Luft auf den Blutdruck des Menschen. *Medizinische Jahrbücher*: 489–497.

von Cyon, E. 1910. Eduard Pflüger. *Pflügers Archiv* 132: 1–19.

von der Goltz, T. Freiherr 1898. Festrede zur Feier des 50jährigen Bestehens der landwirtschaftlichen Akademie Poppelsdorf. *Landwirthschaftliches Jahrbuch* 27: 250.

von der Heide, R. 1918. N. Zuntz. *Landwirthschaftliches Jahrbuch* 51: 329–362.

von Herwarth, H. 1911. *Unser Luftreich – unsere Zukunft*. Berlin: Continent.

von Kleist, E. 1909. *Militär und Luftschiffahrt*. In *Wir Luftschiffer*, edited by K. Bröckelmann, pp. 285–306. Berlin: Ullstein.

von Liebig, G. 1885. Die Wirkung des erhöhten Luftdruckes bei den pneumatischen Kammern bei Asthma. *Deutsche medizinische Wochenschrift*: 292–294.

von Liebig, J. 1874. *Reden und Abhandlungen von Justus Liebig*. Leipzig: Winter.

von Schjerning, O. 1922. *Handbuch der ärztlichen Erfahrungen im Weltkriege 1914/1918, Vol. 7*. Leipzig: Barth.

von Schroetter, H. 1899. *Zur Kenntnis der Bergkrankheit*. Vienna: W. Braumüller.

von Schroetter, H. 1901. Zur Kenntnis der Wirkung bedeutender Luftverdünnung auf den menschlichen Organismus. *Die medicinische Woche und balneologische Zentralzeitung* 38: 404.

von Schroetter, H. 1902a. Über den Einfluss grosser Höhen auf den Organismus und über Ballonfahrten zu physiologischen Zwecken. *Illustrirte Aeronautische Mittheilungen* 6: 150.

von Schroetter, H. 1902b. Über Ballonfahrten zu physiologischen Zwecken. *Monatszeitschrift für Gesundheitspflege* 20: 89–94. Vienna: Adler.

von Schroetter, H. 1902c. Über Höhenkrankheit mit besonderer Berücksichtigung der Verhältnisse im Luftballon. *Wiener medizinische Wochenschrift*: 1294, 1351–54 1417–20.

von Schroetter, H. 1903. Über Höhenkrankheit mit besonderer Berücksichtigung der Verhältnisse im Luftballon. In *Protokoll über die dritte Versammlung der Internationalen Kommission für wissenschaftliche Luftschiffahrt vom 20. bis 25. Mai 1902 zu Berlin*, 102–129. Strassburg i. E.

von Schroetter, H. 1904. Zur Physiologie der Hochfahrten. *Illustrierte Aeronautische Mitteilungen* 8: 14–21.

von Schroetter, H. 1906. Der Sauerstoff in der Prophylaxe und Therapie der Luftdruckerkrankungen. In *Sauerstofftherapie*, edited by M. Michaelis, pp. 155–314. Berlin: A. Hirschwald.

von Schroetter, H. 1909. Hygiene der Aeronautik, *Denkschrift zur ersten internationalen Luftfahrtausstellung Frankfurt a. M. 1909*. Frankfurt.

von Schroetter, H. 1912a. *Hygiene der Aeronautik und Aviatik*. Vienna: W. Braumüller.

von Schroetter, H. 1912b. Bemerkungen zur Physiologie und Therapie der Luftwirkung. *Zeitschrift für Balneologie, Klimatologie und Kurorthygiene* 5: 1–12.

von Schroetter, H. 1913. Gesichtspunkte zur Hygiene und Prophylaxe der Luftfahrt. *Das Österreichische Sanitätswesen* 25: 1429–1444, 1457–1478.

von Schroetter, H. 1919a. Zur Psychologie und Pathologie des Feldfliegers. *Zeitschrift für angewandte Psychologie* 15: 299–300.

von Schroetter, H. 1919b. Bemerkungen zur praktischen Physiologie des Fliegers. *Wiener klinische Wochenschrift* 32: 731–736.

von Schroetter, H. 1921. Klimatische Beobachtungen und Studien anlässlich der Landungsmanöver in Dalmatien, August 1911, nebst Notizen zur Hygiene des Marsches. *Denkschriften der Kaiserlichen Akademie der Wissenschaften in Wien, Mathematisch-Naturwissenschaftliche Klasse* 97: 93–132.

von Schroetter, H. 1925. Zur Kenntnis des Energieverbrauchs beim Maschinenschreiben. *Pflügers Archiv* 207: 323–342.

von Schroetter, H. 1927. Zur Physiologie und Hygiene der Luftfahrt. *Ärztliche Sachverständigen-Zeitung* 33: 339–342.

von Tschudi, –. 1902. Das neue Kasernement des Preussischen Luftschiffer-Bataillons. *Illustrierte Aeronautische Mittheilungen* 6: 61–62.

Vossische Zeitung 1917. *Prof. Dr. Zuntz' 70. Geburtstag*. 6. October 1917.

Waldenburg, L. 1873. Ein transportabler pneumatischer Apparat zur mechanischen Behandlung der Respirations-Krankheiten. *Berliner klinische Wochenschrift* 10: 465–469.

Ward, M. P., J. S. Milledge, and J. B. West. 2000. *High Altitude Medicine and Physiology*. London: Arnold.

Wegener, G. 1913. Die Erweiterung der Herrschaft des Menschen über die Erdoberfläche während der letzten 25 Jahre und der Anteil der Deutschen daran. *Festrede vom 16. Juni 1913 an der Handelshochschule Berlin*. Berlin.

Weil, J. V. 1986. Ventilatory control at high altitude. In *Handbook of Physiology, Respiratory System*, edited by Neil S. Cherniack, John G. Widdicombe and Alfred P. Fishmann, Sec. 3, Vol. II, p. 704. American Physiological Society. Bethesda, MD. Williams & Wilkins Company.

Weitlaner, F. 1910. Die psychologischen, physiologischen und pathologischen Momente beim Simplonflug Chavez'. *Flug- und Motortechnik* 4: 753.

Wenig, O. 1868. Verzeichnis der Professoren und Dozenten der Rheinischen Friedrich-Wilhelms-Universität von 1818–1968. Bonn.

Werner-Bleines, –. 1909. Sauerstoff und Atmungsvorrichtungen für Luftschifffahrt. *Illustrierte Aeronautische Mitteilungen* 13: 1032–1040.

West, J. B. 1981. High Altitude Physiology. *Benchmark Papers in Human Physiology, Vol. 15*. Stroudsburg: Hutchinson Ross Publishing Company.

West, J. B. 1996. Respiratory Physiology: People and Ideas. *American Physiological Society*. New York, NY: Oxford University Press.

West, J. B. 1998. *High Life*. New York, NY: Oxford University Press.

West, J. B., and P. D. Wagner. 1980. Predicted gas exchange on the summit of Mt Everest. *Respiration Physiology* 42: 1–16.

Wigand, A. 1914. Wissenschaftliche Hochfahrten im Freiballon. *Fortschritte der naturwissenschaftlichen Forschung* 10: 203–272.

Williams, John A. 2003. 76th President of APS. *The Physiologist* 46: 45, 51–52, 54.

Winau, R. 1987. *Medizin in Berlin*. Berlin: De Gruyter.

Winau, R. 2005. Schriften und Vorträge 1965–2005, unpaginated. Institute for the History of Medicine, Berlin: Center for Hymanities and Health Science.

Wittke, G. 1970. Storia della fisiologia dell'altitudine con particolare riferimento al lavoro del Prof. N. Zuntz e della sua Scuola. *Medicina dello Sport* 23: 295–298.

Wittke, G. 1982. Bedeutung und Begründung des Tierversuchs in der medizinischen Ausbildung. *Schreibmaschinen-Manuskript des Verfassers zum Vortrag anläßlich der Eröffnung der Zentralen Tierlaboratorien am 27.2.1982 in Berlin*. Berlin.

Wittmack, L. 1906. Die Königliche Landwirtschaftliche Hochschule in Berlin. *Festschrift zur Feier des 25jährigen Bestehens*. Berlin: Paul Parey.

Wrynn, A. M. 2006. "A debt paid off in tears": Science, IOC politics and the debate about high altitude in the 1968 Mexico City Olympics. *International Journal of the History of Sport* 23: 1152–1172.

Zabel, E. 1918. Deutsche Luftfahrt. Rückblicke und Ausblicke. Berlin: Braunbeck.

Zuelzer, W. W. 1981. *Der Fall Nicolai*. Frankfurt: Societäts-Verlag.

Zuntz, L. 1899. Untersuchungen über den Gaswechsel und Energieumsatz des Radfahrers. *Inaugural Dissertation*. Berlin: Schumacher.

Zuntz, L. 1903. Über die Wirkung des Höhenklimas auf den gesunden und kranken Organismus. *Fortschritte der Medizin*: 601–606, 631–634.

Zuntz, L. 1926. N. Zuntz. *Blätter für Volksgesundheitspflege* 26: 201–202.

Index

Printed and bound by CPI Group (UK) Ltd, Croydon, CR0 4YY

03/10/2024

01040410-0003